Contemporary Topics in

Semiconductor Spintronics

Contemporary Topics in Semiconductor Spintronics

Editors

Supriyo Bandyopadhyay
Virginia Commonwealth University, USA

Marc Cahay
University of Cincinnati, USA

Jean-Pierre Leburton
University of Illinois at Urbana-Champaign, USA

World Scientific

NEW JERSEY • LONDON • SINGAPORE • BEIJING • SHANGHAI • HONG KONG • TAIPEI • CHENNAI • TOKYO

Published by

World Scientific Publishing Co. Pte. Ltd.
5 Toh Tuck Link, Singapore 596224
USA office: 27 Warren Street, Suite 401-402, Hackensack, NJ 07601
UK office: 57 Shelton Street, Covent Garden, London WC2H 9HE

British Library Cataloguing-in-Publication Data
A catalogue record for this book is available from the British Library.

CONTEMPORARY TOPICS IN SPINTRONICS

ISBN 978-981-3149-81-6

Desk Editor: Rhaimie Wahap

Preface

Spintronics, which utilizes both of the electron's degrees of freedom (spin and charge), has attracted strong interest in the last three decades because of the promise it holds for novel device functionalities, both classical and quantum-mechanical. This book addresses some of the recent frontier research topics in semiconductor-based spintronics. These include the creation, manipulation, control and detection of spin polarized currents in semiconductor structures by purely electrical means. A daunting challenge is to preserve spin coherence in a device for the longest time or over the longest distance in order to produce reliable spintronic processors. Some progress is now being made to meet this challenge.

In Chapter 1, Balocchi *et al.* present a review of the control of electron spin dynamics in (111)-oriented zinc-blende quantum well semiconductor structures by adjusting the interplay between the Dresselhaus and Rashba spin-splitting contributions. The symmetry-driven origin of spin relaxation is presented and selected experimental results from the literature reviewed. The chapter concludes by offering some new perspectives in the field of spintronics brought about by this new class of spin-based devices.

In Chapter 2, Puttisong *et al.* focus on the identification and optimization of the key physical processes that determine the efficiency of the spin-filtering effect and the dynamic spin transfer between conduction and defect electrons and between electrons and defect nuclei in a semiconductor at room temperature. This chapter gives a brief introduction to defect-engineered spin functionalities in a semiconductor, followed by an in-depth analysis of the key physical parameters that govern this mechanism with the aid of a coupled rate-equation approach. A guideline on how to bring the spin systems to complete polarization is provided at the end of the chapter.

In Chapter 3, Sanchez and Leburton explain various spin effects, such as a carrier density-dependent spin-flip in dilute quasi-one-dimensional (1D) systems using an unrestricted Hartree-Fock approach with a three-dimensional (3D) Coulomb interaction. They show dimensionality is critical in identifying a complex pattern of magnetic phases varying with confinement and magnetic field. In the absence of magnetic field and spin-orbit interaction, they predict the existence of degenerate excited states with opposite spin polarizations above a confinement-dependent 1D concentration threshold. Their model shines new light on the emergence of anomalous structures at 0.3 G_0 and ~0.6–0.7 G_0 in the conductance of semiconductor quantum constrictions at low temperatures.

In Chapter 4, Hossain *et al.* discuss spin transport in one- and quasi-one-dimensional structures where the D'yakonov-Perel' spin relaxation mechanism can be completely eliminated if all spin carriers are restricted to a single subband. They discuss experimental results that demonstrate a 10-fold increase in the spin relaxation time at room temperature in an InSb nanowire compared to bulk or quantum wells of InSb, because 96% of the electrons occupy the lowest subband in the conduction band at room temperature. They also use this effect to demonstrate the working principle of a novel spintronic infrared photodetector. Spin transport in nanowires of other materials such as GaN, Ge and Si are discussed as well.

In Chapter 5, Jones *et al.* use a non-equilibrium Green's function formalism to suggest a set of experiments which can provide indirect evidence of the fine and non-local electrostatic tuning of the onset of spin polarization in a quantum point contact or QPC with two sets of closely spaced in-plane side gates (SGs)in the presence of lateral spin-orbit coupling (LSOC). When calculated as a function of the common sweep voltage V_{sweep} applied to two of the SGs, the conductance plots show several conductance anomalies, including intrinsic bistability below $G_0 = 2e^2/h$. The shape of the conductance anomalies and size of hysteresis loops are different when the biasing conditions on the leftmost and rightmost SGs are swapped. This rectifying behavior is an additional indirect evidence for the onset of spontaneous spin polarization in nanoscale devices made of QPCs.

In Chapter 6, Sheng-Chin Ho reviews the fundamental physics of spin field effect transistors and describes the realization of an all-electric all-semiconductor spin-field effect transistor by using two QPCs as spin injector and detector. The spin polarization in QPCs is created as a result of LSOC when an asymmetric one dimensional confinement is applied. The polarized spins traveling between the injecting and detecting contacts experience a Rashba spin-orbit coupling (controlled by a bias applied to a middle gate) and rotate spins in the channel between the two contacts.

The discovery of large magnetoresistance effects in InMnAs heterojunction diodes and bipolar heterojunction transistors at room temperature has spawned a serious interest in exploiting these devices for spintronic logic. In Chapter 7, Wessels *et al.* describe an emitter coupled transistor (ECT) logic family based on the bipolar magnetic junction transistor (MJT). The logic circuit is predicted to have improved speed, area and power characteristics compared to conventional ECT.

Organic semiconductors typically exhibit much longer spin relaxation time than their inorganic counterparts, but because the carrier mobility in organics is very low, the spin relaxation length is usually smaller than in inorganic semiconductors. Recently, there have been remarkable advances in spin injection and transport (or tunneling) in organic spin valves and magnetic tunnel junctions. Two emerging fields are engineering of organic-inorganic interfaces for high degree of spin polarization and control of spins at the single molecule level. In Chapter 8, Majumdar discusses this exciting field and the opportunities it affords.

In Chapter 9, Kamalakar, *et al.* present two important advances in the field of graphene spintronics: the recent achievement of long distance spin communication in large scale chemical vapor deposited graphene, and the demonstration of enhanced spin injection and spin filtering effects in ferromagnet/h-BN-graphene van der Waals heterostructures. They also discuss how these results impact the present state of the art and open new avenues for future developments.

We hope this book will be useful for practicing researchers in the field, but more importantly, will attract generations of graduate students to pursue advanced studies in the fascinating field of spintronics. This book is by no means comprehensive. To keep abreast of the other many interesting advances in the field of spintronics, the reader may refer to Gordon Research Conferences website, https://www.grc.org/programs.aspx?id=15876.

Supriyo Bandyopadhyay,
Marc Cahay,
and Jean-Pierre Leburton

Contents

Contents

Chapter 1

Spin Manipulation in (111) Quantum Wells

A. Balocchi*, T. Amand and X. Marie

Université de Toulouse, INSA-CNRS-UPS, LPCNO, 135 Ave. de Rangueil, 31077 Toulouse, France

This chapter presents a review of the electron spin dynamics and its control possibilities in (111)-oriented zinc-blende quantum well semiconductor structures, as a result of the adjustable interplay between the Dresselhaus and Rashba spin-splitting contributions. We describe the symmetry-driven origin of the spin relaxation and selected experimental results from state-of-the-art research in these unusually oriented hetero-structures. We finally evoke some new perspectives in spintronics opened by this new class of spin-based devices.

1. Introduction

The manipulation and control of the carriers' spin in semiconductor materials is the cornerstone of future spintronics and spin-based quantum information technology and has been the driving force behind the research on the spin-based processing and devices for the past decades.[1–3] The main challenge is to find materials or systems for which the carriers' spin relaxation and coherence time is long enough to perform information processing or control its propagation. Understanding the mechanisms governing the spin relaxation processes is thus crucial and have been the subject of ample theoretical and experimental works. The dominant mechanisms governing the spin decay in crystals originates form the spin-orbit interaction (SOI).[3–7] This relativistic effect describes the coupling between the electron spin and orbital angular momenta via the effective magnetic field $\vec{B}_{SO}$ appearing in the electron inertial reference frame moving in the presence of the crystal electric field. In general, the spin-orbit interaction plays a fundamental role in the determination of the semiconductor crystal band structure. In the $\vec{k} \cdot \vec{p}$ approach, one can distinguish between different kinds of terms introduced by the SOI. First, $\vec{k}$-independent terms in the hamiltonian, which are e.g. responsible for the splitting between the degenerated heavy and light hole and the split-off levels in $\vec{k}$=0 in semiconductor crystal of O_h or T_d symmetry like {Si, Ge, ...} or {GaAs, InAs, CdTe, ...} respectively. The order of magnitude of this spin-orbit splitting is typically 50 $\sim$ 500 meV.[8]

*andrea.balocchi@insa-toulouse.fr

Second, in crystals with low-enough symmetry, a small $\vec{k}$-dependent term arises, due to the SOI between remote conduction bands or between device sub-bands and the valence bands. This interaction, analogous to an effective magnetic field $\vec{B}_{SO}(\vec{k})$, splits generally the energy levels of otherwise spin degenerate bands (except in peculiar crystal directions) at any given wavevector. The corresponding contribution hamiltonian can be written as $H_{SO}(\vec{k}) = \mathbf{\Omega}_{SO}(\vec{k}) \cdot \hat{\vec{S}} = \frac{\hbar}{2}\mathbf{\Omega}_{SO}(\vec{k}) \cdot \hat{\vec{\sigma}}$, where $\hat{\vec{\sigma}} = (\sigma_x, \sigma_y, \sigma_z)$ is the Pauli matrix vector operator in the $\{\uparrow, \downarrow\}$ basis. As $\vec{\Omega}_{SO}$ is odd with respect to inversion, this symmetry driven SOI exists only in crystals lacking inversion symmetry (for instance in the zinc-blende structure and not in the diamond one as sketched in figure 1 a) and is termed the Bulk Inversion Asymmetry (BIA) or Dresselhaus[9] term[a]. The magnitude and direction of the associated precession vector $\mathbf{\Omega}_{BIA}$ depend on the speed and direction of the electron, such that the spin-orbit interaction results in a $\vec{k}$-dependent spin-splitting of the bands, where $\vec{k}$ is the wavevector of the electron. The SOI spin-splitting occurs for both conduction and valence band states, however, in what follows we will concentrate on the more thoroughly investigated spin-splitting of the conduction band. The magnitude of the BIA spin splitting is of the order of $10^{-4} \sim 10^{-3}$ eV.[2] It is widely accepted that the dominant process governing the spin relaxation in III-V zinc-blende semiconductors for a wide range of temperatures is the mechanism proposed by Dyakonov and Perel (DP) .[3–7,11] According to this model, the spin relaxation process originates from the $\vec{k}$-dependent Larmor precession, in intervals between collisions, of the electron

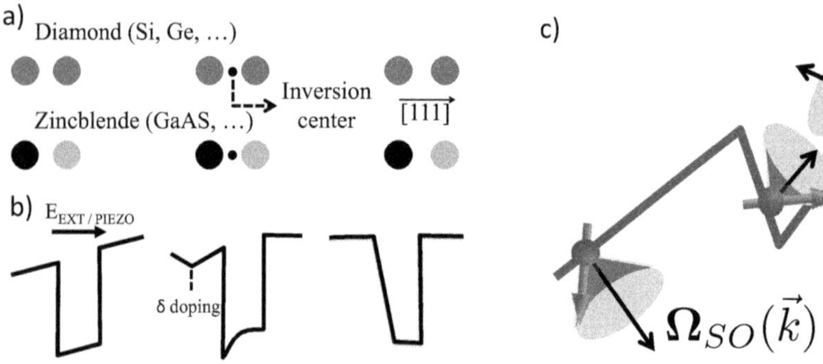

Fig. 1. a) Simplified schematic of the presence (diamond structure) or lack (zinc-blende structure) of an inversion centre along the [111] direction. In the latter, the spin-orbit terms from the Bulk Inversion Asymmetry arise. b) Different possibilities of built-in asymmetries yielding Structural Inversion Asymmetry spin-orbit terms. From left to right the asymmetry is due to an internal or external electric field, an asymmetric doping or again to an asymmetric potential due to a graded alloy interface. c) Schematic representation of the Dyakonov-Perel mechanism in which the electron spin (red arrows) precesses around the $\vec{k}$ dependent spin-orbit effective magnetic field precession vector $\mathbf{\Omega}_{SO}(\vec{k})$ in intervals between collisions.

[a] In solids it is considered that the spin-orbit coupling Dresselhaus term is related to the breaking of crystal inversion symmetry. Recent works report that spin-orbit effects should also be observable in certain centrosymmetric crystals.[10]

spin in the spin-orbit effective magnetic field (figure 1c). The spin relaxation time τ_i^s of a conduction band (CB) electron's spin oriented along a direction i can be simply written in the collision dominated regime as $1/\tau_i^s = \langle \mathbf{\Omega}_\perp^2 \rangle \tau_p^*$, where $\langle \mathbf{\Omega}_\perp^2 \rangle$ is the mean square of the precession vector in the plane perpendicular to the spin direction i and τ_p^* the electron momentum scattering time.

Lowering the dimensionality from 3D to 2D systems naturally modifies not only the band structures but also the symmetry of the system and as a consequence the electron spin relaxation. The choice of the growth direction thus offers a first obvious mean of manipulation of the electron spins. In 2D heterostructures an additional effective magnetic field, $\vec{B}_{SIA}(\vec{k})$, can emerge if the confining potential of the system lacks inversion symmetry as sketched in figure 1 b. This contribution linked to Structural Inversion Asymmetry (SIA) is often referred to as the Rashba term.[12] As such, the corresponding $\mathbf{\Omega}_{SIA}(\vec{k})$ strength can be tailored, for instance, by an internal or external electric field. The enormous interest represented by the 2D systems resides in the fact that the presence of both BIA and SIA contributions provides two independent "control knobs" of the the electron spin relaxation sources. The BIA/SIA adjustable interplay can thus lead to a rich variety of spin-related phenomena.[13–18] It is worth noting that, after being subject of debates, it was recently shown that the Rashba-induced conduction-band spin splitting requires the application of a real electric field in the structure (through asymmetric doping or diode structures), whereas CB potential gradients, which also break the inversion symmetry, yield a negligibly small Rashba splitting.[2,19,20] Finally, Interface Inversion Asymmetry (IIA), due to the anisotropy of the chemical bonds at the QW interfaces, as in non-common-atom quantum wells,[21] and strain are additional sources of conduction band spin splitting.

In this chapter we present recent experimental realizations of the control of the spin relaxation of conduction band electrons obtained through the proper tailoring of the SOI effective fields via the structure growth symmetry and quantum confinement (BIA term) and the applied electric field (SIA term). Spin manipulation by means of electric field is especially attractive as it can be easily implemented in devices by doping (e.g. diode structures). In this scenario, (111)-oriented QWs provides a most propitious configuration as the formally identical expression of the BIA and SIA terms (at least in the first order in the wavevector) allows for the suppression of the conduction band electron spin splitting for any spatial directions by the application of an electric field of proper strength and direction providing a dramatic control of the spin-relaxation and coherence times.

2. Symmetry dependence of the BIA and SIA terms in zinc-blende Quantum Wells

2.1. *The Bulk Inversion Asymmetry contribution*

The Bulk Inversion Asymmetry term, often called the Dresselhaus term, derived by symmetry considerations for zinc-blende bulk semiconductors, at small wavevector writes:[1,9]

$$H_{BIA} = \frac{\hbar}{2}\boldsymbol{\Omega^{3D}_{BIA}(k)}\cdot\boldsymbol{\sigma} = \gamma[k_x(k_y^2 - k_z^2)\boldsymbol{\sigma_x} + k_y(k_z^2 - k_x^2)\boldsymbol{\sigma_y} + k_z(k_x^2 - k_y^2)\boldsymbol{\sigma_z}], \quad (1)$$

where $\boldsymbol{\sigma}$ are the Pauli matrices, $\boldsymbol{\Omega^{3D}_{BIA}(k)}$ represents the Dresselhaus effective magnetic field vector, γ the material-dependent Dresselhaus coefficent and $\vec{e_x} = (1,0,0)$, $\vec{e_y} = (0,1,0)$, $\vec{e_z} = (0,0,1)$ the three unit vectors along the crystallographic axes of the unit cell. Note that $\boldsymbol{\Omega^{3D}_{BIA}(k)}\cdot\boldsymbol{k} = 0$. As the interaction is small, it is usually treated in a perturbative approach.

In quantum wells the quantum confinement effect has to be taken into account.[5] The transition from 3D to 2D is performed by the replacement of the z growth axis momentum component terms k_z and k_z^2 by their quantum-mechanical average values in the considered subband (generally the lowest one), $\langle k_z\rangle = 0$ and $\langle k_z^2\rangle \neq 0$ respectively, the latter being proportional to the electron confinement energy ($\langle k_z^2\rangle = (\pi/L)^2$ for an infinite quantum well of thickness L). In the limit of small $\vec{k}$ vectors or low temperature, i.e. when $(\langle k_z^2\rangle)^{1/2} \gg k_x, k_y$, the weaker cubic terms can be neglected and the Dresselhaus effective magnetic field becomes linear with respect to the transverse $\vec{k}$ components:

$$\boldsymbol{\Omega}^{(1)}_{BIA}(\boldsymbol{k}_\perp) = \frac{2\gamma_{2D}}{\hbar}\langle k_z^2\rangle\,(-k_x, k_y, 0) \qquad \text{for [001] growth axis} \qquad (2a)$$

$$\boldsymbol{\Omega}^{(1)}_{BIA}(\boldsymbol{k}_\perp) = \frac{\gamma_{2D}}{\hbar}\langle k_z^2\rangle\,(0, 0, k_y) \qquad \text{for [110] growth axis} \qquad (2b)$$

$$\boldsymbol{\Omega}^{(1)}_{BIA}(\boldsymbol{k}_\perp) = \frac{4\gamma_{2D}}{\hbar\sqrt{3}}\langle k_z^2\rangle\,(k_y, -k_x, 0) \qquad \text{for [111] growth axis} \qquad (2c)$$

where $\vec{k}_\perp = (k_x, k_y)$ is the in-plane wavevector for each respective axis orientation as defined in table 1 and the superscript (1) indicates the linear approximation [b,c]. The previous equations imply that $\boldsymbol{\Omega}^{(1)}_{BIA}$ is not only a function of the in plane wavevector $\vec{k}_\perp$ but its amplitude and orientation also depend on the growth direction and confinement energy: for a growth along both [001] and [111] the effective magnetic field is in the quantum well plane such that the Dyakonov-Perel spin relaxation exists for spin components along the growth axis. For [110] grown wells the opposite situation will occur.

[b] Using the classical definition of the bulk Dresselhaus constant, the spin precession vector components used in references[22–25] should be multiplied by 2.

[c] The γ_{2D} parameter differs from γ. It depends both on the semiconductor compounds of the heterostructure as well as the QW width. These expressions can also be derived by symmetry considerations on the ensemble { bulk crystal of the QW + confining potential }.

Table 1. Coordinates axis orientation correspondence for the different growth planes. The z-axis defines the growth direction in all cases.

	growth plane		
	(001)	(110)	(111)
x-axis	[100]	[001]	$[11\bar{2}]$
y-axis	[010]	$[1\bar{1}0]$	$[\bar{1}10]$
z-axis	[001]	[110]	[111]

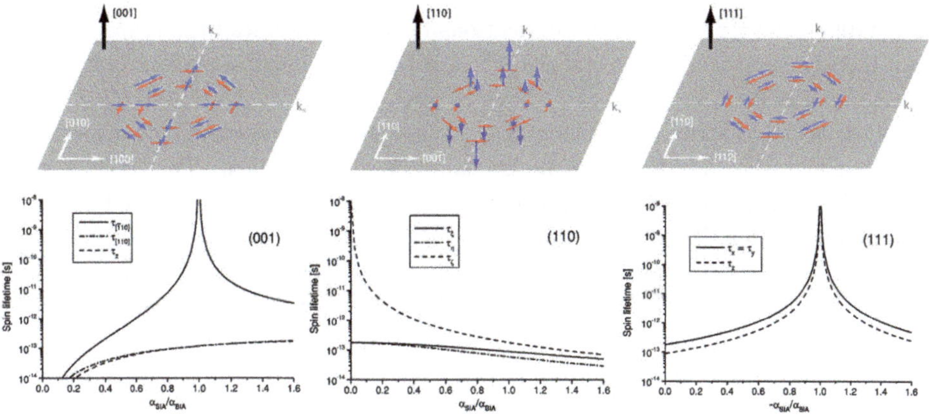

Fig. 2. Top: Representation of the Dresselhaus (blue arrows) and Rashba (red arrows) spin-orbit fields as a function of electron crystal momentum in a zinc-blende quantum well grown along the (a) [001], (b) [110], (c) [111] direction indicated by the black arrow Credits:APS/Alan Stonebraker, https://physics.aps.org/articles/v4/73 September 19, 2011• Physics 4, 73, reference[26]. Bottom: Spin relaxation times for the three spin components as a function of the ratio of the SIA to the BIA strength for [001], [110] and [111]-grown QWs (from left to right). In the case of (111) grown structures, all three components of the spin relaxation tensor present a simultaneous resonance, as opposed to a single one for the other structures. The labels refer to the spin relaxation tensor components according to the axis orientations defined in Table 1 or as defined in reference [27].Reprinted (figure2) with permission from [X. Cartoixà, D. Z.-Y. Ting, Y.-C. Chang, Physical Review B, 71, 045313, (2005)] Copyright (2005) by the American Physical Society.

2.2. *The Structural Inversion Asymmetry contribution*

The form of the SIA contribution, often referred to as Rashba but originally proposed by Vas'ko,[28] is instead independent on the crystallographic orientation of the QW, and if it is induced by and electric field E applied along the growth direction, as it is for most cases, it always lies in the QW plane and writes:

$$\boldsymbol{\Omega}_{SIA}\left(\boldsymbol{k}_{\perp}\right) = \frac{2r_{41}E}{\hbar}\left(k_y, -k_x, 0\right) \tag{3}$$

where $\alpha = 2r_{41}E$ is the"Rashba" coefficient and r_{41} a positive constant,[2] with the x and y axis defined according to table 1. Both BIA and SIA strengths can be controlled, the adjusting parameters being the well width variation of the $\langle k_z^2 \rangle$

term for the former, the applied electric field, for instance, for the latter.[18,19,29,30] Figure 2 top row schematically depicts the orientation of $\mathbf{\Omega}_{BIA}$ and $\mathbf{\Omega}_{SIA}$ as a function of $\vec{k}_{\perp}$ in QWs grown along three different crystallographic directions. Their simultaneous presence can produce interesting interference effects. Figure 2 bottom row presents the spin relaxation times for the spatial components as a function of the ratio of the Dresselhaus and Rashba strengths α_{BIA} and α_{SIA} respectively, for the corresponding different growth crystallographic orientations. As an example of the BIA/SIA interplay in (001) wells, by a proper choice of the adjustable parameters, the total effective precession vector $\mathbf{\Omega}_T^{(1)} = \mathbf{\Omega}_{BIA}^{(1)} + \mathbf{\Omega}_{SIA}$ can be made to vanish, and so does the Dyakonov-Perel mechanism for electrons' wavevectors $\vec{k} \parallel [\bar{1}10]$.[13,15,17,27,31] Succesful applications have demonstrated the realization of a non-ballistic spin-field-effect transistor[16] and the experimental evidence of the persistent spin helix.[18]
In (110) oriented structures, being $\mathbf{\Omega}_{BIA}^{(1)} \parallel z$, spin oriented along the growth axis will be naturally excluded form the Dyakonv-Perel mechanism. The Rashba term, adding an in-plane effective magnetic field component, can then be used as a spin relaxation "switch"[14,29,32] .
A rather unique situation occurs in the case of (111)-oriented heterostructures as the Rashba and Dresselhaus contributions have the same symmetry. In contrast to (100) and (110) QWs, where the DP spin relaxation can vanish for one given spin direction only, it can be suppressed in principle for the three directions in space for (111) QWs, since the conduction bands can become spin degenerate to first order in $\vec{k}$. When an electric field of strength

$$E_c = \frac{2\gamma_{2D}\langle k_z^2 \rangle}{\sqrt{3}r_{41}} \tag{4}$$

is applied perpendicular to the QW plane, the two effective magnetic field compensate each other, i.e. $\mathbf{\Omega}_T^{(1)}$=0, and the DP mechanism is suppressed for all spin components. A new class of spin-based devices relying on this electrical cancellation of the CB spin splitting have been demonstrated: the simultaneous control of the the spin relaxation time in the three spin directions has been recently reported in biased structures up to 250 K.[22–24,33] The role of the non-equivalent $\langle 111 \rangle$ substrate orientations has also been determined.[25] More recently, the gate-control of the electron spin diffusion length has also been reported.[34] The particular spin relaxation characteristics of (111)-oriented quantum wells and a review of related relevant experimental results will be the central topic of this chapter.

2.3. *Higher order terms*

Despite being negligible near the CB edge and at low temperature, the previously neglected Dresselhaus nonlinear terms progressively play the dominant role while the linear ones are compensated by the Rashba contribution in an electrically tuned device. This leads to a limitation of the spin lifetimes (the spin relaxation tensor divergence is removed) but also to new interesting possibilities such as the control of the spin relaxation anisotropy. The role of the non-linear Dresselhaus effective magnetic field $\mathbf{\Omega}^{(3)}_{BIA}$ becomes even more relevant as the temperature increases since higher $\vec{k}$ states are populated.[35] Taking into account these nonlinear terms, the total CB spin splitting is determined by:

$$\mathbf{\Omega}_T = \mathbf{\Omega}^{(1)}_{BIA} + \mathbf{\Omega}^{(3)}_{BIA} + \mathbf{\Omega}_{SIA} \equiv \begin{cases} \Omega_x = \frac{1}{\hbar}\left[\frac{2\gamma_{2D}}{\sqrt{3}}\left(2\langle k_z^2\rangle - \frac{k_\perp^2}{2}\right) + 2r_{41}E\right] k_y \\ \Omega_y = -\frac{1}{\hbar}\left[\frac{2\gamma_{2D}}{\sqrt{3}}\left(2\langle k_z^2\rangle - \frac{k_\perp^2}{2}\right) + 2r_{41}E\right] k_x \\ \Omega_z = \frac{2\gamma_{2D}}{\hbar\sqrt{6}}\left[3k_x^2 k_y - k_y^3\right] \end{cases} \quad (5)$$

A weaker and broader spin relaxation dependence on the effective magnetic field tuning at higher temperatures T is expected due to the impossibility of minimizing the spin splitting for all wavevectors simultaneously. The relative importance of the higher order to linear terms scales as T^2 and therefore it will experience a three order of magnitude increase from T=10 K to T=300 K, drastically curbing the compensation effect. A high temperature re-equilibration of the linear to cubic contributions can ideally be attained in narrower wells if, yet, the larger compensation field can be achieved before electron leakage from the well sets in.

2.4. *Determination of the spin relaxation time: experimental techniques and samples description*

The determination of the electron spin relaxation time in the experiments reviewed below is performed by complementary optical spectroscopic techniques. At their core stand the optical selection rules which, on the base of symmetry and conservation of angular momentum, allow for the photogeneration of conduction band electrons and valence band holes with a definite spin orientation. In time resolved photoluminescence (TRPL) experiments, the samples are excited by 1.5 ps pulses generated by a mode-locked Titanium:Sapphire laser and the photoluminescence (PL) dynamics is then dispersed by a monochromator and detected by a Hamamatsu S1 photocathode streak camera with an overall resolution of 8 ps.[36] In the case of zinc-blende QWs of the GaAs type, a circularly polarized (e.g. σ^+) optical excitation is typically set in quasi-resonant condition with the lowest confined subband pairs. The fundamental photoluminescence transition intensities co-polarized (I^+) and counter-polarized (I^-) with the excitation laser are then recorded and the emission circular polarization degree extracted according to $P_c = (I^+ - I^-)/(I^+ + I^-)$. Symmetrically, the same optical selection rules allow for the determination of the spin relaxation time τ_s by monitoring the P_c dynamics. In case of fast hole spin relaxation, as it is mostly the case in zinc-blende QWs,[3] P_c probes directly the QW electron spin dynamics. The electron lifetime τ_e can instead be extracted from the dynamics of the total PL intensity $I^+ + I^-$.

In time resolved Kerr rotation experiments, the samples are typically excited near normal incidence with degenerate pump and a weaker delayed probe pulses. The pump and probe beams are respectively circularly and linearly polarized, and the electron spin density lifetime T_s (being $1/T_s = 1/\tau_s + 1/\tau_e$) is extracted from the

analysis of the probe beam polarization rotation. Contrary to TRPL experiments, Kerr rotation does not allow for the independent determination of τ_s and τ_e but its superior sensitivity makes it a valuable tool for weak polarisation variations.
The spin diffusion length D_s has been measured by Transient-Spin-Grating (TSG) experiments. This technique consists in a variant of the Kerr set-up in which the pump beam is focused on a phase mask with a period d splitting the beam into its first diffraction orders. The interference between the two pump beams with orthogonal linear polarization yields an excitation spot of constant intensity but with spatially modulated circular polarization. This interference pattern will generate a periodic spin density, called the spin grating. A delayed probe beam with linear polarization is incident on the spin grating and the corresponding diffracted beam is detected as a function of the time delay between the pump and the probe pulses. A heterodyne detection technique through the mixing of the diffracted signal and a reference laser beam split from the pump allows to improve the sensitivity of the experiment.[37,38] The monitoring of the exponentially decaying diffracted probe signal yields the spin diffusion coefficient D_s through the spin-grating decay rate Γ and wavector $\boldsymbol{q}$ through $\Gamma = D_s q^2 + 1/T_s$. The measurement of Γ for different grating wavectors $\boldsymbol{q}$ allows to discriminate between the effects of spin diffusion and spin relaxation, yielding both D_s and T_s.
The investigated samples have a common underlying design consisting of GaAs / AlGaAs multiple quantum wells (MQWs) embedded in the intrinsic part of a diode structure grown on (111)-oriented semi-insulating GaAs substrates using molecular beam epitaxy. The doping profile is n-type on the top and p-type on the bottom (i.e. close to the substrate) for a n-i-p structure and the opposite for a p-i-n one. The samples differ in the size of the QWs and substrate orientation to highlight distinct features of the electron spin control. Their characteristics are summarized in table 2.
The low temperature electron spin properties have been investigated in sample I, a n-i-p structure grown on (111)B substrate misoriented 3° towards$< 111 >$A [39] consisting of L=15 nm wells. This slight misorientation has no sizeable effect on the spin decay but allows to get better structural quality of the samples. The results are compared to those obtained in wider (L=25 nm n-i-p, sample II) and narrower (L=7.5 nm n-i-p, sample III) (111)B MQWs structures. Sample I is also used for the electrical control of the spin diffusion length and as a test ground to validate the theoretical model. Finally, the role of the non-equivalent substrate vertical orientations and the influence of the temperature on the electron spin control have been evidenced by comparing the results of sample I with the data for sample IV and V, respectively consisting of L=12 nm n-i-p (111)A and L=15 nm p-i-n (111)B MQW structures.

Table 2. Characteristics of the investigated samples. L indicates the well width, while A and B characterize the gallium or arsenic rich surface respectively. All samples contain 20 quantum wells. See references for more details. Data and sample characteristics for sample II are taken from the works performed by the group of P.V. Santos.[23,40]

Sample	Substrate	type	L (nm)
I[22]	B	*n-i-p*	15
II[23]	B	*n-i-p*	25
III[22]	B	*n-i-p*	7.5
IV[25]	A	*n-i-p*	12
V[34]	B	*p-i-n*	15

3. Spin dynamics engineering in (111) zinc-blende quantum wells

3.1. *Electrical control of the electron spin*

3.1.1. *Spin relaxation and anisotropy*

Representative measurements of electron spin relaxation dynamics at low temperature in sample I are reported in figure 3, showing the PL circular polarization dynamics for different applied electric fields. The laser excitation energy is here set to 1.56 eV, exciting the electron-heavy hole transition only. The QW spin dynamics in these quasi-resonant excitation conditions corresponds to the electron spin[1,41] since (i) the hole spin relaxation time is very fast (of the order of $\tau_h \approx$ 1ps) and (ii) the exciton spin relaxation time due to the electron-hole exchange interaction is inefficient due to the short τ_h.[7,41,42] Strikingly, a two order of magnitude increase of the electron spin relaxation time is observed when the magnitude of the electric field amplitude increases under a reverse bias. This is what was predicted from the electric field induced Rashba contribution when it progressively compensates the Dresselhaus terms.[27] The electron spin relaxation times along the z-axis are displayed in figure 3b), circles. They rapidly increase to values greater than 10 ns, becoming much longer than the electron lifetime as the electric field is varied from E=-20 kV to E=-55 kV (inset figure 3b). In this quantum wells, compensation of BIA and SIA contributions is obtained for $E \approx$-55 kV/cm. Due to carriers leakage, it has not been possible here to apply larger negative biases to fully explore regions beyond the peak values where the Dyakonov-Perel mechanism should be switch on again due to the overcompensation by the Rashba contribution. As expected from the well width approximate dependence of $\langle k_z^2 \rangle \propto 1/L^2$, weaker compensation fields $E_c \approx$15 kV/cm are reported for sample II,[23] figure 4b), considerably extending the tuning range possibilities, whereas no tuning has been observed for the 7.5 nm wide

well sample III before diode reverse current breakdown is reached[22] as displayed in the inset of figure 3a). It is important to notice that the origin of the field dependence of the spin polarization decay is not due to a reduction of the exciton exchange interaction as reported by Vinattieri *et al.*[43] As a matter of fact, the opposite behaviour, i.e. a decrease of the electron spin relaxation time under a reverse bias is observed in sample V ruling out any possible excitonic contributions. As it will be detailed in section 3.3, due to symmetry reasons, in (111)B *p-i-n* structure, the Dresselhaus and Rashba contributions add up, instead to substract, under a negative bias.
An obstacle to the detection of low noise signals at high applied negative fields is represented by the the quantum confined Stark effect (QCSE), which is proportionally stronger in larger well. The decrease of the spatial overlap between the electron and hole wave functions substantially reduces the radiative recombination rate leading to both weaker and longer photoluminescence dynamics which can be incompatible with the technique adapted for lower applied biases. The original technique proposed by Hernández-Mínguez *et al.*[23,40] overcomes this limitation by combining pulsed illumination and biasing with standard TRPL allowing the determination of the electron spin relaxation time over a wide range of electric fields. The working principle is the following: Laser and bias pulses with the same repetition frequency are here synchronized with each other. The QCSE strongly reduces the recombination of the photoexcited carriers under bias. However, the spin relaxation time at high biases is reconstructed from successive bias pulses of different durations by measuring the PL polarisation degree at the end of voltage pulses when carriers can now rapidly recombine, as reported in figure 4.

In the experiments described so far, the average spin direction is always oriented along the z growth direction as the propagation direction of the excitation laser beam normal to the sample surface photogenerates electron spins only parallel to z. The key feature of (111)-oriented QWs is the possibility of a simultaneous electrical control of all the spin components. In order to demonstrate it, spin polarization along the plane has to be generated. This is achieved by the standard circular polarized optical excitation normal to the plane and by the application of a magnetic field $\boldsymbol{B}_\perp$ perpendicular to the growth axis (namely the Voigt geometry). The electron spin experiences now both an in-plane and out-of-plane relaxation as it precesses around the external magnetic field, as shown as inset of figure 5a). The detected PL circular polarisation degree P_c will consequently oscillate in time reflecting the oscillating orientation of the electron spins. The in-plane spin relaxation time $\tau_\perp^s$ can now be calculated from the decay time τ_{env}^s of the P_c beats envelope according to $1/\tau_{env}^s = (1/\tau_z^s + 1/\tau_\perp^s)/2$, provided that $\left[(\tau_\perp^s)^{-1} - (\tau_z^s)^{-1}\right]/2\omega_L \ll 1$,[44] ω_L being the angular frequency of the Larmor precession. An illustration of such measurements in sample I is reported in figure 5 for selected applied electric fields (red curves) jointly with the corresponding data

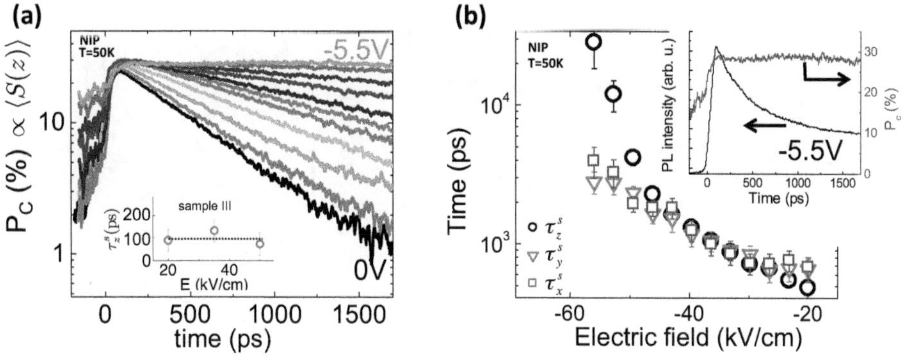

Fig. 3. (a) Electron spin dynamics at $T = 50$ K for different reverse voltages for sample I. Inset: The spin relaxation time for sample III. (b) Dependence of the electron x, y, and z spin relaxation times as a function of the electric field. Inset: PL and P_c dynamics at -5.5 V bias. Both data refer to sample I. Adapted from.[22]

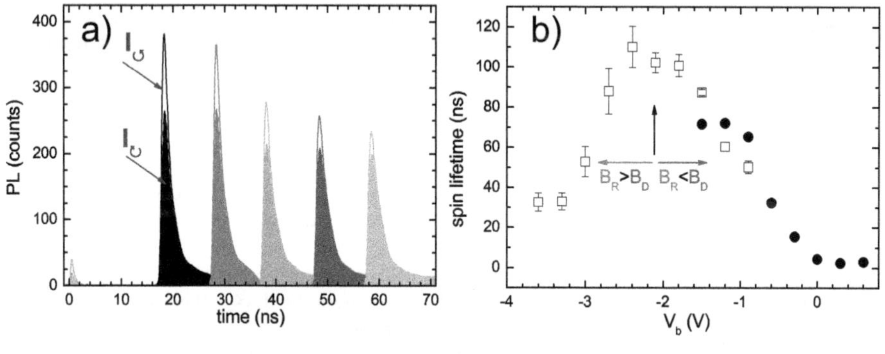

Fig. 4. Sample II. a) Co- and counter-polarized PL traces detected just after the application of a bias pulse with V_b =-2.1 V and durations of 20 ns (black), 30 ns (red), 40 ns (green), 50 ns (blue), and 60 ns (cyan). b) Spin lifetimes as a function of the bias V_b obtained from the measurements with standard Time-resolved photoluminescence (black dots) and by the pulsed bias technique (open squares). Adapted from,[40] licensed under CC-BY.[45]

for $\boldsymbol{B}_{\perp}$=0, black curves. The circular polarization decay time in the presence of the external transverse magnetic field is as well modified by the applied electric field, demonstrating that the DP spin relaxation time in (111) quantum wells can not only be controlled electrically for the S_z electron spin component but also for S_x and S_y. However, the actual electric field dependence of $\tau_{\perp}^s$ is rather different compared to τ_z^s. At V=0 (i.e. when the Rashba term is solely due to the junction built-in electric field and the Dresselhaus term dominates), the magnetic field induced oscillations of the P_c dynamics are systematically above the corresponding values for $\boldsymbol{B} = 0$ yelding $\tau_z^s < \tau_{\perp}^s$, consistently with the usual spin relaxation anisotropy predicted by Dyakonov and Kachorovsky[5] for zinc-blende QWs (i.e. $\tau_{\perp}^s = 2\tau_z^s$). By

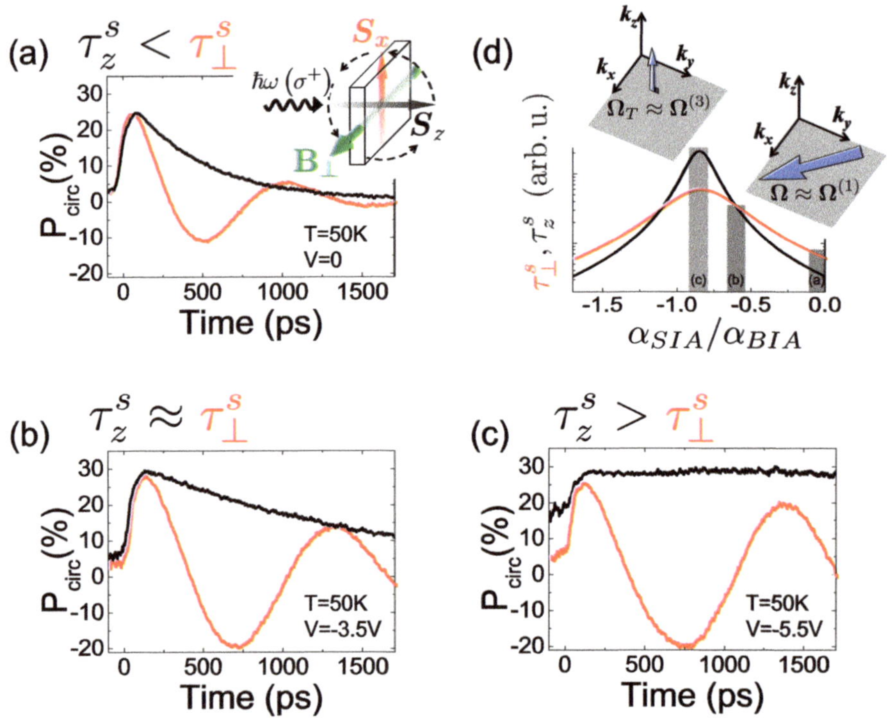

Fig. 5. Comparison between the longitudinal and transverse spin relaxation times τ_z^s,$\tau_\perp^s$ measured through the PL circular polarization dynamics P_c with and without an external magnetic field. (a) $\tau_z^s < \tau_\perp^s$; (b)$\tau_z^s \approx \tau_\perp^s$; (c)$\tau_z^s > \tau_\perp^s$. The inset in (a) shows a schematic representation of the technique used to measure the in plane spin relaxation time $\tau_\perp$: $\boldsymbol{S_z}$ is the photogenerated spin which precesses under the application of an external magnetic field $\boldsymbol{B}_\perp$ generating a periodic in-plane spin polarization ($\boldsymbol{S}_\perp$). (d) The dependence of τ_z^s and$\tau_\perp^s$ as a function of the ratio between the Rashba (α_{SIA}) and Dresselhaus (α_{BIA}) linear contributions to the conduction band splitting. The inset presents a schematic representation of the effective magnetic field precession vector $\boldsymbol{\Omega}$ in the two opposite situations achievable with the application of the external bias: (i) $\alpha_{SIA} \ll \alpha_{BIA}$ or $\alpha_{SIA} \gg \alpha_{BIA}$ and (ii) $\alpha_{SIA} \approx -\alpha_{BIA}$. Adapted from reference [46] figure 1, licensed under CC-BY 3.0.[45]

increasing the external bias, we observe that the anisotropy of the spin relaxation time can be first canceled, $\tau_z^s = \tau_\perp^s$, figure 5(b) and eventually inverted: $\tau_z^s > \tau_\perp^s$, figure 5(c). The origin of this behaviour can be understood as a consequence of the role of the cubic Dresselhaus terms. Equation 5 shows that for a given $\vec{k}$, compensation of the Rashba and Dresselhaus terms can only occur for the QW in-plane components whereas the z component is unaffected by the electric field. As the in-plane precession vector is progressively brought to compensation, electron spins oriented along the z axis will experience longer spin relaxation times compared to the ones oriented in the plane since the latter will precess around a nonzero vector parallel to the growth direction, as depicted in inset of figure 5(d). The black and red curves in figure 5(d) schematically reproduce the spin relaxation times as a function of the relative weight of the Rashba to Dresselhaus linear contributions to the spin orbit Hamiltonian α_{SIA} and α_{BIA} respectively. The three different regimes to which the spin relaxation anisotropy has been adjusted in the experimental data are here shaded. For $\alpha_{SIA} \ll \alpha_{BIA}$ or $\alpha_{SIA} \gg \alpha_{BIA}$ the precession vector is dominated by the linear terms $\mathbf{\Omega} \approx \mathbf{\Omega}^{(1)} = (2/\hbar)\left[(2\gamma_{2D}/\sqrt{3})\langle k_z^2\rangle + r_{41}E\right](k_y, -k_x, 0)$ lying in the QW plane. The situation is inverted as compensation of the in-plane precession vector is achieved: now $\alpha_{SIA} \approx \alpha_{BIA}$ and $\mathbf{\Omega} \approx \mathbf{\Omega}^{(3)}$, being $\mathbf{\Omega}^{(3)} = (2\gamma/\hbar\sqrt{6})\left(3k_x^2 - k_y^2\right)(0, 0, k_y)$ the cubic Dresselhaus term which is now essentially oriented along the growth direction and characterized by terms cubic in $\vec{k}$ unaffected by the electric field.

It is worth noting that the tuning or suppression of the Dyakonov-Perel electron spin relaxation should be possible in many other III-V, II-VI zinc blende or (0001) wurtzite QWs[47] as the control principle relies only on symmetry properties. The possibility of tuning the spin relaxation anisotropy represents also a particular interesting possibility in view of the coherent manipulation of the electron spin.

3.1.2. *Spin diffusion length*

An example of a direct application of the full control of the electron spin is the electrical manipulation of the electron spin diffusion length L_s, which represents the average length over which the memory of the initial spin orientation is retained by the moving electron. In semiconductors, L_s is jointly influenced by the specific transport properties of the device and by the spin density relaxation time. The peculiar electron spin control possibilities existing for (111) zinc blende QWs offer a straightforward mean of manipulating the spin diffusion length. Being simply related to the spin density lifetime T_s and spin diffusion coefficient D_s via the equation $L_s = \sqrt{D_s T_s}$, the electrical control of L_s follows from the manipulation of T_s. Despite simple in concept, the experimental evidence needs the concurrence of two complementary techniques: The measurement of the spin-diffusion constant D_s by using the Transient Spin-Grating technique, and the electron spin lifetime T_s by time-resolved photoluminescence spectroscopy. We present here the results obtained for sample I at T=10 K. In the TSG experiment a spin density interference pattern of wavevector $q = 2\pi/\Lambda$ and wavelength Λ is created by cross polarized beams. A combination of measurements of the grating decay rate $\Gamma = D_s q^2 + 1/T_s$ for different q and biases yields a voltage constant diffusion coefficient $D_s \approx$20 cm^2 s^{-1} as reported in figure 6 a) and b). The accurate determination of T_s by TRPL finally gives the field dependent spin diffusion length through $L_s = \sqrt{D_s T_s}$. Figure 6 c) presents the resulting modification of the diffusion length by application of an external electric field in sample I at T= 10 K. It increases by more than 200% reaching values close to L_s=1.8 μm, limited by the electron lifetime τ_e. To overcome this constraint, n-doped structures could be employed. The electron lifetime being no more an issue, larger manipulation ranges of L_s with applied voltage should be obtained as $T_s \approx \tau_s$.

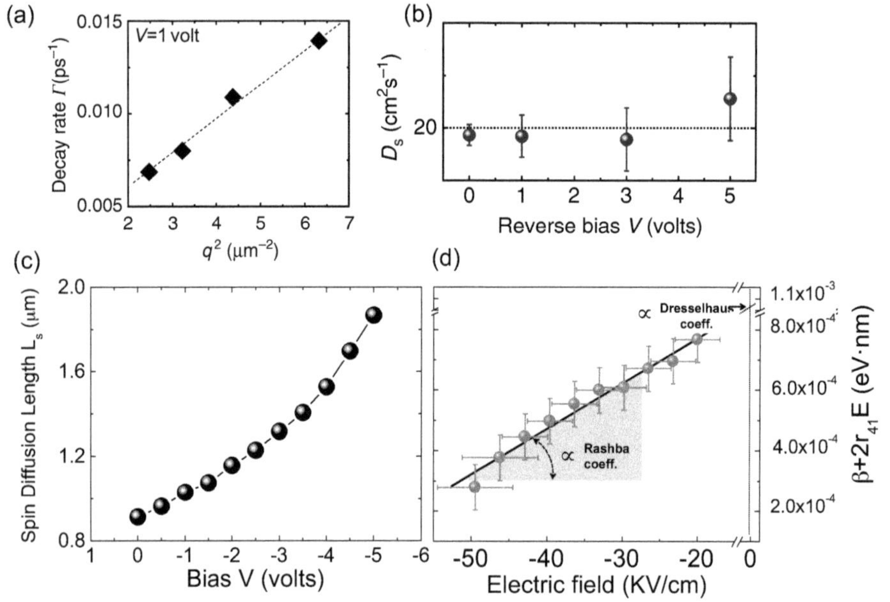

Fig. 6. Sample I. a) The transient spin grating decay rate Γ for a reverse bias V=1 V. b) Spin-diffusion coefficient D_s as a function of the bias voltage V. c) Variation of the spin-diffusion length $L_s = \sqrt{D_s T_s}$ as a function of the reverse bias V applied to the sample. d) Electric field variation of $\beta + 2r_{41}E$. The slope of the linear regression (black line) and its intercept with the ordinate axis are respectively proportional to the Rashba and Dresselhaus coefficients. Adapted from.[34]

3.2. *Determination of the Dresselhaus and Rashba coefficients*

The Dresselhaus and Rashba effective magnetic fields are the key parameters determining the spin relaxation in semiconductors and the previous section has shown that their independent manipulation is crucial to the effective control of the electron spin. The knowledge of their value, still rather scattered in literature, should facilitate the design of novel structures as well as validate theoretical calculations. Equations 4 (or 8 as presented later on) show that once the compensation field is measured, through time resolved photoluminescence spectroscopy for instance, only the Dresselhaus to Rashba coefficients ratio can be determined. An additional measurement technique is necessary for an independent determination. As an illustration, if we consider that in sample I E=-55 kV is the compensation field owing to the very long spin relaxation time measured, equation 4 yields $r_{41}E/\gamma = 5 \times 10^6$ m^{-2}. At this point, the literature value or an independent measurement of either r_{41} or γ is needed to determine the value of the other term. However, despite the diverse experimental and theoretical investigations, the actual values of r_{41} or γ_{2D} are still rather dispersed, ranging from $|\ \gamma_{2D}\ |$= 3 to 27 eV $\mathrm{\AA}^3$[48] and r_{41}= 5 to 10 $e\mathrm{\AA}^2$.[2]

The TSG technique offers a rather neat possibility of independently determine both r_{41} and γ. Recalling that the dominant Dyakonv-Perel electron spin-relaxation time in quantum wells is given by

$$1/\tau_{\perp}^{s} = \langle \boldsymbol{\Omega}_T^2 \rangle \tau_p^*, \tag{6}$$

where $\boldsymbol{\Omega}_T = \boldsymbol{\Omega}_{BIA} + \boldsymbol{\Omega}_{SIA} = \left(\frac{\beta}{\hbar} + \frac{2r_{41}E}{\hbar}\right)(k_y, -k_x, 0)$, calculating the mean square precession vector $\langle k_x^2 \rangle = \langle k_y^2 \rangle = k_\theta^2/2$, where $k_\theta^2 = 2m^* k_B T/\hbar^2$ is the thermal de Broglie wavevector on simple equipartition grounds, equation 6 yields (away from resonance):

$$\beta + 2r_{41}E = \hbar^2/\sqrt{2m^* k_B T \tau_s \tau_p^*}. \tag{7}$$

Equation 7 states a linear dependence of the right hand side on the applied electric field: the intercept is proportional to the Dresselhaus coefficient, the slope to the Rashba one. A direct determination of both terms is thus possible by measuring the electric field variation of the electron spin and momentum relaxation times τ_s and τ_p^*. Figure 6 d) presents the experimental determination of β and r_{41} exploiting equation 7 in spin-diffusion experiments on sample I according to the following reasoning. Considering that in undoped AlGaAs/GaAs wells and under low excitation power, i.e. in non-degenerate conditions and when spin-drag effects are absent,[49] the electron spin-diffusion coefficient is equal to the electron diffusion one (i.e. D_e=D_s[50]), τ_p^* can be readily extracted from the Einstein relation $\tau_p^* = m^*\mu/e = (m^* D_e/k_B T)$, where $m^*$and e are the electron effective mass and charge. The Dresselhaus and Rashba coefficients extracted by a linear regression from the data in figure 6d) are $\gamma_{2D} = 18 \pm 2$ eV $\AA^3$ and $r_{41} = 7 \pm 1$ $e\AA^2$. Although reported values are similar to those presented here, γ_{2D} ≈11–16 eV $\AA^3$ [48,51,52] and r_{41}=5.2 e $\AA^2$,[2] as stated before, no consensus exists yet in literature on their actual values.

3.3. *Substrate orientation dependence*

The results presented in the previous sections have demonstrated that in (111)B-oriented zinc-blende QWs the electron spin relaxation time can be dramatically increased by the application of a negative bias. However, the definition of *negative* bias is here ambiguous, referring simply to the convention applied to diode doping regions and not the underlying crystal structure. Indeed, differently form the (001) case, the zinc-blende crystal possesses two non-equivalent substrate orientations: the (111)A which presents a gallium reach surface and the (111)B or arsenic rich surface (figure 7a). As the spin-orbit effective magnetic field emerges from the breaking of inversion symmetry, it is reasonable to expect a non-equivalence of the electric manipulation on the different A and B substrate orientations. If it is agreed to denominate a diode structure from its doping region starting from the *topmost* surface (i.e. opposite to the substrate), then the preceeding results demonstrate that for (111)B quantum wells a *negative* bias has to be applied in order to *increase* the spin relaxation time (up to the compensation field) in an *n-i-p* structure: the

required electric field has to point opposite the growth direction. The results reported in figure 8 clarify the role played by the growth direction. Figure 8 a) and b) show the time evolution of the Kerr rotation signal for different applied voltages in sample I and sample IV respectively at T=50 K. An opposite behaviour is observed for the Kerr rotation dynamics as the bias is modified. For sample I a clear increase of the electron spin lifetime is measured when the reverse bias varies from +0.4 V to -2.6V, in accordance to the experimental evidence presented in the previous sections. For sample IV, the application of a reverse bias yields the opposite behaviour compared to the (111)B-grown sample I: a decrease of the electron spin lifetime is measured as the voltage varies from +3V to -1V. The results reported in figure 8 c) for sample V complete the scenario showing that for a (111)B *p-i-n* structure a negative bias produces an opposite behaviour compared to sample I.
To give an overview of the observed phenomena, it is here useful to fix the crystal reference frame by setting the z axis along the direction [111] for any sample orientation. For the (111)B *n-i-p* samples, the increase of the electron spin lifetimes requires the application of a reverse bias, i.e., the electrons in the quantum wells must feel an electric field E *opposite* to the growth direction or again, E must be oriented in the same direction as the z axis. Referring to the same coordinate frame for (111)A grown QWs, the observation of an extended spin lifetime requires the application of an external electric field still along the z direction, but corresponding now to an orientation *along* the growth direction. These requirements are a simple consequence of the material dependent relative sign of the Rashba and Dres-

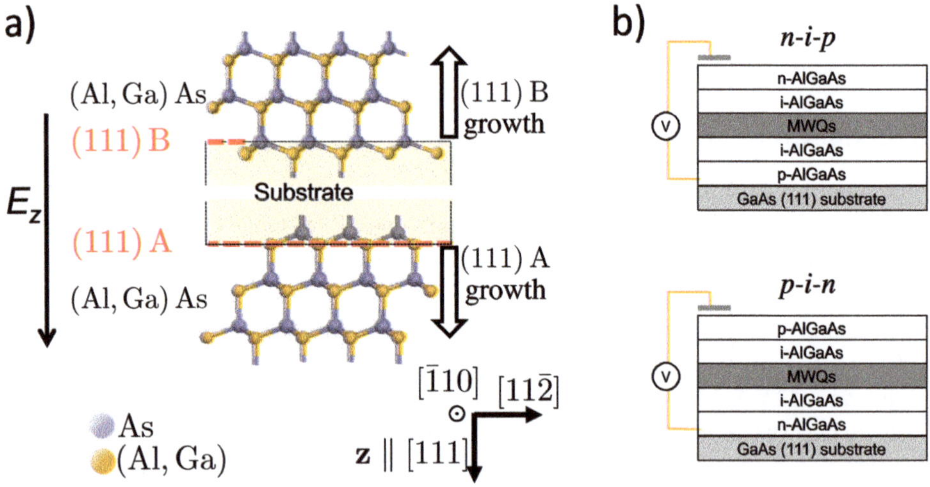

Fig. 7. a) Schematic representation of the crystal structure for samples grown on (111)A and (111)B substrates. The applied electric field E_z is represented here with the direction required to observe an increase of the electron spin relaxation time. b) Illustration of the convention used to identify the structures according to their doping sequence. Reproduced from [Applied Physics Letters 101, 032104 (2012); doi: 10.1063/1.4737007], with the permission of AIP Publishing.[25]

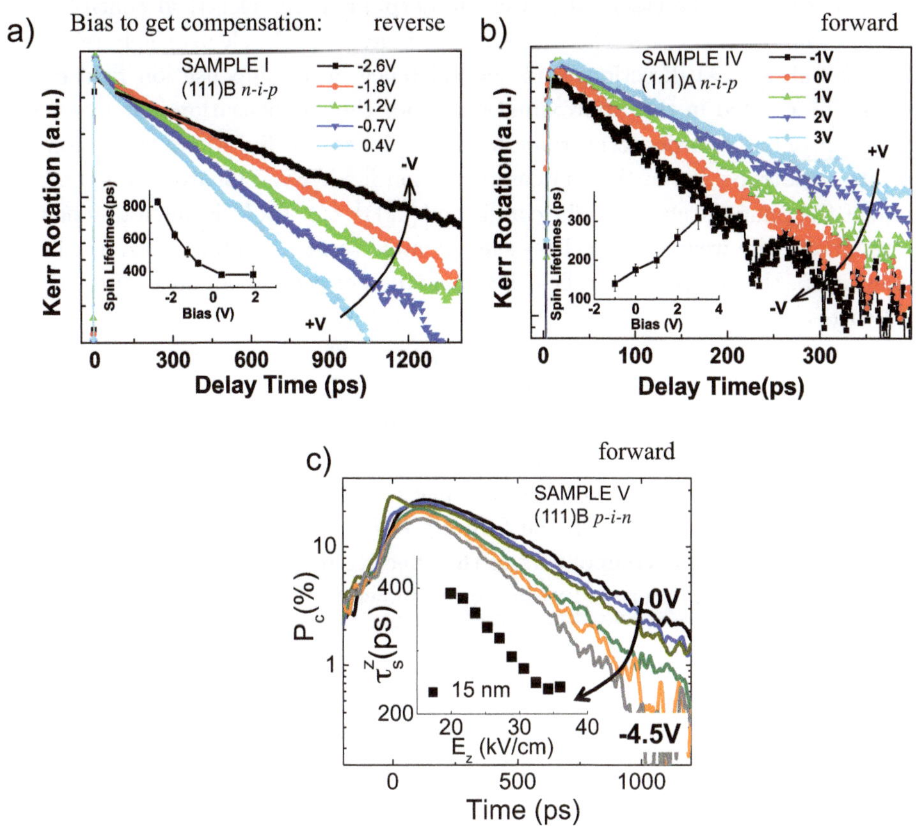

Fig. 8. a) and b) Kerr rotation dynamics at T=50 K for sample I and IV respectively under different applied voltages. Insets in (a) and (b): the extracted electron spin lifetimes. c) Circular polarization dynamics for different reverse bias voltages applied to sample V and the corresponding spin relaxation times (inset). a) and b) Reproduced from [Applied Physics Letters 101, 032104 (2012); doi: 10.1063/1.4737007], with the permission of AIP Publishing.[25] c) adapted from.[22]

selhaus terms. From a crystallographic standpoint, the experiments displayed in figure 8 correspond to the same physical situation since the positive and negative voltages are defined with respect to the standard diode convention and not to the fixed crystallographic orientation. These examples show that, in addition to the control possibilities of the spin-orbit magnetic field via the quantum confinement and applied electric field amplitude, the growth direction represents another degree of freedom which can provide an efficient engineering tool for spin relaxation times in quantum wells.

3.4. *Temperature dependence*

The previous sections have shown that as long as the condition $(\langle k_z^2 \rangle)^{1/2} \gg k_x, k_y$ holds, the linear terms in the expression of the Dresselhaus spin-orbit field are sufficient to grasp the essential physics underpinning the electrical control of the electron spin in (111) QWs: the simultaneous increase of the electron spin relaxation times for all spatial directions and the existence of a compensation field can be readily interpreted in this approximation. Once the compensation is reached, only the effects of higher order terms remain. Examples are the finite spin relaxation time and its anisotropy at the (linear) compensation field. Even far from compensation field, as the temperature rises, the condition $(\langle k_z^2 \rangle)^{1/2} \gg k_x, k_y$ is inevitably no more satisfied and the weight of the Dresselhaus cubic terms is enhanced. According to equations 5, the compensation field becomes temperature dependent (through $\mathbf{k}_\perp$) and writes:

$$E_c = -\frac{\gamma_{2D}}{\sqrt{3}r_{41}}\left(2\langle k_z^2\rangle - \frac{k_\perp^2}{2}\right). \tag{8}$$

When the temperature increases, electrons spread over a larger wavevectors range and at microscopic level a given external electric field cannot yield a perfect compensation of the Dresselhaus and Rashba effective field for all the populated electron states. Recasting equation 8 in the form $k_c^2 \equiv k_\perp^2 = 4\langle k_z^2\rangle + 2r_{41}\sqrt{3}E_c/\gamma_{2D}$, it is found that the in-plane components of the effective magnetic field are strictly zero only along the circumference of radius k_c as shown in figure 9. At higher tempera-

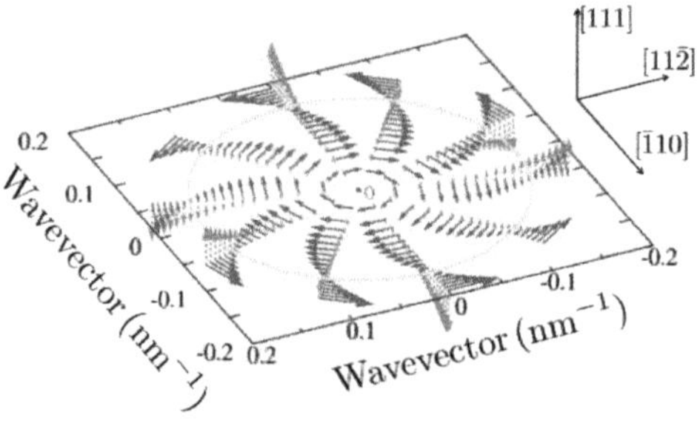

Fig. 9. Schematic of the effective spin-orbit magnetic field components under a fixed value of the electric field showing that strict compensation of the in-plane components (blue arrows) can be achieved only along a circle of radius k_c represented by the yellow line. The out of plane components (red arrows) are uninfluenced by the applied electric field. Adapted from.[53]

tures or higher carrier densities, the compensation effect must progressively weaken. The impact of the temperature on the electrical control of the spin relaxation is as well expected to produce a modified electric field dependence of the spin relaxation time and anisotropy. Still, the actual scenario is more complicated as spin relaxation is determined by the joint effects of inhomogeneous broadening, scattering processes[53] and electron density. Their combined effect is expected to play a more relevant role as the temperature increases. Vurgaftman and Meyer[35] have studied the feasibility of practical compensating devices and showed that temperature strongly affects the gate-voltage dependence of the inhomogeneous broadening and consequently the electron spin relaxation time. Their calculation predicts that the Dyakonov-Perel spin-relaxation rate can be suppressed by a factor of 2000 at T=10 K but only by one order of magnitude at room temperature in InGaAs/AlInAs

QWs. Sun *et al.* have analysed the problem in the framework of the microscopic many-body kinetic spin Bloch equations in (111) n-doped GaAs QWs revealing the complexity of the interplay between the electron density, temperature dependence and impurities concentration.[53]
The experimental evidence of the influence of the temperature dependence on the electron spin tuneability range is summarised in figure 10. Figure 10 a) shows the effect of the temperature on the tuning range in sample II. In this *large* wells τ_z^s suffers already a fivefold decrease of its peak value as the temperature increases from T=33 K to 60 K. The experimental compensation electric field, determined by the field amplitude at the peak of the spin relaxation time, is also found to be affected (inset of figure 10) confirming that, already in this temperature range and these wide wells, the contribution of the Dresselhaus cubic terms cannot be neglected. A more favourable situation can be obtained in narrower wells where the

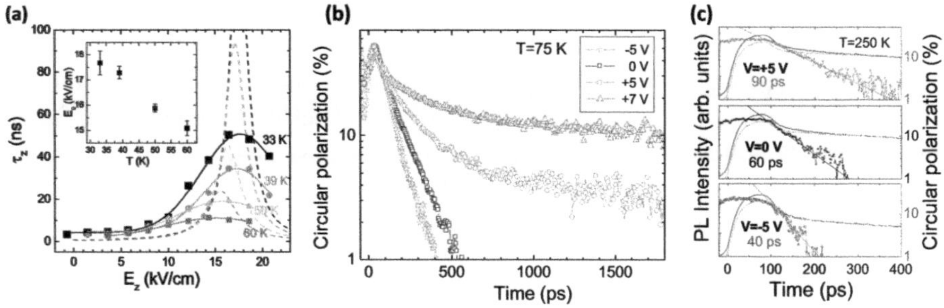

Fig. 10. Effect of the temperature on the eletric field control of the spin relaxation. a) Circular polarization dynamics for different applied voltages at T=75 K from sample II. Adapted from[40] licensed under the CC-BY.[45] b) Time evolution of the circular polarization at T=75 K and c) T=250 K for sample IV. Figures b) and c) Reproduced from [Applied Physics Letters 102, 242408 (2013); doi: 10.1063/1.4811555], with the permission of AIP Publishing.[33]

the conditions $(\langle k_z^2 \rangle)^{1/2} \gg k_x, k_y$ can hold for higher temperatures. Figures 10 b) and c) report the electrical tuning of the electron spin relaxation time along the z axis in sample IV. A clear thirty-fold tuning of the electron spin relaxation time is still observed at T=75 K is these narrow QWs. Remarkably, an electrical control of the spin relaxation time by more than a factor two is even observed at T= 250 K, the highest temperature reported so far in literature. Specifically designed *n-i-p* structures allowing the application of very large electric fields in even narrower well structures without leakage currents[54] should move the proposal of the electrical control of the electron spin relaxation in (111) QWs at room temperature closer to realisation.

3.5. *Model*

The previous sections have highlighted the most prominent features of the full electrical control of the electron spin in <111>-oriented GaAs/AlGaAs quantum wells.

The interpretations of the experimental results have been directly drawn from the considerations on the peculiar form of the total spin-orbit effective magnetic field. These observations seize the essential nature of the physical phenomena yet they remain qualitative. In order to make quantitative estimations of the expected spin relaxation time the total spin-orbit effective field vector square average $\langle \mathbf{\Omega}_{\perp}^2 \rangle$ has to be evaluated for the electron population. This evaluation can be performed at different levels of complexity successively adding physical phenomena giving an increasingly more precise description of the experiments. In this section a brief review of a selected simulation technique will be presented.

3.5.1. *Simple approach*

In a simple approach, the longitudinal and transverse spin relaxation times can be evaluated from the thermal average of the square of the spin-orbit effective precession vector in a infinite quantum well. In undoped non-degenerate semiconductors, under thermal equilibrium condition, the electron energy distribution in the xOy plane can be approximated by the Maxell-Bolzmann function. By performing the angular and thermal averages of the square of the effective magnetic field for each components, the following spin relaxation times are obtained:

$$\begin{aligned} \frac{1}{\tau_x^s} &= \left\langle \overline{\Omega_y^2 + \Omega_z^2} \right\rangle \tau_p^* = \frac{2}{\hbar^2} k_\theta^2 [(\alpha_T - \frac{\gamma_{2D} k_\theta^2}{\sqrt{3}})^2 + \frac{7}{6}(\gamma_{2D}^2 k_\theta^4] \tau_p^* = \frac{1}{\tau_y^s} \\ \frac{1}{\tau_z^s} &= \left\langle \overline{\Omega_x^2 + \Omega_y^2} \right\rangle \tau_p^* = \frac{4}{\hbar^2} k_\theta^2 [(\alpha_T - \frac{\gamma_{2D} k_\theta^2}{\sqrt{3}})^2 + \frac{1}{6}\gamma_{2D}^2 k_\theta^4] \tau_p^* \end{aligned} \tag{9}$$

where $\alpha_T = \frac{2\gamma_{2D}}{\sqrt{3}} \left\langle k_z^2 \right\rangle + r_{41} E$. Finally, the ratio between longitudinal and transverse spin relaxation times yields a relation independent on the momentum scattering time τ_p^*:

$$\frac{\tau_z^s}{\tau_x^s} = \frac{1}{2}[1 + \frac{\gamma_{2D}^2 k_\theta^4}{(\alpha_T - \frac{\gamma_{2D} k_\theta^2}{\sqrt{3}})^2 + \frac{1}{6}\gamma_{2D}^2 k_\theta^4}]. \tag{10}$$

This expression yields τ_z^s/τ_x^s=7/2 when $\alpha_T = \frac{\gamma_{2D} k_\theta^2}{\sqrt{3}}$, i.e. for a compensation electric field $E_c = \frac{2\gamma_{2D}}{\sqrt{3} r_{41}} \left(k_\theta^2/2 - \langle k_z^2 \rangle \right)$, and shows that $\tau_z^s/\tau_x^s \approx 1/2$ as $|\ E \ | \gg \frac{\gamma_{2D}}{\sqrt{3} r_{41}} \langle k_z^2 \rangle$. Despite its simplicity, the model takes into account the cubic terms and predicts an isotropic point where the longitudinal and transverse spin relaxation times are equal.

3.5.2. *Density matrix approach*

In this model the electron spin relaxation is described in a more accurate way. The Dyakonov-Perel mechanism is modelled using the spin density matrix formalism explicitly including the contribution of the the collision processes as originally developed by Averkiev and Golub for < 001 >-oriented QWs.[13,15] A few simplifying assumptions are considered here: only the first confined level in the QW is populated, the energy dispersion is isotropic and the collision between particles are dominated by elastic scattering processes. The dynamics of the spin-density matrix $\boldsymbol{\rho}$ for the spin-orbit Hamiltonian $H_{SO} = H_D + H_R$ writes:

$$\frac{\partial \rho(k)}{\partial t} + \frac{i}{\hbar}[H_{SO}(k), \rho(k)] = I_k(t) \tag{11}$$

where $H_{SO}(k) = H^{(1)} + H^{(3)}$ with

$$\begin{aligned} H^{(1)} &= \left(\frac{2\gamma_{2D}}{\sqrt{3}}\langle k_z\rangle^2 + r_{41}E\right)(k_y\sigma_x - k_x\sigma_y) = \alpha_T(k_y\sigma_x - k_x\sigma_y) \\ H^{(3)} &= \gamma_{2D}\left[\frac{k_\perp^2}{2\sqrt{3}}(k_x\sigma_y - k_y\sigma_x) + \frac{k_y}{\sqrt{6}}\left(3k_x^2 - k_y^2\right)\sigma_z\right] \end{aligned} \tag{12}$$

$I_k(t) = -\sum_{k'} W_{k,k'}(\rho(k) - \rho(k'))$ is the collision integral and $W_{k,k'}$ the electron scattering probability from wavevector state $\vec{k}$ to $\vec{k}'$. The spin-orbit interaction Hamiltonian H_{SO} is considered here as a small perturbation such that the spin relaxation times are much longer than the time needed for the momentum distribution to become isotropic, which is of the order of 1 ps.[55] The density matrix is thus assumed to have no angular dependence. We delineate here only the main steps needed to calculate the spin relaxation times. A detailed analysis can be found in references.[13,15,46] The solution in the two dimensional case is found by expanding H_{SO} in terms of its Fourier angular components:

$$H_{SO}(k) = \sum_{n=-\infty}^{\infty} H_n(k)e^{in\varphi}. \tag{13}$$

The collision integral is also expanded on its Fourier angular components:

$$W_{k_\perp}(\phi - \phi') = \sum_n W_n(k)e^{in(\phi-\phi')} \tag{14}$$

assuming that the scattering process, at a given kinetic energy, depends only on the polar angle difference between initial and final electron wavevector. The characteristic scattering times are thus derived:[56,57]

$$\frac{1}{\tau_l} = \mathsf{W}_0 - \mathsf{W}_l = \int \frac{d\phi}{2\pi} W_{k_\perp}(\phi)\left[1 - \cos(l\phi)\right] \tag{15}$$

Using these notations, the density matrix equations can then be conveniently reformulated. The average spin is defined as $\hat{\vec{S}} = tr\left[\rho(k)\,\vec{\sigma}\right]$. Its time evolution takes now the form:

$$\frac{dS_i}{dt} = -\Gamma_{ij}(k_\perp)S_j \tag{16}$$

where the spin relaxation tensor in a given energy shell $E(k_\perp)$ of fixed $|k_\perp|$ is:

$$\Gamma_{ij} = \frac{1}{\hbar^2}Tr\left(\sum_{n=0}^{\infty} \tau_n[H_{-n}, [H_n, \sigma_j]]\sigma_i\right). \tag{17}$$

It turns out that due to the form of the spin-orbit hamiltonian, the summation contains only n=1 and n=3 terms. The spin relaxation tensor takes the form:

$$\Gamma = \begin{bmatrix} \Gamma_{xx} & 0 & 0 \\ 0 & \Gamma_{yy} & 0 \\ 0 & 0 & \Gamma_{zz} \end{bmatrix} \tag{18}$$

where

$$\Gamma_{xx} = \Gamma_{yy} = \frac{2\tau_1}{\hbar^2}\left[\left(\alpha_T\left|k'_\perp\right| - \frac{\gamma_{2D}\left|k'_\perp\right|^3}{2\sqrt{3}}\right)^2 + \frac{\tau_3}{6\tau_1}\gamma_{2D}^2\left|k'_\perp\right|^6\right] = \Gamma_\perp \tag{19}$$

$$\Gamma_{zz} = \frac{4\tau_1}{\hbar^2}\left(\alpha_T\left|k'_\perp\right| - \frac{\gamma_{2D}\left|k'_\perp\right|^3}{2\sqrt{3}}\right)^2 . \tag{20}$$

The spin relaxation tensor is finally averaged over the electron distribution by performing the thermal average of the corresponding scattering rates in the non degenerate regime and using the relation:[27]

$$\left\langle \tau_1 k^{2n} \right\rangle_\theta = \frac{(n+\nu)!}{(\nu+1)!}\tau_p k_\theta^{2n} \tag{21}$$

where τ_p is the transport mobility time. Most scattering mechanism have a power-law dependence of τ_1 and τ_3 on the energy E, with the same characteristic exponent ν so that $\tau_1 \propto \mathrm{E}^\nu$ and $\tau_3 \propto \xi \mathrm{E}^\nu$, where ξ is a constant independent of E. The ratio $\xi = \tau_3/\tau_1$ is a constant describing the angular characteristic of the scattering mechanism.[57] After performing the calculation of the spin relaxation tensor and its thermal average, the spin relaxation rates parallel and perpendicular to the growth direction are obtained as:

$$\frac{1}{\tau_{zz}} = \frac{4\tau_p}{\hbar^2}k_\theta^2\left\{\alpha_{tot}^2 + \frac{2+\nu}{12}\gamma_{2D}^2 k_\theta^4\right\} = \frac{1}{\tau_z^s} \tag{22}$$

$$\frac{1}{\tau_{xx}} = \frac{1}{\tau_{yy}} = \frac{2\tau_p}{\hbar^2}k_\theta^2\left\{\alpha_{tot}^2 + \frac{2+\nu}{12}\left[1 + 2(3+\nu)\frac{\tau_3}{\tau_1}\right]\gamma_{2D}^2 k_\theta^4\right\} = \frac{1}{\tau_\perp^s} \tag{23}$$

with

$$\alpha_{tot}(E) = \alpha_T - \frac{2+\nu}{2\sqrt{3}}\gamma_{2D}k_\theta^2 . \tag{24}$$

It is worth noting that τ_z^s and $\tau_\perp^s$ are symmetric functions of the electric field centred nearby the new compensation field E_p defined by $\alpha_{tot}(E_p) = 0$ and that within this model exact compensation between the linear Dresselhaus and Rashba fields can be reached at any temperature in a non-degenerate electron gas. Only the cubic terms, which give rise to the k_θ^6 contribution in (22,23) can never be compensated as expected. It is useful to recast equations 22 and 23 in a Lorentzian form:

$$\tau_z^s = \frac{1}{\pi}\frac{A_{zz}\Gamma_{zz}}{(E-E_p)^2 + \Gamma_{zz}^2} \tag{25}$$

$$\tau_\perp^s = \frac{1}{\pi}\frac{A_\perp\Gamma_\perp}{(E-E_p)^2 + \Gamma_\perp^2} \tag{26}$$

with:

$$A_\perp \Gamma_\perp = 2A_{zz}\Gamma_{zz} \tag{27}$$

$$\Gamma_{zz} = \frac{\gamma_{2D}}{r_{41}}\left[\sqrt{\frac{2+\nu}{12}}k_\theta^2\right] \tag{28}$$

$$\Gamma_\perp = \left[1 + 2\,(3+\nu)\,\frac{\tau_3}{\tau_1}\right]^{1/2}\Gamma_{zz} \tag{29}$$

$$E_p = -\frac{\gamma_{2D}}{r_{41}}\left[\frac{1}{\sqrt{3}}\left(2\langle k_z^2\rangle - \frac{2+\nu}{2}k_\theta^2\right)\right] \tag{30}$$

$$A_{zz}\Gamma_{zz} = \frac{\hbar^2\pi}{4\tau_p r_{41}^2 k_\theta^2}. \tag{31}$$

From the previous equations the ratio of the spin relaxation times parallel and perpendicular to the growth direction at the compensation field $E = E_p$ is:

$$\frac{\tau_{z,p}^s}{\tau_{\perp,p}^s} = \frac{1}{2}\left[1 + 2(3+\nu)\frac{\tau_3}{\tau_1}\right] \tag{32}$$

in which the subscript p indicates the peak value. Kainz *et al.*[57] have provided the value of the characteristic exponent ν for different scattering mechanism, as listed in table 3.

Table 3. Characteristic exponent ν in the scattering time for different scattering mechanisms in a 2D electron gas, as well as the ratio $\xi = \tau_3/\tau_1$ and $\tau_{z,p}$ and $\tau_{\perp,p}$.[57]

Scattering mechanism	ν	$\frac{\tau_3}{\tau_1}$	$\frac{\tau_{z,p}}{\tau_{\perp,p}}$
Type I : acoustic phonons (deformation potential) optical phonons (deformation potential) ionized impurities (screened) neutral impurities alloy scattering interface roughness	0	1	$\frac{7}{2}$
Type II : acoustic phonon (polar ,piezoelectric) optical phonon (polar)	1	$\frac{1}{3}$	$\frac{11}{6}$
Type III : ionized impurities (weakly screened)	2	$\frac{1}{9}$	$\frac{19}{18}$

Equation (32) does not allow the independent determination of the Dresselhaus and Rashba coefficients as before. However different interesting features can be extracted for a given QW at a given temperature following these sequential steps: The spin relaxation anisotropy at compensation can be used to determine the dominant

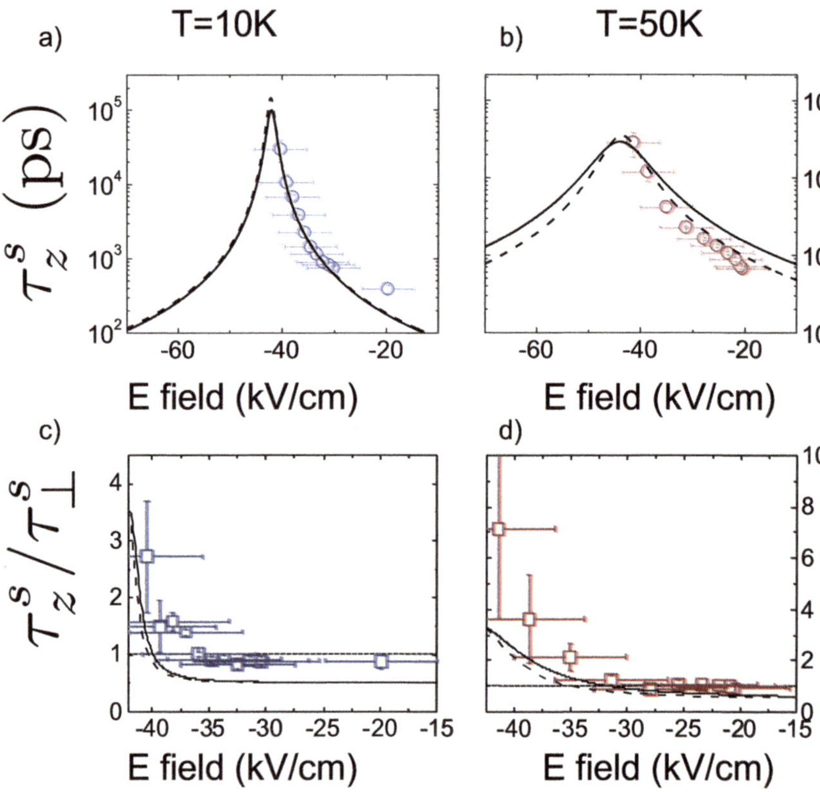

Fig. 11. Comparison between experimental and simulated electric field dependence of the spin relaxation time. a) and b) circles: spin relaxation time parallel to the growth direction τ_z^s as a function of the electric field measured at T=10 and 50 K respectively. The solid lines are fits to the data according to the model presented in the text. c) and d) squares: The measured spin relaxation time ratio $\tau_z^s/\tau_\perp^s$ as a function of the electric field for the same temperatures. The solid lines are the calculated ratios according to the model. The horizontal line is a guide to the eye to mark the isotropic spin relaxation points. In all figures the dashed lines correspond to the same fitting procedure using the compensation field E_p as an additional fitting parameter. After,[46] figure 2, licensed under CC-BY 3.0.[45]

scattering type in the structure with the aid of table 3. Once the class of the scattering processes is determined, E_p and the half width of the spin relaxation curves are solely functions of γ_{2D}/r_{41} for a given temperature. Finally, the momentum scattering time τ_p is fully determined without any mobility measurements from $\tau_{z,p}^s$ (or $\tau_{\perp,p}^s$) if an independent measurement of the Rashba (or Dresselhaus) term is available.

Figure 11 presents the results of this fitting procedure applied to the data obtained for sample I for two different temperatures. The value of the ratio $\tau_z^s/\tau_\perp^s$ at compensation increases up to a maximum greater than 11/6. This implies a dominant

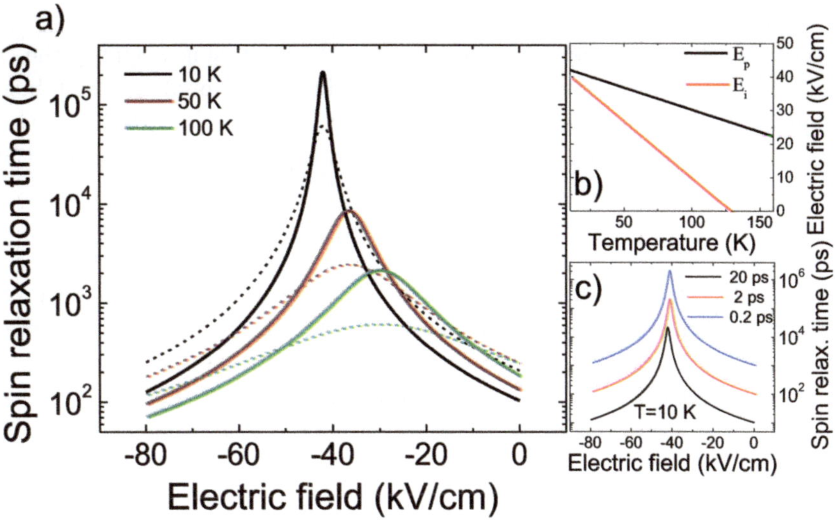

Fig. 12. (a) Calculated temperature dependence of the spin relaxation times. Solid and dashed lines represent τ_z^s and $\tau_\perp^s$ respectively. The parameters are taken from the fit at T=10 K and extrapolated for higher temperatures according to the model. Momentum scattering times are respectively τ_p= 7, 0.2 and 0.2 ps. (b) Temperature dependence of the compensation (E_p) and isotropic (E_i) fields according to equation (33) for ν=0. (c) The motional narrowing effect calculated for T=10 K. After,[46] figure 3, licensed under CC-BY 3.0.[45]

scattering mechanism of type I, thus fixing $\nu=0$ [d] and $\tau_3/\tau_1=1$, reducing the number of fitting parameters. By setting $\langle k_z^2\rangle=2.5\times 10^{16}$ m^{-2} obtained in the framework of a finite QW barrier calculation,[18] $r_{41} = 7\,e\text{Å}^2$,[34] and using τ_p and $\frac{\gamma_{2D}}{r_{41}}$ as only fitting parameters, the procedure gives $\frac{\gamma_{2D}}{r_{41}}\ |_{10K}=2.0\pm0.7$ VÅ, $\frac{\gamma_{2D}}{r_{41}}\ |_{50K}=2.4\pm0.7$ VÅ, thus yielding form sample I: γ_{2D}=14 to 16.8 eV Å^3, consistent with *spds** tight binding calculations yielding γ_{2D}=16 eV Å^3.[58] The corresponding values for the momentum scattering times are finally τ_p^{10K}=7 ps and τ_p^{50K}=195 fs. The proposed model produces a reasonable agreement with the experimental data. Nevertheless, it is clear that the measured spin relaxation time ratio is greater than the maximum possible predicted by this model and this trend is expected to increase with temperature as different processes not considered in this model gain importance. Other possible sources of discrepancy can be due to the spin relaxations mechanisms caused by local variations such as the well confinement potential, inhomogeneities of the electric field and thus of the Rashba field or again the QCSE induced variation

[d] Acoustic phonon and, still more, optical phonon scattering processes are not relevant for temperatures below 100 K since they lead to an at least one order of magnitude larger momentum scattering time τ_p. Being the quantum wells non-intentionally doped and relatively wide, impurities and alloy scattering processes are discarded as well. As a consequence, the most likely scattering process is due to interface roughness.

of the confining potential, and the misalignment of the substrate.
Figure 12 reports the calculated temperature dependences of the spin relaxation times according to the model. Consistently with the increasing weight of the Dresselhaus cubic terms, a weaker spin relaxation modulation is predicted with increasing temperature. To conclude, it is interesting to examine the predicted temperature dependences of the compensation field E_p and isotropic $E_i(\pm)$ fields, related by:

$$E_i(\pm) = E_p(T) \pm \left\{ \frac{2+\nu}{12} \left[(1 + 2\,(3+\nu))\, \frac{\tau_3}{\tau_1} - 2 \right] \right\}^{\frac{1}{2}} \frac{\gamma_{2D}}{r_{41}} k_\theta^2. \tag{33}$$

Both kinds of fields vary linearly with the temperature (neglecting the possible variation of ξ with T in our experimental range). Their respective temperature trends are traced in figure 12 b), showing a more distinct variation of E_i compared to E_c. No dependence on τ_p^* is instead anticipated for both E_i and E_p, whereas the variation of the spin relaxation time is, as expected, due to the Dyakonov-Perel motional narrowing feature (figure 12 c).

4. Perspectives and conclusions

We have shown that exploiting crystal symmetry properties allows an efficient electric field manipulation of the electron spin relation in 2D zinc-blende heterostructure, which is especially attractive as it can offer a practically viable means of integration into devices. In particular, the unique relative orientation of the Dresselhaus and Rashba fields in (111)-oriented quantum wells allows for the simultaneous control of the electron spin relaxation, its anisotropy and diffusion length for the three directions in space. As this demonstration is solely based on symmetry considerations, control of the spin relaxation should be also possible in many other III-V and II-VI zinc-blende nanostructures, as well as and in wurtzite (0001) nanostructures.[47] Although perfect spin-splitting suppression is unattainable due to the presence of cubic terms in the Dresselhaus effective field expression, a two order of magnitude increase of the electron spin relaxation time is demonstrated at cryogenic temperature. The demonstration of a large spin relaxation tuning range at room temperature is a prerequisite for "real" applications, yet the use of specifically designed narrower quantum wells diode structures with low leakage current should in principle allow the realisation of a proof-of-concept room temperature device. These demonstration naturally generates several engaging follow-ups: the transport of long-lived spins in devices by moving potentials created by surface acoustic waves,[40,59,60] or the increase of visibility of the spin transport by Pauli blocking,[61] or again the realization of devices with defined spin relaxation properties in the absence of an applied external field by exploiting the built-in piezoelectric field in strained (111) InGaAs quantum wells. As an example of this last proposal, figure 13 shows the spin relaxation time measured in otherwise identical (100) and (111)B 10 nm InGaAs/GaAs quantum wells for three different In content.[62] While in (100) QWs the the small differences in the electron spin relaxation can be associated to the variation of the quantum confined related Dresselhaus terms, in piezoelectric (111)B structures, the SIA contribution largely determines the spin relaxation times. Finally, these experiments could be extended to wurtzite based systems or again to the demonstration of the control of the hole spin relaxation as predicted by Wang *et al.*[63]

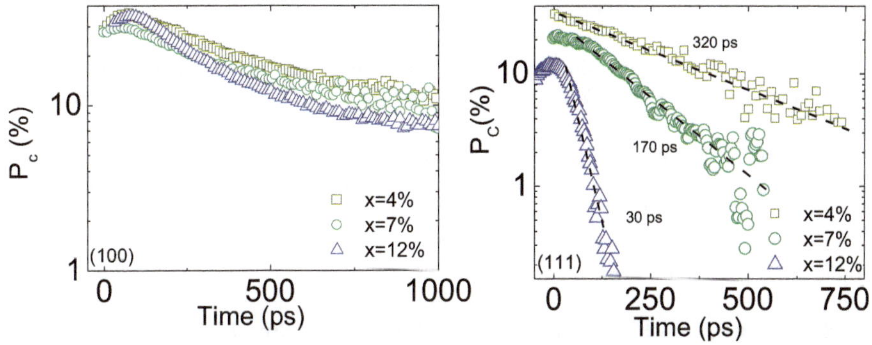

Fig. 13. Electron spin dynamics measured in (100) (left) and (111)B (right) 10 nm InGaAs/GaAs quantum wells with varying indium concentration. Reproduced from [Applied Physics Letters 108, 082103 (2016); doi: 10.1063/1.4942600], with the permission of AIP Publishing.[62]

References

1. F. Meier and B. P. Zakharchenya, *Optical Orientation.* (Elsevier Science Ltd, Amsterdam ; New York : New York, Nov. 1984). ISBN 978-0-444-86741-4.
2. R. Winkler, *Spin-orbit Coupling Effects in Two-Dimensional Electron and Hole Systems.* (Springer, Berlin ; New York, Dec. 2003), 2003 edition edition. ISBN 9783540011873.
3. M. I. Dyakonov, *Spin Physics in Semiconductors.* (Springer Science & Business Media, July 2008). ISBN 978-3-540-78820-1.
4. M. Dyakonov and V. Perel, Spin orientation of electrons associated with the interband absorption of light in semiconductors, *Soviet Physics JETP.* **33**, 1053, (1971).
5. M. Dyakonov and V. Kachorovskii, Spin relaxation of two-dimensional electrons in noncentrosymmetric semiconductors, *Soviet Physics Semiconductors.* **20**, 110, (1986).
6. R. S. Britton, T. Grevatt, A. Malinowski, R. T. Harley, P. Perozzo, A. R. Cameron, and A. Miller, Room temperature spin relaxation in GaAs/AlGaAs multiple quantum wells, *Applied Physics Letters.* **73**(15), 2140–2142 (Oct., 1998). doi: doi:10.1063/1.122403.
7. A. Malinowski, R. S. Britton, T. Grevatt, R. T. Harley, D. A. Ritchie, and M. Y. Simmons, Spin relaxation in GaAs/Al_{x}Ga_{1-x}As quantum wells, *Physical Review B.* **62**(19), 13034–13039 (Nov., 2000). doi: 10.1103/PhysRevB.62.13034.
8. P. Y. Yu and M. Cardona, *Fundamentals of Semiconductors.* Graduate Texts in Physics, (Springer Berlin Heidelberg, Berlin, Heidelberg, 2010).
9. G. Dresselhaus, Spin-orbit coupling effects in zinc blende structures, *Physical Review.* **100**, 580–586, (1955). doi: 10.1103/PhysRev.100.580.

10. X. Zhang, Q. Liu, J.-W. Luo, A. J. Freeman, and A. Zunger, Hidden spin polarization in inversion-symmetric bulk crystals, *Nature Physics.* **10**(5), 387–393, (2014). doi: 10.1038/nphys2933.
11. R. I. Dzhioev, K. V. Kavokin, V. L. Korenev, M. V. Lazarev, B. Y. Meltser, M. N. Stepanova, B. P. Zakharchenya, D. Gammon, and D. S. Katzer, Low-temperature spin relaxation in n-type GaAs, *Phys. Rev. B.* **66**, 245204 (Dec, 2002). doi: 10.1103/PhysRevB.66.245204.
12. Y. A. Bychkov and E. I. Rashba, Oscillatory effects and the magnetic susceptibility of carriers in inversion layers, *Journal of Physics C: Solid State Physics.* **17**(33), 6039 (Nov., 1984).
13. N. Averkiev and L. Golub, Giant spin relaxation anisotropy in zinc-blende heterostructures, *Physical Review B.* **60**(23), 15582–15584 (Dec., 1999). doi: 10.1103/PhysRevB.60.15582.
14. Y. Ohno, R. Terauchi, T. Adachi, F. Matsukura, and H. Ohno, Spin Relaxation in GaAs(110) Quantum Wells, *Physical Review Letters.* **83**(20), 4196–4199 (Nov., 1999). doi: 10.1103/PhysRevLett.83.4196.
15. N. S. Averkiev, L. E. Golub, and M. Willander, Spin relaxation anisotropy in two-dimensional semiconductor systems, *Journal of Physics: Condensed Matter.* **14**(12), R271, (2002).
16. J. Schliemann, J. C. Egues, and D. Loss, Nonballistic Spin-Field-Effect Transistor, *Physical Review Letters.* **90**(14), 146801 (Apr., 2003). doi: 10.1103/PhysRevLett.90.146801.
17. N. S. Averkiev and L. E. Golub, Spin relaxation anisotropy: microscopic mechanisms for 2d systems, *Semiconductor Science and Technology.* **23**(11), 114002, (2008).
18. M. P. Walser, C. Reichl, W. Wegscheider, and G. Salis, Direct mapping of the formation of a persistent spin helix, *Nature Physics.* **8**(10), 757–762, (2012).
19. P. S. Eldridge, W. J. H. Leyland, P. G. Lagoudakis, R. T. Harley, R. T. Phillips, R. Winkler, M. Henini, and D. Taylor, Rashba spin-splitting of electrons in asymmetric quantum wells, *Physical Review B.* **82**(4), 045317 (July, 2010).
20. H. Ye, C. Hu, G. Wang, H. Zhao, H. Tian, X. Zhang, W. Wang, and B. Liu, Anisotropic in-plane spin splitting in an asymmetric (001) GaAs/AlGaAs quantum well, *Nanoscale Research Letters.* **6**, 520, (2011). ISSN 1556-276X. doi: 10.1186/1556-276X-6-520.
21. O. Krebs and P. Voisin, Giant Optical Anisotropy of Semiconductor Heterostructures with No Common Atom and the Quantum-Confined Pockels Effect., *Physical review letters.* **77**(9), 1829–1832 (Aug., 1996). ISSN 1079-7114.
22. A. Balocchi, Q. Duong, P. Renucci, B. Liu, C. Fontaine, T. Amand, D. Lagarde, and X. Marie, Full Electrical Control of the Electron Spin Relaxation in GaAs Quantum Wells, *Physical Review Letters.* **107**(13), 1–5 (Sept., 2011). doi: 10.1103/PhysRevLett.107.136604.
23. A. Hernández-Mínguez, K. Biermann, R. Hey, and P. V. Santos, Electrical suppression of spin relaxation in GaAs(111)b quantum wells, *Physical Review Letters.* **109**(26), 1–5, (2012). doi: 10.1103/PhysRevLett.109.266602.
24. K. Biermann, A. Hernández-Mínguez, R. Hey, and P. V. Santos, Electrically tunable electron spin lifetimes in GaAs(111)B quantum wells, *Journal of Applied Physics.* **112**(8), 083913–083913–7 (Oct., 2012).
25. H. Q. Ye, G. Wang, B. L. Liu, Z. W. Shi, W. X. Wang, C. Fontaine, A. Balocchi, T. Amand, D. Lagarde, P. Renucci, and X. Marie, Growth direction dependence of the electron spin dynamics in {111} GaAs quantum wells, *Applied Physics Letters.* **101**(3), 032104–032104–5 (July, 2012). doi: doi:10.1063/1.4737007.
26. M. E. Flatté, Viewpoint A well-rounded life for an electron spin, *Physics.* **4**, 73, (2011).

doi: 10.1103/Physics.4.73.

27. X. Cartoixà, D. Z.-Y. Ting, and Y.-C. Chang, Suppression of the dyakonov-perel spin-relaxation mechanism for all spin components in [111] zincblende quantum wells, *Physical Review B.* **71**(4), 045313 (Jan., 2005).
28. F. Vas'ko, Spin splitting in the spectrum of two-dimensional electrons due to the surface potential, *[Pis'ma Zh. Eksp. Teor Fiz* **30**, *9, 574 (1979)] JETP Letters.* **30**, 541, (1979).
29. O. Z. Karimov, G. H. John, R. T. Harley, W. H. Lau, M. E. Flatt, M. Henini, and R. Airey, High temperature gate control of quantum well spin memory, *Physical Review Letters.* **91**(24), 246601 (Dec., 2003).
30. V. Lechner, L. E. Golub, P. Olbrich, S. Stachel, D. Schuh, W. Wegscheider, V. V. Bel'kov, and S. D. Ganichev, Tuning of structure inversion asymmetry by the ?-doping position in (001)-grown GaAs quantum wells, *Applied Physics Letters.* 94(24):242109, (2009). doi: http://dx.doi.org/10.1063/1.3156027.
31. S. D. Ganichev and L. E. Golub, Interplay of Rashba/Dresselhaus spin splittings probed by photogalvanic spectroscopy A review, *physica status solidi (b).* **251**(9), 1801–1823 (Sept., 2014). ISSN 1521-3951. doi: 10.1002/pssb.201350261.
32. G. Wang, A. Balocchi, A. V. Poshakinskiy, C. R. Zhu, S. A. Tarasenko, T. Amand, B. L. Liu, and X. Marie, Magnetic field effect on electron spin dynamics in (110) GaAs quantum wells, *New Journal of Physics.* **16**(4), 045008 (Apr., 2014). doi: 10. 1088/1367-2630/16/4/045008.
33. G. Wang, A. Balocchi, D. Lagarde, C. R. Zhu, T. Amand, P. Renucci, Z. W. Shi, W. X. Wang, B. L. Liu, and X. Marie, Temperature dependent electric field control of the electron spin relaxation in (111)A GaAs quantum wells, *Applied Physics Letters.* **102**(24), 242408 (June, 2013). doi: 10.1063/1.4811555.
34. G. Wang, B. L. Liu, A. Balocchi, P. Renucci, C. R. Zhu, T. Amand, C. Fontaine, and X. Marie, Gate control of the electron spin-diffusion length in semiconductor quantum wells, *Nature Communications.* **4**, 1–5, (2013). doi: 10.1038/ncomms3372.
35. I. Vurgaftman and J. R. Meyer, Spin-relaxation suppression by compensation of bulk and structural inversion asymmetries in [111]-oriented quantum wells, *Journal of Applied Physics.* 97(5):053707, (2005). doi: http://dx.doi.org/10.1063/1.1858876.
36. T. A. A. Balocchi and X. Marie, *Time-Resolved Optical Spectroscopy in: Semiconductor Research: Experimental Techniques.* (A. Patané and N. Balkan Editors, Springer-Verlag Berlin and Heidelberg GmbH & Co. K, Heidelberg, Germany ; New York, 2012), 2012 edition. ISBN 978-3-642-23350-0.
37. A. R. Cameron, P. Riblet, and A. Miller, Spin Gratings and the Measurement of Electron Drift Mobility in Multiple Quantum Well Semiconductors, *Physical Review Letters.* pp. 4793–4796, (1996).
38. N. Gedik and J. Orenstein, Absolute phase measurement in heterodyne detection of transient gratings., *Optics letters.* **29**(18), 2109–11 (Sept., 2004).
39. B. J. Garcia, C. Fontaine, and A. Munoz-Yague, Molecular beam epitaxial growth and optical characterization of $GaAs/Al_xGa_{1-x}As$ quantum wells on nominally oriented (111)b GaAs substrates, *Applied Physics Letters.* **63**(19), 2691–2693, (1993).
40. A. Hernández-Mínguez, K. Biermann, R. Hey, and P. V. Santos, Spin transport and spin manipulation in GaAs (110) and (111) quantum wells, *physica status solidi (b).* **251**(9), 1736–1752 (Sept., 2014). doi: 10.1002/pssb.201350202.
41. T. Amand, X. Marie, P. Le Jeune, M. Brousseau, D. Robart, J. Barrau, and R. Planel, Spin quantum beats of 2d excitons, *Phys. Rev. Lett.* **78**, 1355–1358 (Feb, 1997). doi: 10.1103/PhysRevLett.78.1355.
42. M. Dyakonov, X. Marie, T. Amand, P. Le Jeune, D. Robart, M. Brousseau, and J. Barrau, Coherent spin dynamics of excitons in quantum wells, *Phys. Rev. B.* **56**,

10412–10422 (Oct, 1997).
43. A. Vinattieri, J. Shah, T. C. Damen, K. W. Goossen, L. N. Pfeiffer, M. Z. Maialle, and L. J. Sham, Electric field dependence of exciton spin relaxation in GaAs/AlGaAs quantum wells, *Applied Physics Letters.* **63**(23), 3164–3166, (1993). doi: http://dx.doi.org/10.1063/1.110211.
44. V. K. Kalevich, B. P. Zakharchenya, K. V. Kavokin, A. V. Petrov, P. L. Jeune, X. Marie, D. Robart, T. Amand, J. Barrau, and M. Brousseau, Determination of the sign of the conduction-electron g factor in semiconductor quantum wells by means of the hanle effect and spin-quantum-beat techniques, *Physics of the Solid State.* **39**(4), 681–685 (Apr., 1997). doi: 10.1134/1.1129913.
45. URL `http://http://creativecommons.org/licenses/by/3.0/`.
46. A. Balocchi, T. Amand, G. Wang, B. L. Liu, P. Renucci, Q. H. Duong, and X. Marie, Electric field dependence of the spin relaxation anisotropy in (111) GaAs/AlGaAs quantum wells, *New Journal of Physics.* **15**(9), 095016 (Sept., 2013). doi: 10.1088/1367-2630/15/9/095016.
47. N. J. Harmon, W. O. Putikka, and R. Joynt, Prediction of extremely long mobile electron spin lifetimes at room temperature in wurtzite semiconductor quantum wells, *Applied Physics Letters.* **073108**(2011), 11–14, (2012). doi: 10.1063/1.3555628.
48. B. Jusserand, D. Richards, G. Allan, C. Priester, and B. Etienne, Spin orientation at semiconductor heterointerfaces, *Phys. Rev. B.* **51**, 4707–4710 (Feb, 1995). doi: 10.1103/PhysRevB.51.4707.
49. I. D'Amico and G. Vignale, *Phys Rev. B* **62**, 4853 (2000); C. P. Weber, N. Gedik, J. E. Moore, J. Orenstein, J. Stephens, and D. D. Awschalom, Observation of spin Coulomb drag in a two-dimensional electron gas, *Nature.* **437**(7063), 1330–1333 (Oct., 2005). doi: 10.1038/nature04206.
50. P. Eldridge, W. Leyland, P. Lagoudakis, O. Karimov, M. Henini, D. Taylor, R. Phillips, and R. Harley, All-optical measurement of Rashba coefficient in quantum wells, *Physical Review B.* **77**(12), 1–5 (Mar., 2008). doi: 10.1103/PhysRevB.77.125344.
51. D. Richards, B. Jusserand, G. Allan, C. Priester, and B. Etienne, Electron spin-flip raman scattering in asymmetric quantum wells: spin orientation, *Solid State Electronics.* **40**(1-8), 127–131, (1996).
52. W. Leyland, R. Harley, M. Henini, a. Shields, I. Farrer, and D. Ritchie, Oscillatory Dyakonov-Perel spin dynamics in two-dimensional electron gases, *Physical Review B.* **76**(19), 1–6 (Nov., 2007). doi: 10.1103/PhysRevB.76.195305.
53. B. Y. Sun, P. Zhang, and M. W. Wu, Spin relaxation in n-type (111) GaAs quantum wells, *Journal of Applied Physics.* **108**(9), 093709, (2010). doi: 10.1063/1.3504859.
54. A. J. Bennett, R. B. Patel, A. J. Shields, K. Cooper, P. Atkinson, C. A. Nicoll, and D. A. Ritchie, Indistinguishable photons from a diode, *Applied Physics Letters.* 92 (19):193503, (2008). doi: http://dx.doi.org/10.1063/1.2918841.
55. J. L. Oudar, A. Migus, D. Hulin, G. Grillon, J. Etchepare, and A. Antonetti, Femtosecond orientational relaxation of photoexcited carriers in GaAs, *Phys. Rev. Lett.* **53**, 384–387 (Jul, 1984). doi: 10.1103/PhysRevLett.53.384.
56. M. Z. Maialle, E. A. de Andrada e Silva, and L. J. Sham, Exciton spin dynamics in quantum wells, *Phys. Rev. B.* **47**, 15776–15788 (Jun, 1993). doi: 10.1103/PhysRevB.47.15776.
57. J. Kainz and R. Winkler, Temperature dependence of Dyakonov-Perel spin relaxation in zinc-blende semiconductor quantum structures, *Physical Review B.* **195322**, 1–9, (2004). doi: 10.1103/PhysRevB.70.195322.
58. R. Benchamekh, P. Voisin, private communications (2015).
59. T. Sogawa, P. V. Santos, S. K. Zhang, S. Eshlaghi, A. D. Wieck, and K. H. Ploog, Transport and lifetime enhancement of photoexcited spins in GaAs by surface acoustic

waves, *Phys. Rev. Lett.* **87**, 276601 (Dec, 2001). doi: 10.1103/PhysRevLett.87.276601.

60. S. Hermelin, S. Takada, M. Yamamoto, S. Tarucha, A. D. Wieck, L. Saminadayar, C. Bäuerle, and T. Meunier, Electrons surfing on a sound wave as a platform for quantum optics with flying electrons, *Nature.* **477**(7365), 435–438 (Sept., 2011). ISSN 0028-0836. doi: 10.1038/nature10416.
61. F. Cadiz, D. Paget, A. C. H. Rowe, T. Amand, P. Barate, and S. Arscott, Effect of the Pauli principle on photoelectron spin transport in p^+ GaAs, *Phys. Rev. B.* **91**, 165203 (Apr, 2015). doi: 10.1103/PhysRevB.91.165203.
62. S. Azaizia, A. Balocchi, H. Carrre, P. Renucci, T. Amand, A. Arnoult, C. Fontaine, and X. Marie, Control of the electron spin relaxation by the built-in piezoelectric field in ingaas quantum wells, *Applied Physics Letters.* **108**(8), 082103, (2016). doi: http://dx.doi.org/10.1063/1.4942600.
63. L. Wang and M. W. Wu, Hole spin relaxation in p-type (111) GaAs quantum wells, *Physical Review B.* **85**(23), 235308 (June, 2012). ISSN 1098-0121. doi: 10.1103/PhysRevB.85.235308.

Chapter 2

Room Temperature Defect-Engineered Spin Functionalities: Concept and Optimization

Y. Puttisong, I. A. Buyanova and W. M. Chen

Department of Physics, Chemistry and Biology, Linköping University, SE-581 83 Linköping, Sweden

1. Introduction

In the field of spintronics, great research efforts have been endeavored to utilize spin to improve performance of electronic and photonic devices and even enrich them with new functionalities[1-5]. The prerequisites for the success of proposed innovative spin-functionalized devices are to efficiently generate, maintain, manipulate and detect spins. Perhaps the greatest challenge is to achieve these at room temperature (RT), which has proven to be difficult due to thermally-activated spin depolarization and decoherence processes[6,7]. Recently the unconventional concept of exploiting spin-dependent recombination (SDR) via 'spin-filtering defects' in semiconductors has been shown to be capable of generating and maintaining conduction electron spin polarization, up to 43% at RT thanks to the spin-filtering effect associated with the Ga_i^{2+}-interstitial defects in diluted nitride Ga(In)NAs alloys[8-18]. Such high spin polarization is achieved without introducing ferromagnetic elements into a semiconductor which makes this defect-mediated spin-filtering approach attractive for integration of the spin-functional semiconductor with conventional electronic/optoelectronic devices. Furthermore, recent results have demonstrated that efficient RT spin amplifiers[13] and spin detectors[14] can also be realized by exploiting the spin-filtering concept. Such components are largely expected to be the building-blocks for future spintronic devices. In addition, the high spin polarization of the conduction/defect electron spins mediated by the spin-filtering effect has been shown to drive defect nuclear spins out of equilibrium[16], leading to efficient nuclear spin hyperpolarization of defect atoms at RT that is relevant to initializing nuclear spin states for quantum computation utilizing nuclear spin qubits.

To further improve the efficiency of the defect-enabled spin functionalities, e.g. aiming at the complete polarization of electron spins in a semiconductor at RT, a thorough understanding of the physical mechanisms behind this defect-mediated spin-filtering effect is essential. In this chapter, we focus on the identification and optimization of the key physical processes that determine the efficiency of the spin-filtering effect and the dynamic spin transfer between conduction and defect electrons and between electrons and defect nuclei. We shall begin with a brief introduction to the defect-engineered spin functionalities in a semiconductor, followed by an in-depth analysis of the key physical parameters with the aid of a coupled rate-equation approach. A guideline on how to bring the spin systems to complete polarization will be provided at the end of the chapter.

2. SDR and the spin-filtering effect

The defect-engineered spin-filtering effect relies on spin-selective capture of conduction electrons by the paramagnetic spin-filtering defect centers governed by the Pauli's exclusion principle[8,19-21]. The defect centers must have an orbitally non-degenerated ground state to ensure that the capture process of the conduction electrons by the defect centers is only allowed when the conduction and defect electrons have opposite spin orientations (Fig. 1a). Through the efficient recombination of the captured electrons with spin-unpolarized holes via the defect states, a non-vanishing spin polarization (P_e) of the conduction electrons can be transferred to the defect electron spin polarization (P_C) until a dynamic balance between the conduction and defect electron spins is reached. This can be illustrated by the simplest example shown in Fig. 1a-d by assuming P_e=100% (all conduction electrons having the spin-up orientation) and unpolarized holes. In thermal equilibrium before the SDR is initiated, the defect electron spin is unpolarized such that only one half of the paramagnetic defects can capture the spin-up conduction electrons resulting in two spin-paired-off electrons at each of these defects. One of these two electrons will subsequently recombine with an unpolarized hole captured from the valence band, returning the defect to the paramagnetic state with an equal probability of having a spin-up and spin-down electron. The consequence of this recombination is that P_C increases from 0% to 50% in one cycle and repetitions of the same SDR process can drive P_C towards 100% (see Fig. 1b-d). Once the defect electrons are spin polarized, the conduction electrons that lose their spin orientation due to spin relaxation will be depleted from the conduction band by the spin polarized defects, activating the spin-filtering effect. This will eliminate the consequence of conduction electron spin relaxation and will maintain P_e as long as the defect centers keep their spin polarization. Fortunately, a defect electron generally has a longer spin relaxation time than that of a free electron such that this condition can easily be satisfied.

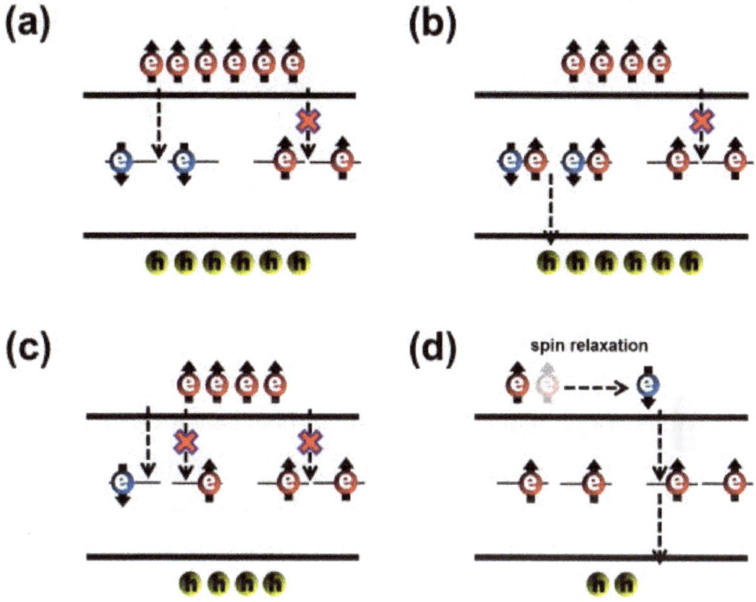

Fig. 1. a SDR via orbitally non-degenerated, spin unpolarized paramagnetic defect centers, **b** Spin-selective capture of the conduction electrons by the centers with an electron of opposite spin to form the spin paired-off charge state of the centers, and their subsequent hole capture returning to the paramagnetic charge state. **c** Dynamic polarization via SDR leads to spin polarization of 75% for the defect electrons in one cycle of SDR, and further higher in the subsequent SDR cycles. **d** Complete spin polarization of the paramagnetic defect centers is established via SDR which activates the spin-filtering effect. The figure was taken from Linköping studies in science and technology. Dissertation No.1607 reprinted with permission.

In reality, however, there are many parameters that determine the spin-filtering efficiency. To give further insights into the physical principle underlying the spin-dependent processes, the coupled conduction-defect electron system is modeled based on spin-dependent rate equations describing the populations of the conduction and defect electrons. We start with a set of simple rate equations shown below:

$$\frac{dn_{\pm}}{dt} = G_{\pm} - \gamma_e n_{\pm} N_{\mp} - \frac{(n_{\pm} - n_{\mp})}{2\tau_S} - \gamma_e^{'} n_{\pm} M_0,$$

$$\frac{dN_{\pm}}{dt} = -\gamma_e n_{\mp} N_{\pm} + \frac{1}{2}\gamma_h p N_{\uparrow\downarrow} - \frac{(N_{\pm} - N_{\mp})}{2\tau_{SC}},$$
$$\frac{dM_0}{dt} = \gamma_h' p M_1 - \gamma_e'(n_+ + n_-) M_0, \qquad (1)$$
$$\frac{dp}{dt} = G - \gamma_h p N_{\uparrow\downarrow} - \gamma_h' p M_1,$$

where G is the free carrier generation rate; $n_{\pm}$ and $N_{\pm}$ are the populations of the conduction electrons and the paramagnetic defect centers in the spin-up (+) and spin down (-) configurations. p denotes the population of free holes that are assumed to be spin unpolarized due to ultra-fast spin relaxation processes induced by spin-orbit coupling[6]. $N_{\uparrow\downarrow}$ is the population of the defect centers with two spin paired-off electrons. The total defect concentration is defined as $N_{SDR} = N_+ + N_- + N_{\uparrow\downarrow}$. $\gamma_{e,h}$ is the electron or hole capturing rate of the defect centers. The $\gamma_e n_{\pm} N_{\mp}$ term secures the spin-dependent capture process in which only the conduction electrons with their spin orientation anti-parallel to that of the defect electrons can be captured. The spin relaxation processes of the free and defect electrons are governed by the terms $(n_{\pm} - n_{\mp})/2\tau_S$ and $(N_{\pm} - N_{\mp})/2\tau_{SC}$ with the spin relaxation times τ_S and τ_{SC} for the free and defect electrons, respectively.

We also include a competing spin-independent defect in the carrier recombination, which can be described by the Shockley-Read-Hall (SRH) model[22, 23]. The relevant defect centers can only be occupied by one electron, changing from empty centers with population M_0 to filled centers with population M_1. The total population of such a non-SDR center is $M_{non-SDR} = M_0 + M_1$. The corresponding capture rate is denoted by $\gamma_{e,h}'$.

We choose the parameter values known from the Ga_i^{2+} - interstitial defects in the diluted nitride Ga(In)NAs alloys. The typical values are τ_S = 150 ps, τ_{SC}= 1500 ps, $\gamma_e / \gamma_h = 4$ and $N_{SDR} = 1\times10^{16}$ cm^{-3} [8,10-15]. The initial spin polarization of the conduction electrons is an input parameter defined by $P_e^i = (G_+ - G_-)/(G_+ + G_-)$. The simulations are performed under the steady state condition. We shall note that the term related to the band-to-band (BB) radiative transition is not included in the model, as the BB recombination time is generally much longer than the typical recombination time via defects and thus does not significantly contribute to the SDR. Once the populations of various states are obtained from the rate-equation analysis, we can calculate spin polarization for the free and defect electrons by the relationships of $P_e = (n_+ - n_-)/(n_+ + n_-)$ and $P_C = (N_+ - N_-)/(N_+ + N_-)$.

3. Dynamic spin polarization and spin amplification

First, we show that the spin-filtering effect leads to an amplification of non-vanishing P_e^i to a much higher degree by facilitating dynamic spin polarization between the conduction and defect electrons. This condition can be achieved when the population of the conduction electrons is comparable with that of the defect centers. This is to guarantee that most of the defect centers are participating in the recombination events. To demonstrate this, we calculate the dependence of P_e and P_C on G (thus on the concentration of free carriers). Note that here we set $M_{non-SDR} = 0$ to study the dynamics of the spin filtering centers alone. Also, all the spin-filtering defect centers are set to be in the paramagnetic state before the recombination occurs. (The more complicated case where these constraints are relaxed will be presented in the later part of the chapter.)

As can be seen from Fig. 2a, P_e and P_C first monotonously increase with increasing G, until they reach the maximum values followed by a decrease at even higher G. At low G, the spin-filtering effect is clearly insignificant, as P_e approximately keeps the same value as P_e^i and P_C close to zero. This is because there is not sufficient number of spin-polarized conduction electrons to lead to a sizable P_C in order to trigger the spin-filtering effect. In this case, P_e is determined by the spin relaxation of the conduction electrons and can be described by the simple relation $P_e = P_e^i/(1 + \tau_c/\tau_s)$ where $\tau_c \sim 1/\gamma_e N_C \sim 1$ ps.

The dynamic spin polarization improves as G increases, leading to a redistribution between N_+ and N_- (thus an increase of P_C) as shown in Fig. 2a and 2b. At the optimal value of G, both P_e and P_C approach 54%, corresponding to an amplification of the initial spin polarization of the conduction electrons by 11 times! The spin amplification can be understood as follows. As the recombination now proceeds with a sufficient number of the conduction electrons, their spin polarization P_e is transferred to P_C. During the next recombination round, the recombination of the conduction electrons having the same spin orientation as the defect electrons slows down whereas the recombination for the anti-parallel configuration between the conduction and defect electron spins speeds up to increase the net polarization of P_e. As P_e increases, the net spin transfer to the defects is further enhanced. Therefore, there is a positive feedback in dynamic spin polarization transfer between the conduction and defect center electrons until the steady-state condition is reached.

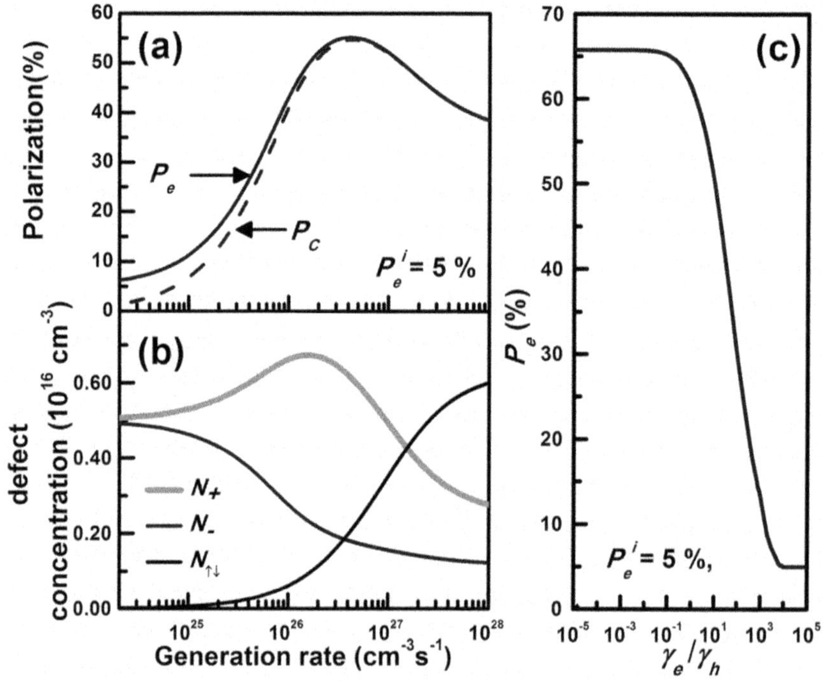

Fig. 2. a P_e (the solid line) and P_C (the dash line) as a function G. **b** The concentrations of the defects with difference spin/charge states, plotted as a function of G. **c** The maximum P_e as a function of the ratio between γ_e and γ_h.

4. Effect of charge state transformation

Here, we attempt to understand the physical parameters that determine the maximum value of P_e and what actually happens at high G where P_e starts to decline. First, let us compare the results of Fig. 2a and 2b. We can see that as P_e and P_C approach the saturation values, $N_{\uparrow\downarrow}$ increases. In other words, more and more defects are occupied by two spin paired-off electrons at a high injection level of carriers. The spin filtering efficiency is, thus, reduced as the net population of the defects in the paramagnetic charge state decreases.

Straight forwardly, by reducing the likelihood of the formation of the spin paired-off charge state, the spin-filtering efficiency should be improved. The first parameter to be considered is the relative rate of the electron and hole capture by the defects. The simulations shown in Fig.2a and 2b were obtained by setting $\gamma_e / \gamma_h = 4$. $\gamma_e / \gamma_h > 1$ implies that the electrons captured by the defect centers are less likely to be recombined with holes. Therefore, this condition promotes the formation of the spin paired-off charge state. As seen in Fig.2c when the ratio γ_e / γ_h keeps increasing, the defects no longer function as the spin-filtering centers. $\gamma_e / \gamma_h < 1$ is thus a preferable condition for the spin-filtering, since it ensures that most of the defect centers remain in the paramagnetic charge state even under high carrier injection. Optimizing this parameter could be difficult, however, since the electron and hole capture rates are intrinsic parameters of the given defects.

5. Effect of doping on spin-filtering efficiency

In general, the assumption that all spin-filtering centers are in the paramagnetic state in thermal equilibrium is oversimplified. The redistribution between the charge states $N_{\pm}$ and $N_{\uparrow\downarrow}$ is determined by the Fermi level of the host material following the relation:

$$N_{\uparrow\downarrow} \Big/ (N_+ + N_-) = \exp\left(\frac{E_F - E_1}{k_B T}\right) \tag{2}$$

where E_F is the Fermi energy. E_1 is the charge-associated energy by adding one electron to the center transforming $N_{\pm}$ to $N_{\uparrow\downarrow}$. k_B is Boltzmann's constant and T is temperature. As E_1 is currently unknown, we shall assume two cases, $E_F >> E_1$ and $E_F << E_1$. This could be referred to as the strong n-type and p-type doping of the host material. The distribution of the charge states before the carrier injection is

$$\begin{aligned} &\text{for p-type}: (N_+ + N_-) = N_{SDR}, \\ &\text{for n-type}: N_{\uparrow\downarrow} = N_{SDR}, \end{aligned} \tag{3}$$

This corresponds to the following charge neutrality equations:

$$\text{for p-type: } p = n_+ + n_- + N_{\uparrow\downarrow} + M_1 - n_0 + p_0,$$
$$\text{for n-type: } p = n_+ + n_- - N_+ - N_- - M_0 - n_0 + p_0, \tag{4}$$

where $n_0(p_0) = \pm n_i \exp((E_F - E_i)/k_B T)$. $n_0(p_0)$ is the free electron (hole) concentration in thermal equilibrium, n_i is the intrinsic carrier concentration at room temperature, set to $n_i = 1.967 \times 10^8 \text{cm}^{-3}$ that corresponds to GaNAs with the energy gap of 1.22 eV and the electron effective mass reported in Ref. 24. E_i is the intrinsic Fermi energy.

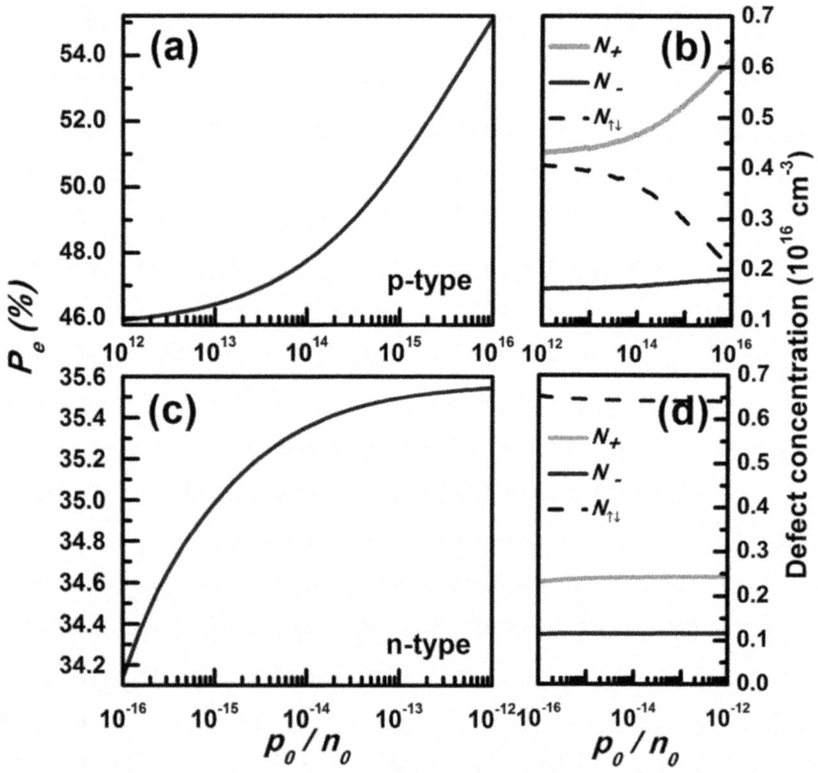

Fig. 3. P_e, $N_\pm$, and $N_{\uparrow\downarrow}$ as a function of p_0/n_0 assuming p-type doping (**a** and **b**) and n-type doping (**c** and **d**) of the host material.

The simulation of P_e by varying p_0/n_0 under the p-type and n-type conditions (constraints set by Eqs.3 and 4) are displayed in Fig. 3a-d. Here, we use $\gamma_e/\gamma_h = 4$. A preferential condition for the spin-filter effect is found in the p-type host, as being reflected by overall stronger P_e at the optimum G as compared with the n-type condition under the same P_e^i = 5%. This can be explained by a reduction of the population of the charge state with two spin paired-off electrons. Under the n-type condition, the dominant defect charge state before the carrier injection is readily the spin paired-off one. Therefore, the centers are not fulfilling the condition of the spin-filtering centers from the start. Carrier injection by the help of SDR could transform $N_{\uparrow\downarrow}$ to $N_{\pm}$. However, this process is clearly ineffective with $\gamma_e / \gamma_h = 4$. This is in sharp contrast to the p-type condition, where the centers are readily in the paramagnetic state that is suitable for SDR. High p_0 (thus, p-doping concentration) is shown to lead to better spin-filtering performance. This is similar to a larger γ_h that helps to reduce the population of the two-electron trap state governed by the term $0.5\gamma_h pN_{\uparrow\downarrow}$ in Eqs. (1). However, a change of n_0 does not significantly contribute to the spin-filtering efficiency in the n-type host as shown in Fig. 3(c) where only a slight improvement of P_e by 1% was found by varying p_0 / n_0 over four orders of magnitude.

6. Effect of competing spin-independent recombination centers

The presence of competing spin-independent recombination via defects degrades the spin-filtering performance[25]. We demonstrate this in Fig. 4 under both p-type and n-type conditions. Here, we simulate P_e by varying the relative ratio $N_{SDR} / M_{non-SDR}$. Both centers are set to have the same electron and hole capture rates. When $M_{non-SDR} >> N_{SDR}$, the recombination is dominated by the non-SDR centers where there is no spin filtering effect. The maximum P_e in this case is thus $P_e \sim P_e^i$. In practice, both types of the recombination centers are very likely co-exist in a material. Therefore, a sufficient number of the SDR-active centers is essential to make the SDR channel dominant. We find that $N_{SDR} / M_{non-SDR} > 10^3$ is the optimum ratio for efficient spin-filtering.

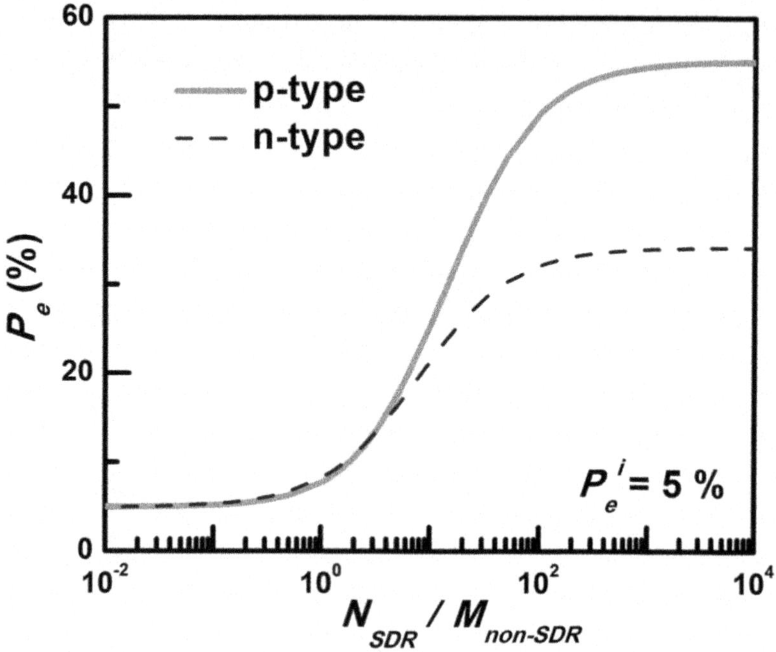

Fig. 4. P_e as a function of $N_{SDR}/M_{non-SDR}$ under both p-type and n-type conditions. N_{SDR} is set at a constant value of $1\text{x}10^{16}$ cm^{-3}.

7. Effect of the defect concentration

Since the spin-filtering process is controlled by efficient spin-dependent carrier recombination via the spin-filtering defects, it implies that the recombination rate of the conduction electrons, which is closely linked to the total concentration of the spin-filtering defect, is another key factor that determines the efficiency of the spin-filtering effect. In order to achieve efficient SDR, the capture time of the conduction electrons by the paramagnetic defect center ($\sim 1/\gamma_e N_{SDR}$) should be much shorter than the conduction electron spin relaxation time τ_S. Here, we compute P_e as a function of $\gamma_e N_{SDR} \tau_S$, i.e. the ratio between the two time constants, and the result is shown in Fig. 5a. Note that we only vary N_{SDR} while keeping γ_e and τ_S constant. Clearly, when τ_S is shorter than the effective electron capture time (i.e. $\gamma_e N_{SDR} \tau_S \ll 1$), the spin-filtering effect is not effective under any conditions regardless of n- or p-type, P_e^i= 5 or 50%. Increasing N_{SDR} shortens the effective capture time, and thus improves spin polarization transfer resulting in better spin-filtering performance.

In addition, we shall show that a high concentration of the spin-filtering centers also leads to better recombination dynamics that reduces the likelihood of the defects in the unwanted charge state under a fixed level of carrier injection. This is shown in Fig. 5b and 5c, where we plot P_e and $N_{\uparrow\downarrow} / N_{SDR}$ as a function of G with the following three choices of N_{SDR}: N_{SDR}=0.5* N_0, N_0 and 5* N_0 (N_0=10^{16} cm^{-3}). As it can be seen, a higher concentration of N_{SDR} pushes the saturation point of the maximum P_e towards a higher G. Consequently, this gives more room for the system to establish the dynamic polarization without being prevented by overfill of the defects by electrons, shown by the reduction of the $N_{\uparrow\downarrow} / N_{SDR}$ ratio at higher N_{SDR} under a fixed G .

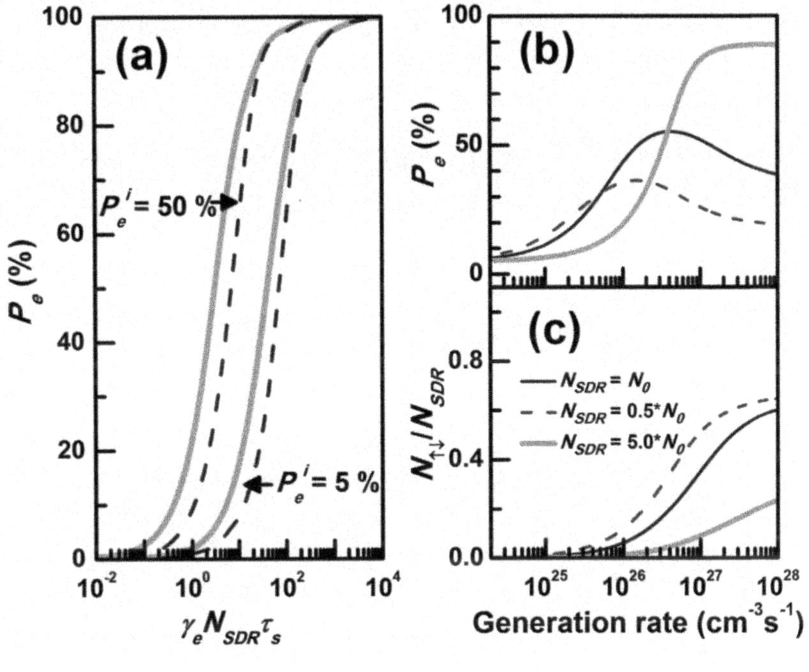

Fig. 5. a P_e as the function of the ratio between τ_S and the effective electron capture time (1/ $\gamma_e N_{SDR}$) computed for P_e^i = 5% and 50% under both p-type (the solid line) and n-type condition (the dash line) of the host material. **b** and **c** P_e and $N_{\uparrow\downarrow}$ (scaled with N_{SDR}) as a function of G with three different concentrations of N_{SDR} ($N_0 = 10^{16}$ cm^{-3} , $M_{non-SDR} = 0$ cm^{-3} and P_e^i =5%).

The simulations shown in Fig.5a show that, by properly tuning N_{SDR}, complete spin polarization of the electrons could be achieved even with a small initial spin injection of 5%. The requirement for a large N_{SDR} can understandably be relaxed with a higher spin injection efficiency as shown for the case of P_e^i = 50%. Furthermore, the complete spin polarization can be achieved even under the n-type condition that was previously shown to be generally unfavorable for the spin-filtering. On the other hand, turning the n-type material into an efficient spin filter by introducing sufficiently large N_{SDR} can be useful when n-doping is necessary for certain applications.

8. Hyperfine interaction and dynamic nuclear polarization

So far, we have considered only the electron spins. The nuclear spins can be involved in the presence of a strong hyperfine interaction (HFI) between electron and nuclear spins. This is the case for the Ga_i spin-filtering defects in Ga(In)NAs alloys in which recent studies have shown a significant contribution of HFI in the spin filtering efficiency[26, 27]. The effective spin Hamiltonian describing the HFI between the electron and nuclear spins of a defect reads:

$$\vec{H} = \vec{S} \cdot A \cdot \vec{I} = A\hat{S}_Z\hat{I}_Z + A\left(\hat{S}_+\hat{I}_- + \hat{S}_-\hat{I}_+\right)/2 \qquad (5)$$

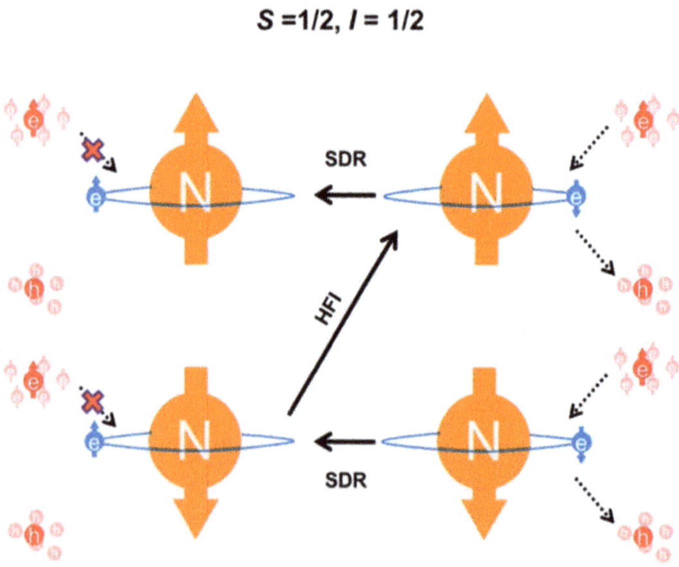

Fig. 6 A schematic diagram for the DNP induced by the combined effect of SDR and HFI, taken as an example S =1/2 and I = 3/2. The figure was taken from Dissertation No.1607, Linköping studies in science and technology. Reprinted with permission.

where $\vec{S}$ is the defect electron spin and A is the HFI parameter. $\hat{S}_Z$ and $\hat{I}_Z$ are the electron and nuclear spin opcrators along the quantization axis Z, respectively. The operator $\hat{S}_{\pm}$ ($\hat{I}_{\pm}$) is the rising/lowering electron (nuclear) spin operator. The second term of spin Hamiltonian leads to a mutual flip-flop between the electron and nuclear spin of the defect. In the case of spin-polarized defect electrons, the flip-flop process can drive nuclear spins out-of-equilibrium. Combining with the SDR process, the injection of spin-polarized free electrons can trigger dynamic nuclear polarization (DNP) resulting in a sizable nuclear spin polarization. The process is schematically presented in Fig. 6.

On the other hand, the spin flip-flop process is also an important source of spin relaxation of the defect electrons. This is evident from the direct correlation between the spin relaxation rate ($1/\tau_A$) and A^2 following the Fermi golden rule of the spin transitions.[27] Taking into account the nuclear spin states, Eq.(1) can be generalized by replacing the defect concentration $N_{\pm}$ by $N_{\pm,m_I}$ where $m_I = I, I-1, \ldots, -I+1, -I$. The defect electron spin relaxation term can now be separated into two parts. One is associated with electron spin flips without involving the nuclear spin, with the characteristic electron spin relaxation time denoted by T_{1SC}. This refers to spin-lattice relaxation of the defect electron governed by a spin-orbit interaction (SOI). The other involves a mutual flip-flop process between the defect electron spin and the nuclear spin, with a spin flip-flop time τ_A. For the Ga_i-interstitial defect in Ga(In)NAs with I = 3/2, the modified rate equations have the following form:

$$\frac{dn_\uparrow}{dt} = G_\uparrow - \gamma_e n_\uparrow \left(N_{\downarrow,-3/2} + N_{\downarrow,-1/2} + N_{\downarrow,+1/2} + N_{\downarrow,+3/2}\right) - \frac{n_\uparrow - n_\downarrow}{2\tau_S},$$
$$\frac{dn_\downarrow}{dt} = G_\downarrow - \gamma_e n_\downarrow \left(N_{\uparrow,-3/2} + N_{\uparrow,-1/2} + N_{\uparrow,+1/2} + N_{\uparrow,+3/2}\right) - \frac{n_\downarrow - n_\uparrow}{2\tau_S},$$
$$\frac{dp}{dt} = (G_\uparrow + G_\downarrow) - \gamma_h p\left(N_{\uparrow\downarrow,-3/2} + N_{\uparrow\downarrow,-1/2} + N_{\uparrow\downarrow,+1/2} + N_{\uparrow\downarrow,+3/2}\right),$$

$$\begin{aligned}
\frac{dN_{\downarrow,-3/2}}{dt} &= \frac{1}{2}\gamma_h p N_{\uparrow\downarrow,-3/2} - \gamma_e n_\uparrow N_{\downarrow,-3/2} - \left(\frac{N_{\downarrow,-3/2} - N_{\uparrow,-3/2}}{2T_{1SC}}\right), \\
\frac{dN_{\downarrow,-1/2}}{dt} &= \frac{1}{2}\gamma_h p N_{\uparrow\downarrow,-1/2} - \gamma_e n_\uparrow N_{\downarrow,-1/2} \\
&\quad - \left(\frac{N_{\downarrow,-1/2} - N_{\uparrow,-1/2}}{2T_{1SC}}\right) - \left(\frac{N_{\downarrow,-1/2} - N_{\uparrow,-3/2}}{2\tau_A}\right), \\
\frac{dN_{\downarrow,+1/2}}{dt} &= \frac{1}{2}\gamma_h p N_{\uparrow\downarrow,+1/2} - \gamma_e n_\uparrow N_{\downarrow,+1/2} \\
&\quad - \left(\frac{N_{\downarrow,+1/2} - N_{\uparrow,+1/2}}{2T_{1SC}}\right) - \left(\frac{N_{\downarrow,+1/2} - N_{\uparrow,-1/2}}{2\tau_A}\right), \\
\frac{dN_{\downarrow,+3/2}}{dt} &= \frac{1}{2}\gamma_h p N_{\uparrow\downarrow,+3/2} - \gamma_e n_\uparrow N_{\downarrow,+3/2} \\
&\quad - \left(\frac{N_{\downarrow,+3/2} - N_{\uparrow,+3/2}}{2T_{1SC}}\right) - \left(\frac{N_{\downarrow,+3/2} - N_{\uparrow,+1/2}}{2\tau_A}\right), \\
\frac{dN_{\uparrow,-3/2}}{dt} &= \frac{1}{2}\gamma_h p N_{\uparrow\downarrow,-3/2} - \gamma_e n_\downarrow N_{\uparrow,-3/2} \\
&\quad - \left(\frac{N_{\uparrow,-3/2} - N_{\downarrow,-3/2}}{2T_{1SC}}\right) - \left(\frac{N_{\uparrow,-3/2} - N_{\downarrow,-1/2}}{2\tau_A}\right), \\
\frac{dN_{\uparrow,-1/2}}{dt} &= \frac{1}{2}\gamma_h p N_{\uparrow\downarrow,-1/2} - \gamma_e n_\downarrow N_{\uparrow,-1/2} \\
&\quad - \left(\frac{N_{\uparrow,-1/2} - N_{\downarrow,-1/2}}{2T_{1SC}}\right) - \left(\frac{N_{\uparrow,-1/2} - N_{\downarrow,+1/2}}{2\tau_A}\right), \\
\frac{dN_{\uparrow,+1/2}}{dt} &= \frac{1}{2}\gamma_h p N_{\uparrow\downarrow,+1/2} - \gamma_e n_\downarrow N_{\uparrow,+1/2} \\
&\quad - \left(\frac{N_{\uparrow,+1/2} - N_{\downarrow,+1/2}}{2T_{1SC}}\right) - \left(\frac{N_{\uparrow,+1/2} - N_{\downarrow,+3/2}}{2\tau_A}\right), \\
\frac{dN_{\uparrow,+3/2}}{dt} &= \frac{1}{2}\gamma_h p N_{\uparrow\downarrow,+3/2} - \gamma_e n_\downarrow N_{\uparrow,+3/2} - \left(\frac{N_{\uparrow,+3/2} - N_{\downarrow,+3/2}}{2T_{1SC}}\right),
\end{aligned} \tag{6}$$

$$\frac{dN_{\uparrow\downarrow,-3/2}}{dt} = \gamma_e n_\downarrow N_{\uparrow,-3/2} + \gamma_e n_\uparrow N_{\downarrow,-3/2} - \gamma_h p N_{\uparrow\downarrow,-3/2} - \left(\frac{3N_{\uparrow\downarrow,-3/2} - N_{\uparrow\downarrow,-1/2} - N_{\uparrow\downarrow,+1/2} - N_{\uparrow\downarrow,+3/2}}{2\tau_N}\right),$$

$$\frac{dN_{\uparrow\downarrow,-1/2}}{dt} = \gamma_e n_\downarrow N_{\uparrow,-1/2} + \gamma_e n_\uparrow N_{\downarrow,-1/2} - \gamma_h p N_{\uparrow\downarrow,-1/2} - \left(\frac{3N_{\uparrow\downarrow,-1/2} - N_{\uparrow\downarrow,-3/2} - N_{\uparrow\downarrow,+1/2} - N_{\uparrow\downarrow,+3/2}}{2\tau_N}\right),$$

$$\frac{dN_{\uparrow\downarrow,+1/2}}{dt} = \gamma_e n_\downarrow N_{\uparrow,+1/2} + \gamma_e n_\uparrow N_{\downarrow,+1/2} - \gamma_h p N_{\uparrow\downarrow,+1/2} - \left(\frac{3N_{\uparrow\downarrow,+1/2} - N_{\uparrow\downarrow,-3/2} - N_{\uparrow\downarrow,-1/2} - N_{\uparrow\downarrow,+3/2}}{2\tau_N}\right),$$

$$\frac{dN_{\uparrow\downarrow,+3/2}}{dt} = \gamma_e n_\downarrow N_{\uparrow,+3/2} + \gamma_e n_\uparrow N_{\downarrow,+3/2} - \gamma_h p N_{\uparrow\downarrow,+3/2} - \left(\frac{3N_{\uparrow\downarrow,+3/2} - N_{\uparrow\downarrow,-3/2} - N_{\uparrow\downarrow,-1/2} - N_{\uparrow\downarrow,+1/2}}{2\tau_N}\right),$$

$$N_{SDR} = N_{\downarrow,-3/2} + N_{\downarrow,-1/2} + N_{\downarrow,+1/2} + N_{\downarrow,+3/2} + N_{\uparrow,-3/2} + N_{\uparrow,-1/2} + N_{\uparrow,+1/2} + N_{\uparrow,+3/2} + N_{\uparrow\downarrow,-3/2} + N_{\uparrow\downarrow,-1/2} + N_{\uparrow\downarrow,+1/2} + N_{\uparrow\downarrow,+3/2},$$

$$p = n_\uparrow + n_\downarrow + N_{\uparrow\downarrow,-3/2} + N_{\uparrow\downarrow,-1/2} + N_{\uparrow\downarrow,+1/2} + N_{\uparrow\downarrow,+3/2} - n_0 + p_0.$$

The nuclear spin polarization of the defect atom can thus be calculated by the equation:

$$P_N = \frac{\sum_{m_I=-3/2}^{+3/2} (N_{\uparrow,m_I} + N_{\downarrow,m_I}) m_I}{\frac{3}{2}\sum_{m_I=-3/2}^{+3/2} (N_{\uparrow,m_I} + N_{\downarrow,m_I})} \tag{7}$$

Here, the nuclear spin relaxation time τ_N has been included in Eqs.(6) to describe nuclear spin relaxation processes that do not originate from the electron-nuclear spin flip-flops. Possible physical mechanisms for such processes include a nuclear spin dipole-dipole interaction between the defect atom and its surrounding lattice atoms, resulting in a nuclear spin loss within τ_N that is expected to be much larger than τ_A.

We shall note that Eqs.(6) assume the pure defect electron and nuclear spin states. This is strictly true only under a strong external magnetic field when the Zeeman interaction overcomes the hyperfine interaction. At zero-field, the pure spin representation may not be applicable as the hyperfine interaction induces strong mixing between the electron and nuclear spins. A full description of such a coupled spin system requires a density matrix approach through master equations which requires numerical computation. This full description can be found elsewhere[29,30]. Here, we use the pure spin treatments as the first approximation to illustrate the physical principle. We find that with a sufficiently high number of the conduction electrons, i.e. when the electron capture time is sufficiently faster than the time evolution of the eigenstates induced by hyperfine mixing, the effective equations (6) provides a good approximation in calculating the electron and nuclear spin polarization[16].

The effect of DNP on the spin-filtering effect under the injection of weakly spin-polarized conduction electrons (i.e., P_e^i=5%) is represented in Fig. 7(a). For comparison, we calculate P_C as a function of G with Eqs.(6) by assuming a weak HFI-induced spin relaxation ($\tau_A \to \infty$) and a strong SOI-induced spin relaxation (T_{1SC}= 1500 ps). This prohibits the electron-nuclear spin flip-flops. As a result, the simulation returns to the same condition as that in Eqs. (1) and as expected the same results are found. For the Ga_i^{2+} spin filtering defects, however, the HFI is found to be the dominant factor governed the spin relaxation and electron-nuclear coupling[26, 27]. Therefore, we set $T_{1SC} \to \infty$ and $\tau_A = 1500$ ps. The simulation under this condition is shown by the dashed lines in Fig. 7a and 7b.

With the DNP feedback, P_C (and thus P_e) improves, indicating better spin-filtering performance due to DNP. As shown in Fig. 7b, the DNP process polarizes the defect nuclear spin. The combined effect of the spin-polarized nucleus and SDR leads to preferred population of the state with m_S=1/2 and m_I= 3/2 under the spin injection of the spin-up conduction electrons. For the state $|m_S = 1/2, m_I = 3/2\rangle$ the electron-nuclear spin flip-flops are prohibited, which in turn means that spin depolarization of the defect electrons induced by the HFI is deactivated. Therefore, by driving the population towards $N_{\uparrow,+3/2}$ the effective spin depolarization rate of the defect centers will slow down resulting in better spin filtering performance.

As one would expect, dynamic spin polarization of the conduction electrons can be improved even with a low value of G when the defect atoms have no nuclear spin, reaching the maximum performance of the spin filters as shown by the simulation in Fig. 7a. However, we shall point out that for simplicity the simulation was made for the case with $T_{1SC} \to \infty$ and $\tau_A \to \infty$. Though such condition is extreme, this simulation nevertheless provides a guideline that nuclear spin-free spin-filtering defects can be exploited to improve the dynamic range of the spin-filtering effect at different levels of carrier injection and the efficiency of the spin-filtering effect in general.

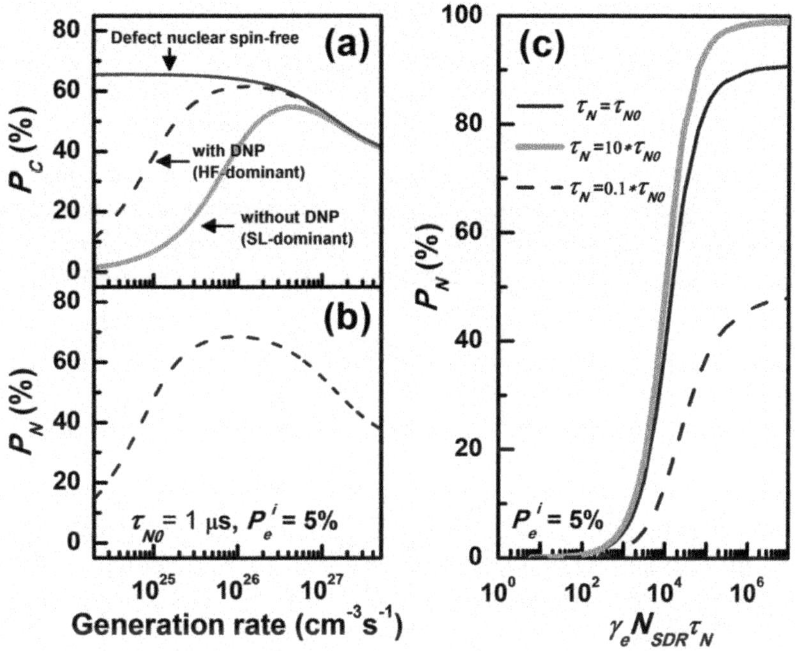

Fig. 7. a Simulated P_C as a function of G when the electron spin relaxation mechanism is dominated by the SOI, or the HFI, or neither (negligible spin relaxation). **b** P_N as a function of G when the HFI-induced electron spin relaxation dominates. **c** P_N as a function of the ratio between τ_N and the effective electron capture time ($1/\gamma_e N_{SDR}$). Three different values of τ_N have been considered. All simulations have been done under the p-type condition of the host material.

9. Conditions for complete polarization of the defect nuclear spin

Finally, we show that the complete nuclear spin polarization of the defect centers can be reached with a set of suitable parameters. As shown above, increasing

N_{SDR} can lead to nearly 100% of P_e and P_C. When the HFI-induced spin relaxation is taken into account, the nuclear spin lifetime τ_N plays a key role in achieving complete nuclear spin polarization of the defect atoms. This can be illustrated by the simulations that, if $\tau_N = \tau_{N0} = 1$ μs, complete polarization of P_N cannot be reached though overall P_N improves as N_{SDR} increases. When τ_N increases by one order of magnitude, the complete nuclear spin polarization can be achieved. As expected, the opposite situation is true when we set τ_N = $0.1 * \tau_{N0}$. In this case, the maximum P_N substantially reduces to a much lower value.

10. Conclusion

In this chapter, we have provided a brief introduction to the interesting concept of the defect-engineered spin-filtering effect in a semiconductor that is capable of generating high spin polarization of the conduction electrons at RT. Our recent experimental results have shown that, by exploiting the SDR via the Ga_i^{2+} defects in Ga(In)NAs, up to 43% conduction-electron spin polarization can be generated at RT. Comparing with conventional GaAs where the achievable spin polarization at RT is about 2%, this represents more than 20 times spin amplification gain[13]. Such a model system can be utilized as an efficient spin detector, by using circular polarization of the emitting light from the BB recombination of the spin amplified conduction electrons as the spin-detection signal. Alternatively, the spin signals can also be converted into electrical signals via the carrier gain due to the spin-blockade of the non-radiative carrier recombination via the spin-polarized defect centers[9,19]. Moreover, the hyperfine interaction between the electron and nuclear spins of the defects can lead to dynamic nuclear spin polarization at RT that is stabilized by the combined effects of the SDR and the HFI.

Based on the rate equations analysis, we have identified several key parameters that control the spin-filtering efficiency. The transition to the spin paired-off charge state of the defects was shown to degrade the spin-filtering efficiency, which must therefore be avoided. The presence of other competing spin-independent recombination centers was also shown to undermine the role of the spin-filtering defects. A direct and perhaps the most effective strategy to approach complete spin polarization of the conduction electrons is to introduce a sufficient number of the spin-filtering defects. For the currently employed Ga(In)NAs alloys, the spin-filtering Ga_i defect was found to be a common grown-in defect regardless of the employed growth methods and post-growth treatments[9]. Moreover, previous results have shown that the total concentration of the Ga_i - interstitial defects scales with the amount of N introduced in the GaN_xAs_{1-x} alloy[8, 31], opening the possibility of controlling N_{SDR} by varying nitrogen composition of the alloy.

The joint action of the HFI and SDR was demonstrated to lead to the establishment of the dynamic balance among P_e, P_C and P_N. We showed that, in the case of strong HFI-induced electron-nuclear spin flip-flops, the resulting nuclear spin polarization of the defect atoms can lead to the stabilization of P_e and P_C. This condition is fulfilled for the Ga_i - interstitial defects in Ga(In)NAs. When near 100% of P_e is required, the searching for alternative spin-filtering defects with zero nuclear spin is beneficial as it improves the dynamic range of the maximum gain in spin amplification even under a low level of carrier injection.

On the other hand, the application such as quantum computation requires the transfer of spin information to nuclear spins. We showed that, once the electron-nuclear spin system is optimized to achieve 100% for P_e and P_C, 100% of P_N can be reached by eliminating the nuclear spin relaxation of the defect centers induced by the nuclear spin dipole-dipole interaction with the host lattice environment. In this case, a nuclear spin-free host semiconductor would be beneficial.

Acknowledgements

We would like to thank X.J. Wang, C.W. Tu, L. Geelhaar, H. Riechert, A.J. Ptak, X. Marie, J.C. Harmand, V.K. Kalevich and A.Yu Egorov for fruitful collaborations. This work was supported by Linköping University through the Professor Contracts, the Swedish Research Council (Grant No. 621-2011-4254), the Linköping Linnaeus Initiative for Novel Functional Materials (LiLi-NFM) supported by the Swedish Research Council (Contract number 2008-6582), and the Swedish Government Strategic Research Area in Materials Science on Functional Materials at Linköping University (Faculty Grant SFO-Mat-LiU # 2009-00971).

References

1. S. A. Wolf, D. D. Awschalom, R. A. Buhrman, J. M. Daughton, S. von Molnár, M. L. Roukes, A. Y. Chtchelkanova, and D. M. Treger, Spintronics: A Spin-Based Electronics Vision for the Future, *Science* ***294,*** *1488 (2001).*
2. I. Žutić, J. Fabian, and S. Das Sarma, *Spintronics: Fundamentals and applications*, *Rev. Mod. Phys.* ***76,*** *323 (2004).*
3. D. D. Awaschalom and M. F. Flatté, Challenges for semiconductor spintronics, *Nat. Phys.* ***3,*** *153 (2007).*
4. T. Dietl, D.D. Awschalom, M. Kaminska and H. Ohno, *Spintronics, Semiconductors and Semimetals Vol.82*, Academic Press, New York, (2008).
5. W. M. Chen and I. A. Buyanova, *Handbook of Spintronic Semiconductors*, Pan Stanford, Singapore (2010).
6. F. Meier and B. P. Zakharchenya, *Optical Orientation,* North-Holland, Amsterdam (1984).
7. M. I. Dyakonov, *Spin Physics in Semiconductors*, Springer, Berlin Heidelberg (2008).
8. X. J. Wang, I. A. Buyanova, F. Chao, D. Zhao, D. Lagarde, A. Balocchi, X. Maries, C. W. Tu, J. C. Harmand, and W. M. Chen, Room-temperature defect-engineered spin filter based on a non-magnetic semiconductor, *Nature Mater.* ***8,*** *198 (2009).*
9. X. J. Wang, Y. Puttisong, C. W. Tu, A. J. Ptak, V. K. Kalevich, A. Yu Egorov, L. Geelhaar, H. Riechert, W. M. Chen, and I. A. Buyanova, Dominant recombination centers in Ga (In) NAs alloys: Ga interstitials, *Appl. Phys. Lett.* ***95,*** *241904 (2009).*
10. V. K. Kalevich, E. L. Ivchenko, M. M. Afanasiev, A. Yu. Shiryaev, A. Yu. Egorov, V. M. Ustinov, and Y. Masumoto, Spin-dependent recombination in GaAsN solid solutions, *JETP Lett.* ***82,*** *455 (2005).*
11. D. Lagarde, L. Lombez, X. Marie, A. Balocchi, T. Amand, V. K. Kalevich, A. Shiryaev, E. Ivechenko and A. Egorov, Electron spin dynamics in GaAsN and InGaAsN structures, *Phys. Stat. Sol. (a)* ***204,*** *208 (2007).*
12. Y. Puttisong, X. J. Wang, I. A. Buyanova, H. Carrére, F. Zhao, A. Balocchi, X. Marie, C. W. Tu, and W. M. Chen, Electron spin filtering by thin GaNAs/GaAs multiquantum wells, *Appl. Phys. Lett.* ***96,*** *052104 (2010).*
13. Y. Puttisong, I. A. Buyanova, A. J. Ptak, C. W. Tu, L. Geelhaar, H. Riechert, and W. M. Chen, Room-Temperature Electron Spin Amplifier Based on Ga(In)NAs Alloys, *Adv. Mate.* ***25,*** *738 (2013).*
14. Y. Puttisong, I. A. Buyanova, L. Geelhaar, H. Riechert, and W. M. Chen, Efficient room-temperature spin detector based on GaNAs, *J. Appl. Phys.* ***111,*** *07C303 (2012).*

15. Y. Puttisong, X. J. Wang, I. A. Buyanova, C. W. Tu, L. Geelhaar, H. Riechert, and W. M. Chen, Room-temperature spin injcction and spin loss across a GaNAs/GaAs interface, *Appl. Phys. Lett.* ***98**, 012112 (2011).*
16. Y. Puttisong, X. J. Wang, I. A. Buyanova, L. Geelhaar, H. Riechert, A. J. Ptak, C. W. Tu and W. M. Chen, Efficient room-temperature nuclear spin hyperpolarization of a defect atom in a semiconductor, *Nat. Comm.* ***4**, 1751 (2013).*
17. Y. Puttisong, X. J. Wang, I. A. Buyanova and W. M. Chen, Effect of hyperfine-induced spin mixing on the defect-enabled spin blockade and spin filtering in GaNAs, *Phys. Rev. B* ***87***, 125202 (2013).
18. Y. Puttisong, D. Dagnelund, I. A. Buyanova, C. W. Tu, A. Polimeni, M. Capizzi and W.M. Chen, Room temperature spin filtering effect in GaNAs: Role of hydrogen, *Appl. Phys. Lett.* ***99**, 152109 (2011).*
19. C. Weisbuch and G. Lampel, Spin-dependent recombination and optical spin orientation in semiconductors. *Solid. State. Commun.* ***14**, 141 (1974).*
20. R. C. Miller, W. T. Tsang & W. A. Nordland. Spin-dependent recombination in GaAs. *Phys. Rev. B* ***21**, 1569–1575 (1980).*
21. D. Paget, Optical-pumping study of spin-dependent recombination in GaAs. *Phys. Rev. B* ***30**, 931–946 (1984).*
22. W. Shockley and W. T. Read, Statistics of the Recombinations of Holes and Electrons, *Phys. Rev.* ***87**, 835 (1952).*
23. R. N. Hall, Electron-Hole Recombination in Germanium, *Phys. Rev.* ***87**, 387 (1952).*
24. P. N. Hai, W. M. Chen, I. A. Buyanova, H. P. Xin and C. W. Tu, Direct determination of electron effective mass in GaNAs/GaAs quantum wells, *Appl. Phys. Lett. 77, 1843 (2000).*
25. A. Kunold, A. Balocchi, F. Zhao, T. Amand, N. Ben Abdallah, J. C. Harmand, and X. Marie, Giant spin-dependent photo-conductivity in GaAsN dilute nitride semiconductor, *Phys. Rev.* ***B** 83, 165202 (2011).*
26. Y. Puttisong, X. J. Wang, I. A. Buyanova, W. M. Chen, Effect of hyperfine-induced spin mixing on the defect-enabled spin blockade and spin filtering in GaNAs, *Phys. Rev. B* ***87**, 125202 (2013).*
27. Y. Puttisong, I. A. Buyanova and W. M. Chen, Limiting factor of defect-engineered spin-filtering effect at room temperature, Phys. Rev. B **89**, *195412* (2014).
28. A. Abragam and B. Bleaney, *Electron Paramagnetic Resonance of Transition ions*, Dover Publications, inc., New York (1986).
29. C. Sandoval-Santana, A. Balocchi, T. Amand, J. C. Harmand, A. Kunold and X. Marie, Room-temperature optical manipulation of nuclear spin polarization in GaAsN, *Phys. Rev. B* ***90**, 115205(2013).*
30. E. L. Ivchenko, L. A. Bakaleinikov, and V. K. Kalevich, Spin-dependent recombination and hyperfine interaction at deep defects, *Phys. Rev. B* ***91**, 205202 (2015).*

31. N. Q. Thinh, I. A. Buyanova and W. M. Chen, Formation of nonradiative defects in molecular beam epitaxial GaN_xAs_{1-x} studied by optically detected magnetic resonance, *Appl. Phys. Lett.* ***79***, *3089 (2001).*

Chapter 3

Onset of Spin Polarization and Anomalous Conductance in One-Dimensional Channels*

Alfredo X. Sánchez[+] and Jean-Pierre Leburton[+,‡]

[+]*Department of Physics and Beckman Institute, University of Illinois at Urbana-Champaign, Urbana, Illinois 61801, USA*

[‡]*Department of Electrical and Computer Engineering, University of Illinois at Urbana-Champaign, Urbana, Illinois 61801, USA*

jleburto@illinois.edu

* This chapter is based on the following articles: A. X. Sánchez and J.-P. Leburton, Temperature modulation of the transmission barrier in quantum point contacts, *Phys. Rev. B.* **88**, 075305 (2013)[29]; A. X. Sánchez and J.-P. Leburton. Tunable magnetic phases in quasi-one-dimensional systems, arXiv preprint, http://arxiv.org/abs/1502.01402 (2015)[28]; A. X. Sánchez and J.-P. Leburton, Charge conservation and conductance anomaly in constricted geometries, in preparation as of this writing (2016).

1. Introduction

Electronic transport in one-dimensional (1D) channels – such as quantum wires (QWRs) and quantum point contacts (QPCs) – remains the subject of active research due to its unique characteristics, some of which remain unexplained.[1-3] Nanostructures like QWRs and QPCs are characterized by two-dimensional (2D) confinement exhibiting electronic modes transverse to the 1D motion of charge carriers. The existence of these transverse modes has important consequences on the interaction between carriers and crystal dynamics,[4] as well as amongst carriers themselves, uncovering a vast array of interesting characteristics. In the former case, carrier scattering undergoes size effects[5-8] that modify the transport properties with important technological consequences for device electronics.[9] In the latter case, the 1D alignment of interacting particles enhances many-body interactions (Hartree, exchange and correlation), giving rise to noteworthy phenomena, such as the formation of Wigner localization[10,11] or Luttinger liquids,[12,13] and the demonstration of separate spin and charge excitations in QWRs.[14,15]

Semiconductor quantum point contacts, in particular, have attracted large interest ever since they were first realized in 1988.[16,17] A common QPC configuration consists of a GaAs/AlGaAs heterostructure with a split metallic gate on top of it. When a negative voltage is applied to these split gates, the electron gas directly underneath is depleted, leaving only a narrow channel between a pair of two-dimensional electron-gas (2DEG) reservoirs acting as source and drain. Experiments have shown that the differential conductance through the QPC is quantized in units of $G_0 \equiv 2e^2/h$.[16-20] This is well understood as a result of the quantization of the energy levels, or subbands in the 1D channel, because of the lateral confinement induced by the gates; then, as each new 1D energy subband is populated, a discrete step appears in the conductance.[21,22] However, additional features emerge in the rise toward the first conductance plateau – such as an anomalous shoulder around $0.6-0.7G_0$ [2,18-20,23] and a "kink" observed around $0.3G_0$ [24-26] – which are still not completely explained.[2,27]

In this chapter, we present a comprehensive theoretical model for many-body-mediated transport properties of a quasi-1D channel in the absence of spin-orbit coupling, with which we predict the emergence of a hierarchy of spin-polarized states. Our model is based on a three-dimensional (3D) non-restricted Hartree-Fock formalism in a quasi-1D geometry that accounts for variations in confinement strength, magnetic field and temperature, and self-consistently relates the spin-dependent concentration to an effective 1D potential and to the Fermi level.[28,29] We show that, above a carrier concentration threshold, a quantum wire allows for the existence of a pair of spin-polarized configurations, even in the absence of a magnetic field.[28] The threshold varies with magnetic field and confinement strength but, crucially, in a 1D channel of finite length it is temperature-independent due to the capacitive effect of the split-gate system. For the calculation of the transmission coefficient and the conductance through a constriction, we use the transfer matrix method[30] under the assumption that the confinement along the 1D channel is slowly-varying (adiabatic).[29] We apply our theory to the quantum transmission through a QPC to show that the anomaly in the QPC conductance emerges at the onset of spin polarization at a conductance value around $\sim 0.6-0.7G_0$ regardless of temperature, as the polarization threshold is controlled by the QPC capacitor gate voltage. There is an additional kink at $\sim 0.3G_0$ [24-26] due to the singularity in the 1D carrier density of states when the 1D channel is pinched off, but this kink is smeared out with temperature.[29] Our model accounts for all experimental features of the conductance anomaly, and provides predictive capabilities on the manifestation of this phenomenon for various QPC biases and confinements.

This chapter is structured as follows. In section 2 we present a general single-particle quantum-mechanical formalism that reduces the 3D Schrödinger equation for a long and narrow constriction to an effective 1D energy-momentum relation. In section 3, we apply this formalism to a wire with parabolic confinement potentials, and we then show that spontaneous spin-polarized configurations emerge above a concentration threshold that depends both on the confinement strength and on the applied magnetic field. In section 4, we study the transport properties of a QPC ignoring any potential spin polarization effects, showing the presence of a small kink in the conductance at ~$0.3G_0$ that is related to the singularity of the 1D carrier density of states. Finally, in section 5, we incorporate spin polarization and show that the onset of degenerate spin-polarized states leads to the appearance of a conductance shoulder at $0.7G_0$, with temperature affecting the width of the shoulder but not the gate voltage at which the shoulder emerges.

2. 3D variational model of a narrow constriction

2.1. *General single-particle picture of a long 1D channel*

We first consider the case of a very long and narrow constriction, i.e., a GaAs quantum wire that can be effectively treated as 1D. We neglect the spin-orbit (Rashba) interaction $\hat{H}_{SO} = \alpha(\hat{\boldsymbol{\sigma}} \times \hat{\mathbf{p}})_z / \hbar$, because its contribution to the effective potential is much smaller compared to other materials. Indeed, in GaAs, the Rashba coupling constant for an electric field F_z in the z-direction is

$\alpha = \left(0.052\,\mathrm{e}\cdot\mathrm{nm}^2\right)F_z$, which is much less than in InAs ($\left(1.17\,\mathrm{e}\cdot\mathrm{nm}^2\right)F_z$) and in InSb ($\left(5.23\,\mathrm{e}\cdot\mathrm{nm}^2\right)F_z$).[31]

In this 1D constriction, electrons are free to move along the x–direction but are confined along the y– and z–directions by confinement potentials represented as $U_y(y)$ and $U_z(z)$, respectively. In a single-particle picture, the 3D Schrödinger equation for an electron of effective mass m^* reads

$$-\frac{\hbar^2}{2m^*}\nabla^2\psi_{\{i\}}(\vec{r})+\left[U_y(y)+U_z(z)+U_x+U_{el}(\vec{r})\right]\psi_{\{i\}}(\vec{r}) \\ +\hat{U}_{xc}\left[\psi_{\{j\}}(\vec{r})\right]=E_{\{i\}}\psi_{\{i\}}(\vec{r}) \tag{1}$$

where the eigenenergies $E_{\{i\}}$ are associated with the quantum numbers $\{i\}=\{i_x,i_y,i_z,\sigma_i\}$, σ being the electron spin. U_{el} and $\hat{U}_{xc}$ are the Hartree and exchange-correlation interaction terms, respectively. An additional term U_x is introduced in Eq. (1) to account for potential barriers in finite 1D channels, although for now (in the case of a long quantum wire) it takes on a constant value U_0.

Given that U_x is constant in the quantum wire, the y- and z-dependence in Eq. (1) can be separated by writing the wave function as a plane wave traveling along x multiplied by a function of y and z, i.e.,

$$\psi_{\{i_x,i_y,i_z,\sigma_i\}}=\frac{e^{ik_xx}}{\sqrt{L_x}}\Lambda_{i_y,i_z}(y,z), \tag{2}$$

where L_x is the length of the wire. Then, the x-dependence is integrated away, and Eq. (1) yields

$$\left\{-\frac{\hbar^2}{2m^*}\left(\partial_y^2+\partial_z^2\right)+U_y(y)+U_z(z)+U_x+U_{el}(y,z)\right\}\Lambda_{i_y,i_z}(y,z) \\ +\hat{U}_{xc}\left[\Lambda_{j_y,j_z}(y,z;k_x,\sigma_j)\right]=\left\{E(k_x,i_y,i_z,\sigma_i)-\frac{\hbar^2k_x^2}{2m^*}\right\}\Lambda_{i_y,i_z}(y,z) \tag{3}$$

We take the expectation value of the left-hand side of Eq. (3) and define two energies: E_{yz}, which includes the kinetic energy and confinement along the y- and z-directions; and U_{ee}, which includes electron-electron interactions:

$$E_{yz}\equiv\left\langle-\frac{\hbar^2}{2m^*}\left(\partial_y^2+\partial_z^2\right)\right\rangle+\left\langle U_y\right\rangle+\left\langle U_z\right\rangle \tag{4}$$

$$U_{ee}\equiv\left\langle U_{el}\right\rangle+\left\langle\hat{U}_{xc}\right\rangle \tag{5}$$

Then, we define an effective 1D potential

$$U_{eff}(k_x;i_y,i_z,\sigma_i)=U_x+E_{yz}(i_y,i_z)+U_{ee}(k_x;i_y,i_z,\sigma_i) \tag{6}$$

With this, the 1D electron energy reads

$$E(k_x;i_y,i_z,\sigma_i)=\frac{\hbar^2k_x^2}{2m^*}+U_{eff}(k_x;i_y,i_z,\sigma_i) \tag{7}$$

2.2. *Extreme quantum limit and spin dependence*

We now focus on the extreme quantum limit, i.e. when only the lowest-energy subband is occupied. In that case, the energy-momentum relation (Eq. (7)) reduces to:

$$E_\sigma(k)=\frac{\hbar^2k^2}{2m^*}+E_{yz}+U_{el}(n)+U_{xc}^{\sigma}(k;n_\sigma) \tag{8}$$

The exchange-correlation term depends on the spin-dependent quantities k_σ (wavevector) and n_σ (carrier concentration per unit length); meanwhile, the Hartree term U_{el} depends only on the total carrier concentration $n=n_\uparrow+n_\downarrow$.

In Eq. (8), when each term for spin-up electrons is equal to the same term with spin-down electrons, a spin-unpolarized configuration is realized, i.e. $E_\uparrow(k_\uparrow)=E_\downarrow(k_\downarrow)$ only if $k_\uparrow=k_\downarrow$ and $n_\uparrow=n_\downarrow$. However, it may still be possible for the energy $E_\uparrow(k_\uparrow)$ to be equal to $E_\downarrow(k_\downarrow)$ when $k_\uparrow\neq k_\downarrow$ and $n_\uparrow\neq n_\downarrow$. In that situation, spin-polarized configurations emerge, satisfying the constraint

$$\frac{\hbar^2}{2m^*}(k_\uparrow^2-k_\downarrow^2)+\left[U_{xc}^{\uparrow}(k_\uparrow;n_\uparrow)-U_{xc}^{\downarrow}(k_\downarrow;n_\downarrow)\right]=0 \tag{9}$$

For the remainder of this chapter we neglect the correlation energy U_{corr} because its value, estimated from Perdew and Wang's local-density theory for the exchange and correlation terms,[32,33] is a fraction (~25-30%) of the exchange energy U_{exch} for the densities and confinement strengths of interest. Moreover, the correlation term is most relevant for situations in which the 1D channel results from structural confinement (i.e. non-electrostatic) and simultaneously remains uniform over the channel length.[34,35] However, in the vast majority of cases, the 1D channel is due to electrostatic confinement by negative split-gate biases.[16-20,36,37] Furthermore, in non-uniform QWRs and QPCs, the confinement strength varies along the 1D channel where the position-dependent carrier concentration is modulated by the dual-gate capacitor, thereby preventing localization-induced Wigner crystallization[34,38] along the QWR. Therefore, one can safely assume that effects of correlation[32,33] will just be a correction to the model presented in the next sections.

3. Spin-polarized configurations in a quasi-one-dimensional system[28]

3.1. *Unrestricted Hartree-Fock model of a quantum wire with an applied magnetic field*

For this section we consider a wire in which the confinement is modeled as a superposition of two parabolic wells (see Figure 1), i.e.

$$U_y(y)+U_z(z)\equiv U_{yz}(y,z)=\frac{1}{2}m^*\omega_y^2y^2+\frac{1}{2}m^*\omega_z^2z^2 \tag{10}$$

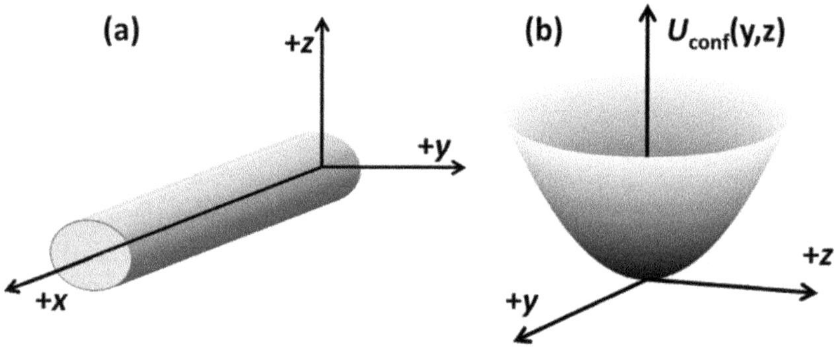

Figure 1. (a) Schematic representation of a quantum wire. (b) Parabolic confinement potentials.

Additionally, the wire is placed in a (constant) magnetic field $\vec{B}=B_x\hat{x}$ parallel to the axis of the wire, with the gauge set to $\vec{A}=B_xy\hat{z}$. In this case, the single particle Schrödinger equation (Eq. (1)), neglecting correlation, picks up additional terms:

$$\begin{aligned}&-\frac{\hbar^2}{2m^*}\nabla^2\psi_{\{i\}}(\vec{r})+\left[U_{yz}(y,z)+U_x+U_{el}(\vec{r})\right]\psi_{\{i\}}(\vec{r})+\hat{U}_{exch}\left[\psi_{\{j\}}(\vec{r})\right]\\&\qquad+\left[\frac{1}{2}m^*\omega_B^2y^2+i\hbar\omega_B y\frac{\partial}{\partial z}+g\mu_B B_x\sigma_i\right]=E_{\{i\}}\psi_{\{i\}}(\vec{r})\end{aligned} \tag{11}$$

Here, $\omega_B=eB_x/m^*$ is the cyclotron frequency (e is the electron charge), $U_B=g\mu_B B_x\sigma$ is the Zeeman splitting term, g is the effective electron g-factor, $\mu_B=e\hbar/2m^*$ is the effective Bohr magneton in the wire, and σ is equal to $+1/2$

or $-1/2$ for spin-up or spin-down electrons, respectively. The electron-electron interaction terms read:

$$U_{el}(\vec{r})\psi_{\{i\}}(\vec{r}) = \sum_{\{j\}} \int d^3\vec{r}' \left|\psi_{\{j\}}(\vec{r}')\right|^2 U_{Coul}(\vec{r},\vec{r}')\psi_{\{i\}}(\vec{r}) \tag{12}$$

$$\hat{U}_{exch}\left[\psi_{\{j\}}(\vec{r})\right] = -\sum_{\{j\}} \delta_{\sigma_i\sigma_j} \int d^3\vec{r}'\, \psi^*_{\{j\}}(\vec{r}')U_{Coul}(\vec{r},\vec{r}')\psi_{\{i\}}(\vec{r}')\psi_{\{j\}}(\vec{r}) \tag{13}$$

U_{Coul}, the Coulomb interaction, is given by

$$U_{Coul}(\vec{r},\vec{r}') = \frac{e^2}{4\pi\epsilon\left|\vec{r}'-\vec{r}\right|} \tag{14}$$

(ϵ is the dielectric constant of the material.) We take the expectation value of the left-hand side of Eq. (11) in the extreme quantum limit, using the following trial wave function:

$$\psi(\vec{r};k_x) = \frac{1}{\sqrt{L_x}} e^{ik_x x}\left(\frac{a^{1/2}}{\pi^{1/4}} e^{-a^2y^2/2}\right)\left(\frac{b^{1/2}}{\pi^{1/4}} e^{-b^2z^2/2}\right), \tag{15}$$

where $a = \sqrt{m^*\omega_y/\hbar}$ and $b = \sqrt{m^*\omega_z/\hbar}$. This wave function corresponds to the exact ground state of Eq. (11) when electron-electron interactions are ignored and the magnetic field is zero. Then, we obtain the energy as a function of k_x:

$$\begin{aligned} E(k_x,\sigma) &= \frac{\hbar^2k_x^2}{2m^*} + \frac{1}{2}\hbar\omega_y + \frac{1}{2}\hbar\omega_z + \frac{1}{4}\left(\frac{\omega_B}{\omega_y}\right)\hbar\omega_B \\ &\quad + g\mu_B B_x\sigma + U_{el}[n] + U_{exch}(k_x,\sigma) \end{aligned} \tag{16}$$

The expectation values U_{el} and U_{exch} are

$$U_{el}[n] = \frac{e^2}{32\pi^2\epsilon}\zeta_{ab}(0)\sum_\sigma \int_{-\infty}^{+\infty} dp_x\, f_T\left[E(p_x,\sigma)\right] = \frac{e^2\zeta_{ab}(0)}{16\pi\epsilon}n \tag{17}$$

$$U_{exch}(k_x,\sigma) = -\frac{e^2}{32\pi^2\epsilon}\int_{-\infty}^{+\infty} dp_x\, \zeta_{ab}(p_x-k_x)f_T\left[E(p_x,\sigma)\right] \tag{18}$$

Here, n is the total 1D electron density in the wire,

$$n = \sum_\sigma n_\sigma(T) = \frac{1}{2\pi}\sum_\sigma \int_{-\infty}^{+\infty} dp_x\, f_T\left[E(p_x,\sigma)\right] \tag{19}$$

$f_T[E]=1/\{1+\exp[(E-E_F(T))/k_BT]\}$ is the Fermi-Dirac distribution for a (spin-independent) Fermi level $E_F(T)$ at a temperature T. $\zeta_{ab}(p)$ is a dimensionless overlap function which depends on the confinement strength and the particular choice of wave function. For the wave function in Eq. (15), $\zeta_{ab}(p)$ is

$$\zeta_{ab}(p)=8\left(\frac{a}{b}\right)\int_0^{+\infty} dt \frac{t\exp(-t^2/2)}{\sqrt{\left(\frac{p}{b}\right)^2+t^2}\sqrt{\left(\frac{p}{b}\right)^2+\left(\frac{a}{b}\right)^2 t^2}} \tag{20}$$

This function is shown in Fig. 2 for different values of a and b. It is a monotonically-decreasing function of p which increases with increasing b (dotted line) and with decreasing a/b ratio (dashed line).

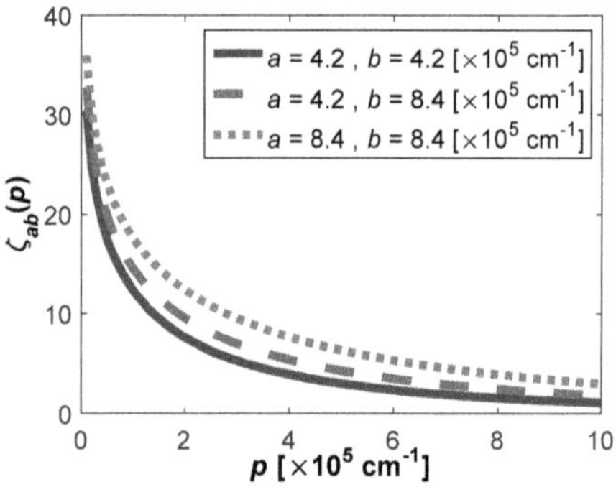

Figure 2. $\zeta_{ab}(p)$ vs. p for different values of a and b.

In general, in order to calculate the spin-dependent concentrations n_σ ($n_\uparrow$ and $n_\downarrow$) and energies $E(k_x,\sigma)$, one must solve the pair of integral equations described by Eq. (16) together with Eq. (17)-(20) for $E(k_x,\uparrow)$ and $E(k_x,\downarrow)$. For the particular case of zero temperature, $f_T(E(k,\sigma))=\theta(E_F-E(k,\sigma))=\theta(k_{F(\sigma)}-|k|)$, where $k_{F(\sigma)}$ is the spin-dependent Fermi wavevector such that $E(k_{F(\sigma)})=E_F$. Then, the exchange term can be rewritten as:

$$U_{exch}(k_x,\sigma)=-\frac{q^2}{32\pi^2\epsilon}\int_{-k_{F(\sigma)}+k_x}^{+k_{F(\sigma)}+k_x} dp\, \zeta_{ab}(p) \tag{21}$$

3.2. *Emergence of spin-polarized regimes*

At zero temperature, when the spin-dependent electron concentrations n_σ are positive, they are associated with the wavevectors $k_{F(\sigma)}$ through the relation $k_{F(\sigma)}=\pi n_\sigma$, so that $E_F=E(k_{F(\sigma)})=E_F[n_\sigma]$. Then, we use Eq. (16) together with

$E_F\left[n_\sigma\right]=E_F\left[n_{-\sigma}\right]$ to obtain a relationship between the total concentration n and n_σ, which we write in terms of $\Delta n_\sigma \equiv n_\sigma - n/2$:

$$\frac{\pi^2}{2}n\Delta n_\sigma+\sigma\frac{egB_x}{2\hbar}-\frac{e^2m^*}{64\pi^2\epsilon\hbar^2}\int_{2\pi(n_0/2-\Delta n_\sigma)}^{2\pi(n_0/2+\Delta n_\sigma)}dp\,\zeta_{ab}\left(p\right)\equiv F\left(\Delta n_\sigma;n\right)=0 \quad (22)$$

This equation holds provided that $0\le\left|\Delta n_\sigma\right|\le n/2$, because of the constraint $0\le n_\sigma\le n$. If Eq. (22) yields the solution $\Delta n_{\uparrow(\downarrow)}>n/2$, the identity $k_{F,\downarrow(\uparrow)}=\pi n_{\downarrow(\uparrow)}$ is not valid anymore, since $n_{\downarrow(\uparrow)}$ would then be negative; therefore, it must be set to zero in order to still satisfy Eq. (16). In such a case, $n_{\uparrow(\downarrow)}=n$, which corresponds to full spin polarization.

In the case of non-interacting electrons (i.e. $\zeta_{ab}\left(p\right)\to 0$), Eq. (22) has a unique solution that is independent of the confinement strength, given by:

$$n_\sigma=\frac{n}{2}\left[1-\left(\frac{n_B}{n}\right)^2\operatorname{sign}\left(gB_x\sigma\right)\right] \qquad \left(n_B\equiv\sqrt{\frac{e\left|gB_x\right|}{\pi^2\hbar}}\right), \quad (23)$$

provided that $n\ge n_B$. When $n<n_B$, the last equation would yield $n_{\uparrow(\downarrow)}<0$ for $\operatorname{sign}\left(gB_x\right)>(<)0$; then, $n_{\uparrow(\downarrow)}$ must be set to zero, and the wire exhibits full spin-down(up) polarization ($n_{\downarrow(\uparrow)}=n$).

In the general case ($\zeta_{ab}\left(p\right)\neq 0$), and if the total concentration is low, Eq. (22) yields only one solution for Δn_σ in terms of n, corresponding to a specific spin configuration. However, once n exceeds a certain threshold value n^{onset}, there are two additional solutions, for a total of three distinct spin configurations. One way to see this is by exapnding the integral in Eq. (22) as a series in Δn_σ, in the limit $\left|\Delta n_\sigma\right|\ll n$. Then, up to third order in Δn_σ, the resulting equation is:

$$\begin{aligned}\frac{\pi^2}{2}n\Delta n_\sigma+\sigma\frac{egB_x}{2\hbar}-\frac{e^2m^*}{16\pi^2\epsilon\hbar^2}\Big\{&\left(\pi\Delta n_\sigma\right)\zeta_{ab}\left(\pi n\right)+\\ &+\frac{2}{3}\left(\pi\Delta n_\sigma\right)^3\zeta''_{ab}\left(\pi n\right)+\ldots\Big\} = 0\end{aligned} \quad (24)$$

Here, $\zeta''_{ab}\left(p\right)=\left[d^2\zeta_{ab}\left(p\right)/dp^2\right]$. Equation (24) is cubic in Δn_σ, and consequently it may have one or three solutions depending on the values of n, B_x, $\zeta_{ab}\left(\pi n\right)$, and constants.

In Figure 3 we plot $F\left(\Delta n_\uparrow;n\right)$ (the left-hand side of Eq. (22)) versus $\Delta n_\uparrow$ for several values of n. The solutions of Eq. (22) correspond to the crossings of the curves with the horizontal axis (dotted line). If the concentration is below n^{onset} (dashed line), there is only a single positive solution for $\Delta n_\uparrow$. However, when $n>n^{\text{onset}}$ (dash-dotted line), two additional negative solutions for $\Delta n_\uparrow$ exist. For $n=n^{\text{onset}}$ (solid line), the slope of $F\left(\Delta n_\uparrow;n\right)$ is zero at the value of $\Delta n_\uparrow$ for which the negative solution emerges. Hence, n^{onset} and the corresponding Δn_σ are found

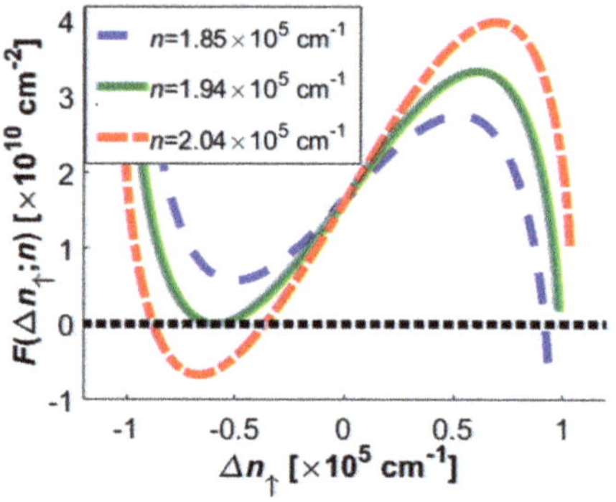

Figure 3. $F(\Delta n_\uparrow;n)$ vs. $\Delta n_\uparrow$ for different concentrations n. $g = 0.44$, $m^* = 0.067m_0$, $\epsilon = 12.9\,\epsilon_0$, $\hbar\omega_y = \hbar\omega_z = 2.0$ meV, and $B_x = 1$ T.

by solving the pair of equations $F(\Delta n_\uparrow;n)=0$ and $\partial F(\Delta n_\uparrow;n)/\partial\Delta n_\sigma = 0$.

To illustrate this model, we consider a GaAs nanowire with $g = 0.44$, $m^* = 0.067m_0$ and $\epsilon = 12.9\,\epsilon_0$,[39] and set the transverse confinement strength to $\hbar\omega_y = \hbar\omega_z = 2.0$ meV. In Figure 4(a) we show the polarization $P = (n_\uparrow - n_\downarrow)/n$ for interacting and non-interacting electrons, as a function of the total concentration n, for a magnetic field $B_x = 1$ T. The solid line in this figure represents the spin polarization for non-interacting electrons, In this case, the Zeeman interaction $U_B = g\mu_B B_x\sigma$ lowers the potential energy of spin-down electrons; therefore, for a given Fermi energy, $n_\downarrow$ exceeds $n_\uparrow$ and P is negative. Partial spin polarization ($-1 < P < 0$) occurs for concentrations $n \geq n_B = 0.817\times10^5\ cm^{-1}$, with P approaching zero as n increases to infinity. For $n < n_B$, there is full spin polarization and $P = -1$. Meanwhile, as n increases to infinity, P slowly drops to zero. This is because the kinetic energy term $T_{x(\sigma)} = \hbar^2 k_{F(\sigma)}^2/2m^* = \hbar^2\pi^2 n_\sigma^2/2m^*$ predominates at high concentrations relative to the concentration-independent Zeeman splitting term U_B.

All three non-solid lines in Figure 4 show the three possible solutions of Eq. (22) for interacting electrons. One of them is present for all concentration values and has $P > 0$ (dashed line); we call it the "↑" or "up" configuration. In this case, the wire is partially spin-polarized in the range $4.51\times10^3\ cm^{-1} \leq n \leq 1.92\times10^5\ cm^{-1}$, with a minimum $P = 0.32$ at $n = 0.54\times10^5\ cm^{-1}$. The concentration values at which the wire becomes fully polarized are calculated by setting $\Delta n_\uparrow = n/2$ in Eq. (22). The exchange interaction U_{exch} lowers the zero-point energy of the 1D energy subband and makes the spin polarization opposite to the non-interacting case. U_{exch} dominates over the kinetic energy T_x at very low n and, since E_F is spin-independent, $n_\uparrow$ has to increase so that $U_{exch(\uparrow)}$ can compensate for the Zeeman splitting.

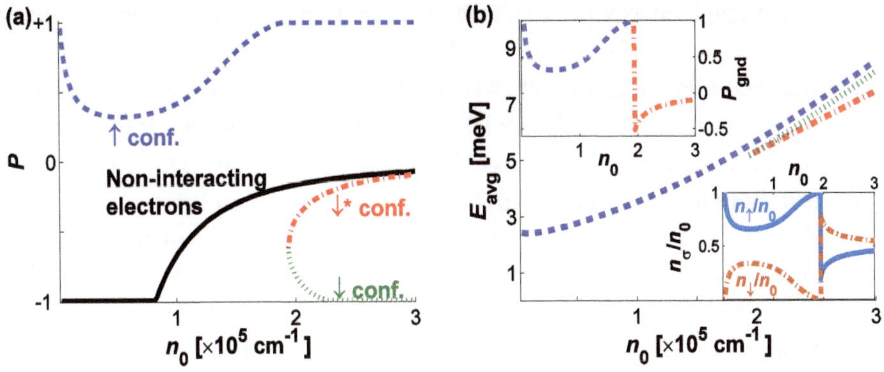

Figure 4. (a) Polarization and (b) average energy per electron as a function of the total concentration at $B_x = 1$ T and $T = 0$ K for non-interacting electrons (solid line) and the three possible polarization regimes for interacting electrons: ↑ (dashed), ↓ (dotted) and $\downarrow^*$ (dash-dotted). $\hbar\omega_y = \hbar\omega_z = 2.0\ meV$. Insets: (top-left) polarization of the ground state vs. total concentration; (bottom-right) fractional concentrations of spin-up and spin-down electrons in the ground state.[28]

When $n \geq n^{onset}\left(=1.94\times10^5\ cm^{-1}\right)$, two new spin configurations emerge, both with $P<0$: "$\downarrow$" or "down" (dotted line in Figure 4(a)), and "$\downarrow^*$" or down-star (dash-dotted line). At the onset concentration, P = -0.62 in both cases, but the two configurations diverge with increasing n. In the "$\downarrow$" case, the polarization becomes stronger (more negative) until full polarization (P = -1) is achieved for $n > n^{full} = 2.27\times10^5\ cm^{-1}$. On the other hand, in the "$\downarrow^*$" case, spin polarization weakens and tends to zero as $n\to\infty$, approaching the behavior of non-interacting electrons. As we will show in Figure 7, the values of n^{onset} and n^{full}, together with their corresponding energies, depend on the confinement strength in the wire.

The interplay between the Zeeman splitting U_B, the exchange interaction U_{exch} and the kinetic energy T_x (in order to keep $E_F\left(n_\uparrow\right)=E_F\left(n_\downarrow\right)$) explains the emergence of the three distinct spin configurations. We illustrate this by looking at the high-concentration behavior (when $P=1$, $P=-1$ and $P\to0$ for the $\uparrow$, $\downarrow$ and $\downarrow^*$ configurations, respectively): (1) if $n_\downarrow < n_\uparrow$, then the combined contribution $U_{exch(\downarrow)} + U_{B(\downarrow)}$ is prevented from becoming too negative; this corresponds to spin-up polarization ($\uparrow$ configuration). (2) If, instead, $n_\downarrow$ is greater than $n_\uparrow$ and $P\to-1$ at high concentrations ($\downarrow$ configuration), the spin-down kinetic energy term $T_{x(\downarrow)}$ increases while the spin-up term $T_{x(\uparrow)}$ decreases, thus lowering $E_F\left(n_\uparrow\right)$ to make it match $E_F\left(n_\downarrow\right)$. (3) If $n_\downarrow > n_\uparrow$ but $n_\downarrow - n_\uparrow \to 0$ as the concentration increases (i.e. $P\to0$, the $\downarrow^*$ configuration), then the kinetic-energy and exchange terms will balance each other out and will both dominate over the concentration-independent U_B.

Out of the three configurations, $\uparrow$, $\downarrow$ and $\downarrow^*$, the one that corresponds to the ground state of the system will be the one with the lowest energy. In Figure 4(b) we show the average energy per electron as a function of n for all three regimes. For $n > n^{onset}$, the $\downarrow^*$ configuration has the lowest energy, and as n increases the energy difference relative to the other two solutions continues to increase. At the threshold concentration $n^{onset} = 1.94 \times 10^5\ cm^{-1}$, the average energy for both the $\downarrow$ and $\downarrow^*$ configurations is $E_{avg} = 5.2\ meV$, but when $n = 3 \times 10^5\ cm^{-1}$, the separation between them has grown up to $0.7\ meV$. The $\uparrow$ solution has the highest energy, which is $0.4\ meV$ higher than for the $\downarrow$ solution regardless of concentration. However, for $n < n^{onset}$, the system must assume the $\uparrow$ configuration, since $\downarrow^*$ is forbidden. As a result, at n^{onset}, the ground state flips abruptly from a positive to a negative polarization. This process is shown on the insets of Figure 4(b). The top-left inset displays the ground-state polarization P_{gnd}, which drops from +1 ($\uparrow$ regime, dashed line) to -0.6 ($\downarrow^*$ regime, dash-dotted line) at n^{onset}. The bottom-right inset shows the ratio n_σ / n for the ground state of the system. Below the onset concentration, the system is in the $\uparrow$ configuration, so $n_\uparrow$ is greater than $n_\downarrow$; the minimum $n_\uparrow \approx 0.7n$ occurs when $n = 0.54 \times 10^5\ cm^{-1}$, whereas $n_\uparrow$ is maximum ($n_\uparrow = n$) when $n \le 4.51 \times 10^3\ cm^{-1}$ or $1.92 \times 10^5\ cm^{-1} \le n < n^{onset}$. At n^{onset}, $n_\downarrow$ becomes greater than $n_\uparrow$ as the system shifts to the $\downarrow^*$ configuration; $n_\downarrow$ is maximum (≈0.75n) at the spin-flip point, but drops toward 0.5n as n increases.

Figure 5(a) illustrates the possible spin-polarization regimes for different concentrations and magnetic fields, separated into five regions. In regions (I) and (II), only the $\uparrow$ configuration is present, with either partial or full polarization, respectively. When $B_x > 4.22\ T$, the $\uparrow$ configuration is fully polarized, independently of the concentration. The line separating regions (I) and (II) from regions (III) and (IV) represents the (magnetic field-dependent) minimum concentration n^{onset} for which the $\downarrow$ and $\downarrow^*$ will emerge. In regions (III) and (IV), the $\uparrow$ configuration is either partially or fully polarized, respectively. Finally, in region (V) on the right end of the diagram, both the $\uparrow$ and $\downarrow$ configurations are fully polarized. The curve separating regions (IV) and (V) corresponds to the minimum concentration n^{full} for full polarization of the $\downarrow$ solution.

The emergence of different polarization regimes in Figure 5 is consistent with direct measurements of the spin polarization in QPCs[40] in which the polarization varies as the concentration and magnetic field are change. Additionally, the abrupt reversal of the polarization predicted by our model (Figure 4(b)) also explains observations of a sudden rearrangement of the spin-up and spin-down energy levels under a strong in-plane magnetic field,[41] which were previously attributed to an exchange-mediated magnetic phase transition.[41-43]

Figure 5(b) shows the evolution of the different spin polarization regimes in the wire with vanishing magnetic fields. As $B_x \to 0$, the $\uparrow$ solution collapses into an unpolarized regime below n^{onset}, and the same happens to the $\downarrow^*$ solution above

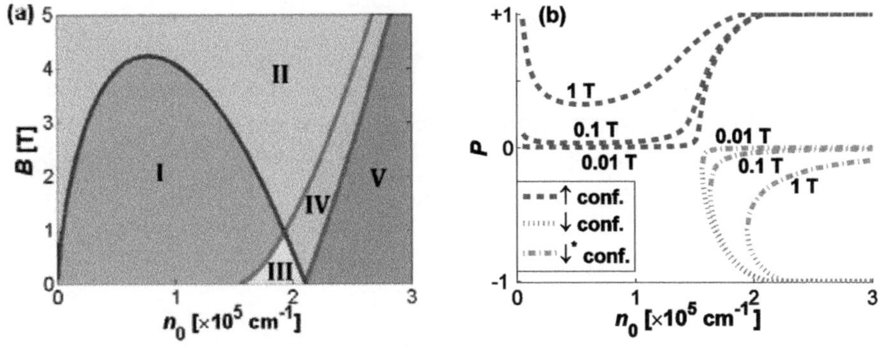

Figure 5. (a) Diagram of the distinct spin-polarized phases as a function of total concentration n and magnetic field B_x: (I) ↑ configuration only, partially polarized; (II) ↑ configuration only, fully polarized; (III) ↑ (partial), ↓ (partial), ↓* (partial); (IV) ↑ (full), ↓ (partial), ↓* (partial); (V) ↑ (full), ↓ (full), ↓* (partial). (b) Polarization vs. total concentration for different magnetic field strengths. $\hbar\omega_x = \hbar\omega_y = 2$ meV and $T = 0$ K in all cases.[28]

n^{onset}. Thus, in the absence of a magnetic field, an unpolarized configuration co-exists with two spin-polarized regimes (the ↑ and ↓ solutions above n^{onset}) which are symmetric relative to each other. Meanwhile, as the magnetic field decreases, the concentration threshold n^{onset} drops to a lower value that, at $B_x = 0$, is common for up- and down-spin polarization.

3.3. *Spin polarization in the absence of magnetic field*

When the magnetic field B_x vanishes, Eq. (24) gives up to three approximate solutions for $\Delta n_\sigma = n_\sigma - n_0/2$ in terms of the total concentration n: the trivial solution $\Delta n_\sigma = 0$, which corresponds to a spin-unpolarized configuration; and a pair of non-zero solutions, both associated with spin-polarized regimes, given by:

$$\Delta n_\sigma \approx \pm\sqrt{\frac{3}{\zeta''_{ab}(\pi n)}\left[a^* n - \frac{\zeta_{ab}(\pi n)}{2\pi^2}\right]} \quad , \tag{25}$$

where a^* is the effective Bohr radius,

$$a^* \equiv \frac{4\pi\epsilon\hbar^2}{e^2 m^*} \tag{26}$$

The non-trivial solutions exist only if the argument inside the square root is non-negative. Therefore, there is a minimum concentration $n^{onset}(B_x = 0)$ for the emergence of the spin-polarized configurations, which satisfies the identity:

$$\frac{\zeta_{ab}\left(\pi n^{onset}_{(B_x=0)}\right)}{\pi n^{onset}_{(B_x=0)}} = 2\pi a^* \tag{27}$$

The onset concentration n^{onset} thus depends on the specific material (via a^*) and on the confinement strength (through parameters a and b).

As the concentration increases, so does the polarization. Full polarization is reached when $n = n^{full}$, where n^{full} is obtained by setting $\Delta n_\sigma = n/2$ in Eq. (22):

$$\left(n^{full}\right)^2 = \frac{e^2 m^*}{16\pi^4 \epsilon \hbar^2} \int_0^{2\pi n^{full}} dp\, \zeta_{ab}(p) \tag{28}$$

The spin-polarized concentrations $n_\uparrow$ and $n_\downarrow$ when $B_x = 0$ are plotted in Figure 6(a) for the ↓ configuration ($n_\downarrow > n_\uparrow$). The solid line represents the unpolarized solution, $n_\uparrow = n_\downarrow = n/2$, while the dashed curves correspond to the spin-polarized regimes that are present for concentrations above the threshold $n^{onset} = 1.54 \times 10^5\ cm^{-1}$. Above this threshold, $n_\downarrow$ continues to increase with n until the wire is fully polarized ($n_\downarrow = n$, $n_\uparrow = 0$) for $n \geq n^{full}$ ($= 2.1 \times 10^5\ cm^{-1}$). The inset shows the variation of the polarization P for this configuration, from 0 at $n \leq n^{onset}$ to -1 at $n \geq n^{full}$.

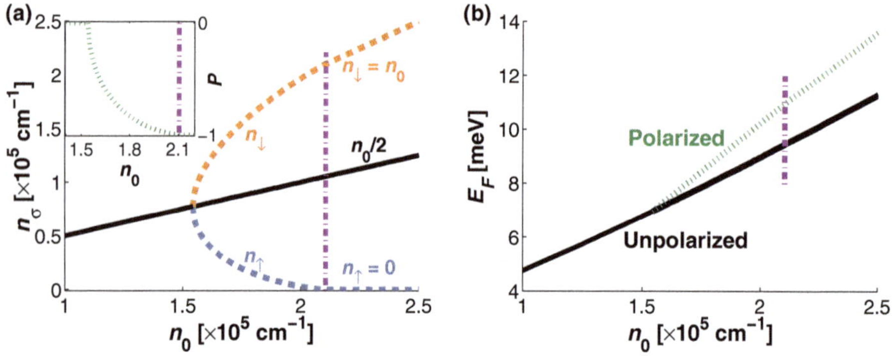

Figure 6. (a) Spin-polarized concentrations vs. total concentration n for the ↓ configuration when $Bx = 0$. Inset: polarization P of the ↓ configuration vs. n. (b) Fermi energy vs. n for both the polarized (solid) and unpolarized (dotted) regimes. The vertical dash-dotted line represents the concentration for full polarization.

Both spin-polarized regimes (↑ and ↓) sit at the same Fermi energy, which is higher than that of the unpolarized configuration. This is illustrated in Figure 6(b), a plot of the Fermi energy as a function of the total concentration. The Fermi energy at the onset of polarization is 7 meV for all configurations, but as n increases beyond n^{onset} the energy of the polarized regimes rises more rapidly, so that the difference in energies between both cases is about 1.5 meV at $n = n^{full}$. Therefore, the ground state of the electrons in the wire at zero temperature will be unpolarized, in agreement with the Lieb-Mattis theorem.[44] At higher temperatures, though, the system can be thermally excited to one of the spin-polarized configurations – especially for concentrations close to n^{onset}, for which the energy difference between the unpolarized and polarized regimes is smaller.

3.4. *Effects of confinement variations*

As mentioned earlier, the concentrations at the onset of polarization and at full polarization (n^{onset} and n^{full}, respectively) also depend on the strength of the transverse confinement along the *y*- and *z*-directions, i.e. on the shape and size of the cross-section of the wire. This dependence arises solely because of the exchange interaction, which manifests itself through the overlap function $\zeta_{ab}(p)$ in Equation (22).

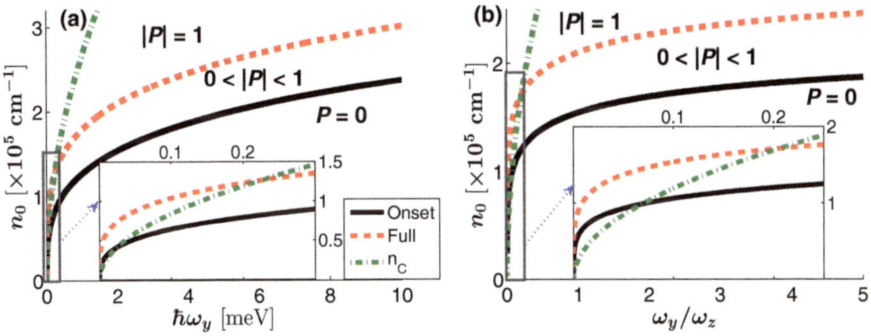

Figure 7. Concentration *n* at the onset of polarization (solid line) and at full polarization (dashed line) as a function of: (a) $\hbar\omega_y$ when $\hbar\omega_y = \hbar\omega_z \equiv \hbar\omega$; (b) ω_y / ω_z for $\hbar\omega_z = 2\,meV$. In both cases, $Bx = 0$ and $T = 0$. Insets: close-ups of regions $\hbar\omega < 0.3\,meV$ (in (a)) and $\omega_y / \omega_z < 0.25$ (in (b)); the dashed line represents the maximum concentration n_C for which only one subband is occupied.

Figure 7(a) shows the variation of n^{onset} (solid line) and n^{full} (dashed line) as a function of the confinement energy $\hbar\omega_y$ for a wire with a circular cross-section ($\hbar\omega_y = \hbar\omega_z \equiv \hbar\omega$) at $B_x = 0$. Both n^{onset} and n^{full} increase when $\hbar\omega$ increases, although the rate of increase slows downs and becomes almost constant for $\hbar\omega > 2\,meV$. The separation between n^{onset} and n^{full} (which corresponds to the concentration range with partial spin polarization) also becomes wider with a stronger confinement. These results show that in wider wires, i.e. for weaker confinements, there is a lower concentration for which polarization is possible. This is true as long as the system remains in the extreme quantum limit, which requires the concentration to be smaller than a critical value n_C given by:

$$n_C = \frac{2}{\pi\hbar}\sqrt{2m^*\left(\hbar\omega_y\right)} \tag{29}$$

$\hbar\omega_y$ is the energy above the bottom of the first subband at which the second subband will start to become filled (taking $\omega_y \leq \omega_z$). The inset of Figure 7(a) shows the n^{onset} and n^{full} curves, together with n_C, for small confinement strengths. The system falls outside the scope of the model ($n^{onset} > n_C$) when $\hbar\omega \lesssim 0.02\,meV$.

In Figure 7(b), the effects of asymmetric confinement on n^{onset} and n^{full} are portrayed for different confinement strength ratios ω_y / ω_z when $\hbar\omega_z = 2\,meV$. Both n^{onset} and n^{full} increase with the ratio, as a result of the growth of the overlap function ζ_{ab} with stronger confinement. The increase of n^{onset} and n^{full} is very rapid for $\omega_y < \omega_z$ and much slower for $\omega_y > \omega_z$. As in Figure 7(a), the spin-polarized regimes are accessible unless $n^{onset} > n_C$; the inset of Figure 7(b) shows that this limit is reached when $\omega_y / \omega_z \lesssim 0.068$, or $\hbar\omega_y \lesssim 0.14\,meV$.

4. Transport properties of a modulation-doped GaAs QPC[29]

4.1. *Description of a QPC structure*

We now apply our theory to the electronic transport through quantum point contacts, to explain their anomalous conductance features at $0.3G_0$ and ~0.6–$0.7G_0$. A quantum point contact is commonly realized in a GaAs/AlGaAs heterojunction consisting of a layer of delta-doped or modulation-doped AlGaAs on top of a GaAs substrate. The interface of the heterojunction is 200–300 nm under the surface. Figure 8 presents a schematic representation of a split-gate QPC with the *z*–axis perpendicular to the heterojunction interface, and larger positive *z*–values corresponding to points deeper in the GaAs substrate. The *x*–direction extends from the source to the drain electrode, while the *y*–axis runs between the gates. The center of the QPC lies at $(x,y)=(0,0)$. Electrons from the 2DEG on the source side of the structure flow into the QPC and are then collected at the drain. A negative potential applied to the split-gate electrodes depletes the 2DEG underneath m and confines the electrons laterally, along the *y*–direction.

Figure 9 shows the different contributions to the potential energy of electrons in the QPC. A depletion potential U_z, which results from the ionized GaAs acceptors, confines charge carriers into a 2DEG underneath the GaAs/AlGaAs hetero-interface, in the $x-y$ plane. Tunneling of electrons into AlGaAs ($z<0$) is prevented by the tall barrier at the interface. Electron energies are limited to discrete

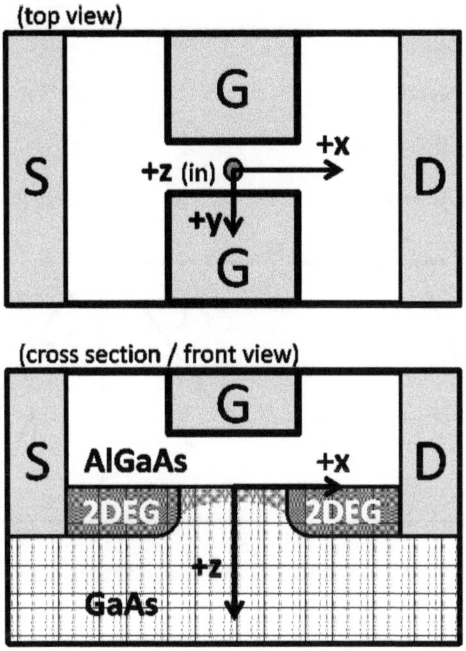

Figure 8. Schematic representation of a quantum point contact. S, D and G correspond to source, drain and gate, respectively.[29]

subbands; at zero temperature, only those subbands with energies below the Fermi energy E_F are occupied. Additionally, a saddle potential $U_{QPC}(x,y)$ is present in the QPC as a result of the finite width of the negatively-biased gate electrodes, and is made up of two contributions: a lateral potential energy well U_{well}, and a smooth potential barrier U_x running along the x-direction. The potential U_{well}, which for a QPC we model as parabolic (see Figure 9(a)), constrains electrons to a narrow channel between the gates along the y-direction, and splits each of the 2DEG subbands (whose energies far away from the QPC, i.e., $x \to -\infty$, are E_{z0}, E_{z1},...) into a new collection of discrete energy subbands with energies E_{y0}, E_{y1},... above those of the 2DEG subbands. Because of the saddle potential, the charge density at the QPC drops, which consequently leads to a reduction of the effect of electron-electron interactions and to a downward shift of the 2DEG subband energies. The energy E_{z0} of the lowest 2DEG subband (Figure 9(b)) is added to the (smallest) energy increase E_{y0} due to the lateral confinement and to the QPC potential barrier U_x, resulting in an effective 1D potential U_{eff}, as shown in Figure 9(c). We neglect the image potential, given that the difference between the dielectric constants of GaAs ($\kappa = 12.9$) and $Al_xGa_{1-x}As$ ($\kappa = 12.0$ for $x = 0.3$) is very small.

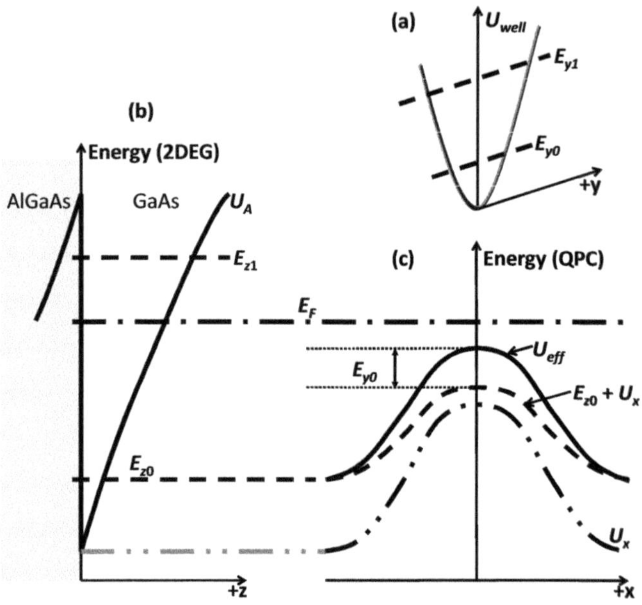

Figure 9. Energy levels and confinement potentials in the QPC. (a) Gate-induced confinement potential U_{well}, showing the energy sub-levels E_{y0}, E_{y1}. (b) Fermi energy and subbands in the 2DEG. (c) Effective potential, showing the individual contributions: energy E_{z0} of the lowest 2DEG subband, potential barrier U_x, and energy shift E_{y0} due to the parabolic confinement.[29]

4.2. *Hartree-Fock model of a long QPC*

We use the results from section 3 to a very long QPC, effectively a quantum wire. For now, we treat the quantum point contact as spin-unpolarized, i.e. $n_\uparrow = n_\downarrow = n/2$. We also allow for the electron concentration in the structure to vary with temperature; thus, we disregard the capacitive effect of the split gates, which would ensure that the concentration remains temperature-independent (i.e., that charge is conserved) over a large range of gate voltages.

In the absence of an applied split-gate voltage, the (2DEG) confinement potential in GaAs due to the depletion from ionized acceptors depends on the acceptor density N_A and the width of the depletion region $z_d = \sqrt{2\epsilon E_C / e^2 N_A}$ ($E_C = 1.52\, eV$ being the bottom of the conduction band in the bulk of GaAs). This potential is well approximated by[45]

$$U_z(z) = \frac{e^2 N_A}{\epsilon} z \left(z_d - \frac{z}{2} \right) \tag{30}$$

This potential satisfies the boundary conditions $U_z(z=0)=0$ and $[\partial U_z/\partial z](z=z_d)=0$ (i.e. the electric field is zero at $z=z_d$). Meanwhile, the parabolic lateral confinement potential induced by the gates is $U_y(y)=m^*\omega_y^2y^2/2$.

The Hartree and exchange terms U_{el} and U_{exch} are still given by Equations (12) and (13), although in this case the Coulomb interaction takes the form

$$U_{Coul}(\vec{r},\vec{r}')=\frac{e^2}{4\pi\epsilon}\left(\frac{1}{|\vec{r}'-\vec{r}|}-\frac{1}{\sqrt{(x'-x)^2+(y'-y)^2+(z'+z)^2}}\right) \tag{31}$$

The second term on the right-hand side corresponds to mirror charges placed on the AlGaAs side of the interface. With this expression for U_{Coul}, $U_{el}(\vec{r})$ satisfies boundary conditions resembling those of U_z, i.e., $U_{el}(z=0)=0$ and $[\partial U_{el}/\partial z](z=z_d)\approx[\partial U_{el}/\partial z](z\to\infty)=0$. (Additionally, this will ensure that the expectation value of U_{el} stays finite.)

We now find the 1D electron energy (Eq. (7)) using the following trial wave function:

$$\psi(\vec{r};k_x)=\frac{1}{\sqrt{L_x}}e^{ik_xx}\left(\frac{a^{1/2}}{\pi^{1/4}}e^{-a^2y^2/2}\right)\left(\frac{b^{3/2}}{2^{1/2}}ze^{-bz/2}\right) \tag{32}$$

This wave function is the product of a plane wave propagating along the x-direction, the ground state of the parabolic potential U_y, and an approximation of the ground state of the depletion potential U_z. For this wave function, the expectation values for the kinetic and potential energy terms are:

$$\left\langle-\frac{\hbar^2}{2m^*}(\partial_y{}^2+\partial_z{}^2)\right\rangle\equiv\langle T_{yz}\rangle=\frac{\hbar^2}{8m^*}(2a^2+b^2) \tag{33}$$

$$\langle U_y\rangle+\langle U_z\rangle=\frac{m^*\omega_y{}^2}{4a^2}+\frac{3e^2N_Az_d}{\epsilon b}\left(1-\frac{2}{z_db}\right) \tag{34}$$

As for the interaction terms U_{el} and U_{exch}, the expectation values are still given by Equations (17) and (18), except that now the overlap function is

$$\zeta_{ab}(p_x)=b\int_0^{+\infty}dp_y\exp\left(-\frac{p_y{}^2}{2a^2}\right)\frac{3P^4+18bP^3+44b^2P^2+54b^3P+33b^4}{(P+b)^6} \tag{35}$$

($P \equiv \sqrt{p_x^2 + p_y^2}$) The parameters a and b are not held fixed as in Section 4, but are instead determined by the variational method, i.e. by minimizing the average effective energy per electron U_{avg}, given by:

$$U_{avg} = E_{yz} + \frac{1}{2}U_{el} + J_{avg} \tag{36}$$

J_{avg} is the average exchange energy per electron:

$$J_{avg} = \frac{1}{2\pi n}\int_{-\infty}^{+\infty} dk_x\, U_{exch}(k_x) f_T\left[E(k_x)\right] \tag{37}$$

As a form of checking the validity of the model, we point out that the expressions for $\langle -(\hbar^2/2m^*)\partial_z^2\rangle \equiv \langle \hat{T}_z\rangle$ and $\langle U_z\rangle$ in the zero-lateral confinement case ($\omega_y = 0$) reduce to their 2DEG counterparts, which are given by $\langle \hat{T}_{z,2D}\rangle = \hbar^2 b^2/8m^*$ and $\langle U_{z,2D}\rangle = (3e^2 N_A z_d/\epsilon b)(1-2/bz_d)$. We also compare $\langle U_{el}\rangle$ (Eq. (17)) to the corresponding expression for a 2DEG, $\langle U_{el,2D}\rangle = 33e^2 n_{2D}/16\epsilon b$, by first expanding $\zeta_{ab}(0)$ in a series:

$$\zeta_{ab}(0) = \frac{33\sqrt{\pi}}{\sqrt{2}}\frac{a}{b} - 144\left(\frac{a}{b}\right)^2 + \frac{413\sqrt{\pi}}{\sqrt{2}}\left(\frac{a}{b}\right)^3 + \ldots \tag{38}$$

Then, to first order in a/b, $\langle U_{el}\rangle = 33e^2 na/16\sqrt{2\pi}\epsilon b$, which equals $\langle U_{el,2D}\rangle$ if n_{2D} is identified with $na/\sqrt{2\pi}$.

As mentioned earlier, in general one must solve an integral equation to obtain the energies $E(k_x)$ and the total concentration n for an arbitrary temperature. At zero temperature, this integral equation will reduce to a nonlinear equation for n in terms of E_F (provided that $n>0$):

$$\begin{aligned} E_F &= \frac{\pi^2\hbar^2 n^2}{8m^*} + U_x + E_{yz} \\ &+ \frac{e^2}{32\pi^2\epsilon}\left[2\pi n\zeta_{ab}(0) - \int_0^{\pi n} dk_x\, \zeta_{ab}(k_x)\right] \end{aligned} \tag{39}$$

Some additional simplifications can be made at very high and very low electron concentrations. At very high concentrations, the exchange term is negligible relative to the Hartree term. In this situation, by setting $E_F = E(k_F) = E(\pi n/2)$, we can obtain an explicit solution for n at $T = 0$:

$$n[T=0] \approx 2\sqrt{\frac{2m^*}{\pi^2\hbar^2}}\theta(E_F - U_x - E_{yz}) \times \left\{ -\sqrt{\frac{2m^*}{\pi^2\hbar^2}}\left(\frac{e^2\zeta_{ab}(0)}{16\pi\epsilon}\right) + \sqrt{\frac{2m^*}{\pi^2\hbar^2}\left(\frac{e^2\zeta_{ab}(0)}{16\pi\epsilon}\right)^2 + (E_F - U_x - E_{yz})} \right\} \tag{40}$$

For $T > 0$, n can be obtained by solving the following expression numerically ($F_{-1/2}$ is the Fermi-Dirac integral of order $-1/2$):

$$n[T] = \sqrt{\frac{2m^* k_B T}{\pi^2\hbar^2}} F_{-1/2}\left[\frac{1}{k_B T}\left(E_F - U_x - E_{yz} - \frac{e^2\zeta_{ab}(0)}{16\pi\epsilon} n[T]\right)\right] \tag{41}$$

Meanwhile, at very low concentrations, the integral in Eq. (39) is approximately equal to $\pi n \zeta_{ab}(0)$, since $\zeta_{ab}(k) = \zeta_{ab}(0) + O(k^2)$. Therefore, in this case, $\langle U_{exch}\rangle \approx -\langle U_{el}\rangle / 2$ and, by substituting $\zeta_{ab}(0)$ with $\zeta_{ab}(0)/2$ in Equations (40) and (41), they can be used to get an approximate solution for the concentration in the presence of the exchange interaction.

Figure 10(a) and (b) show the variational parameters a and b, normalized to their zero-concentration values $a_0 = \sqrt{m^*\omega_y / \hbar}$ and $b_0 \approx (12 m^* e^2 N_A z_d / \hbar^2 \epsilon)^{1/3}$, as functions of n for different confinement strengths $\hbar\omega_y$. We see that, as n increases, a decreases relative to a_0 (so the characteristic length of the wavefunction along y, which is proportional to $1/a$, increases). This indicates that electron-electron interactions force the electrons to spread out over a wider region along the y-axis, thus counteracting the effects of the gate-induced lateral confinement. As shown in the inset of Figure 10(a), a_0 itself increases with increasing lateral confinement strength. Meanwhile, when n increases, b decreases relative to b_0 (see Figure

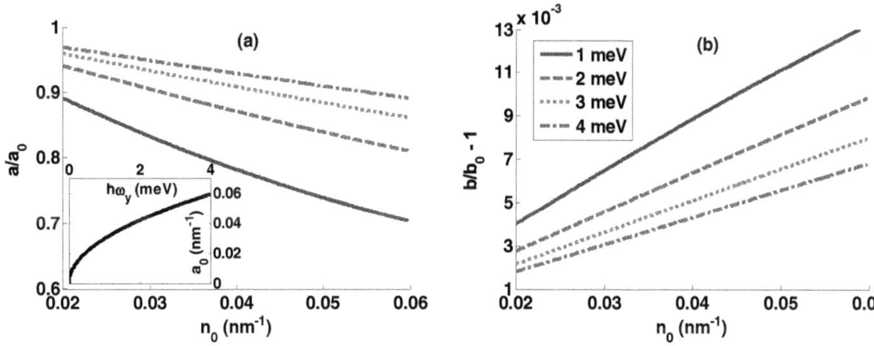

Figure 10. (a) a/a_0 and (b) $b/b_0 - 1$ vs. electron concentration n for different confinement strengths $\hbar\omega_y$ in the wire, from 1 meV (solid line) to 4 meV (dot-dashed line) at $T = 0$. $N_A = 10^{14}$ cm^{-3}. Inset in (a): a_0 vs. $\hbar\omega_y$. For (b), $b_0 = 0.190$ nm^{-1}.

10(b)), meaning that electrons are confined to a thinner surface near the heterojunction interface. Both parameters vary more significantly for smaller $\hbar\omega_y$, but a is more sensitive to changes in electron density than b: when $\hbar\omega_y = 1\,meV$, a decreases by about 25% while b increases just by 1%.

Temperature changes have only a minor effect on a and b, as shown in Figure 11(a) and (b). Even at 1 K, a and b do not vary significantly relative to their zero-temperature values. At most, for very low concentrations, a decreases by less than 0.5% relative to a_0, while b increases by less than 0.05% relative to b_0. At electron concentrations above 0.01 nm^{-1}, the effect of temperature on the variational parameters is negligible.

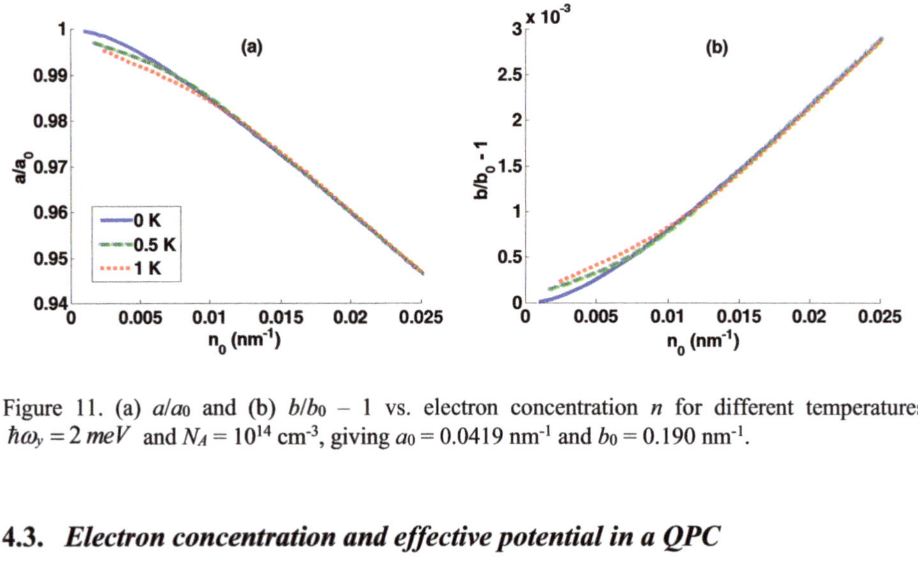

Figure 11. (a) a/a_0 and (b) $b/b_0 - 1$ vs. electron concentration n for different temperatures. $\hbar\omega_y = 2\,meV$ and $N_A = 10^{14}$ cm^{-3}, giving $a_0 = 0.0419$ nm^{-1} and $b_0 = 0.190$ nm^{-1}.

4.3. *Electron concentration and effective potential in a QPC*

In a QPC of finite length, the potential term U_x is no longer constant, but instead depends on x. Additionally, the lateral confinement U_y will now drop to zero for large x. Therefore, the Schrödinger equation (Eq. (1)) is no longer separable. Nevertheless, when the potential barrier U_x varies slowly and smoothly relative to the confinement potentials U_y and U_z, an adiabatic approximation can be used in Eq. (1) to find local energy levels $E(k_x, x)$, a local electron concentration $n(x)$ and x-dependent parameters a and b.

Initially, we set $U_x(x) = U_0 \operatorname{sech}^2(x/x_0)$. The QPC barrier height U_0 gets taller as the gate voltage becomes more negative, while x_0 increases together with the length of the QPC. Given that the confinement length in the y-direction $(1/a)$ is of a few tens of nanometers, the adiabatic approximation should remain valid for x_0 up to $\sim 100\,nm$. This limit on x_0 also ensures that transport through the QPC is ballistic, since the mean free path in GaAs is of the order of several microns.[16,17] Meanwhile, the lateral confinement now takes the form

$U_y(x,y) = m^*[\omega_y(x)]^2 y^2/2$, where we take $\omega_y(x) = \omega_{y,\max}\sqrt{U_x(x)/U_0}$. These expressions are then used to calculate the effective potential $U_{eff}(k_x, x)$.

As for the Fermi level, it is determined by the electron density n_{2D} in the 2DEG far away from the QPC, where the effective potential reduces to $U_{eff}(x \to -\infty) = \langle \hat{T}_z \rangle + \langle U_z \rangle + U_{ee}$. Then,

$$E_F - U_{eff}(x \to -\infty) = \frac{\pi \hbar^2 n_{2D}}{m^*} \tag{42}$$

Figure 12(a) and (b) show the profile of the effective potential U_{eff} (when the particle energy is $E = E(k_F) = E_F$) for two different values of the maximum effective potential $U_{eff,max}$, at $T = 0$. We take $n_{2D} = 8.4 \times 10^{10}\ cm^{-2}$ in order to obtain a Fermi energy of 3 meV above U_{eff} (from Eq. (42)). The solid and dashed lines correspond to the effective potential when the exchange interaction is included or ignored, respectively. The separation between the dotted and solid (or dashed) lines represent the contribution U_{ee} of electron-electron interactions; this contribution

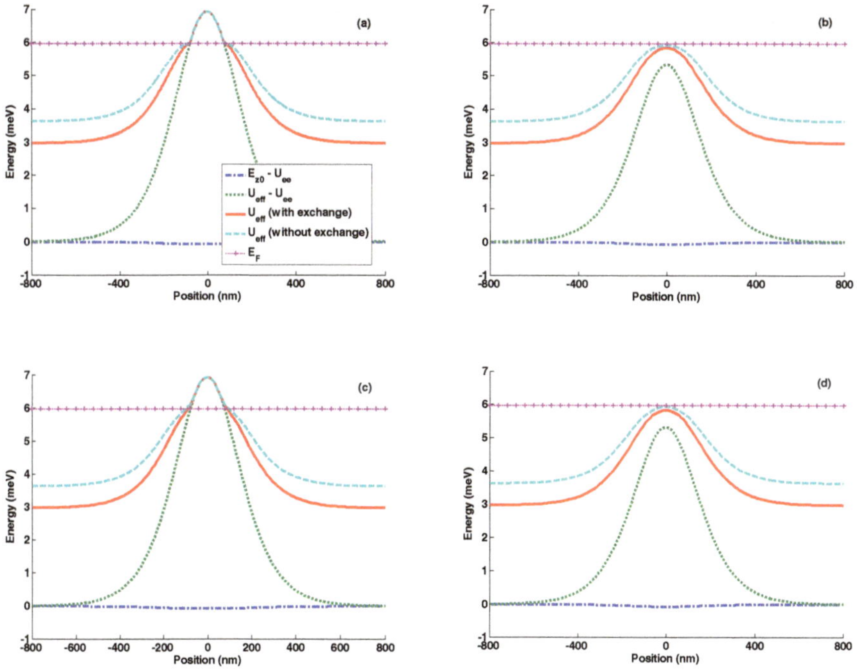

Figure 12. Effective potential vs. position when: (a), (c) $U_{\text{eff,max}} > E_F$ (U_0 = 7 meV); (b), (d) $U_{\text{eff,max}} < E_F$ (U_0 = 5.4 meV). For (a) and (b), T = 0 K; for (c) and (d), T = 1 K. All energies are measured relative to $(E_{z0} - U_{ee})_{x\to-\infty}$. $\hbar\omega_{y,max} = 2\ meV$, $N_A = 10^{14}$ cm^{-3}, $n_{2D} = 8.4\times10^{10}$ cm^{-2} and x_0 = 200 nm.

is more significant far away from the QPC, where the barrier height U_x drops to zero.

As U_x increases near $x=0$, fewer electron states are populated, so U_{ee} becomes smaller and U_{eff} increases less rapidly. Figure 12(a) shows a kink or shoulder in the effective potential at the point where $U_{eff} = E_F$, resulting from the onset of Coulomb interactions. When $U_{eff} > E_F$, the 1D channel is depleted and $U_{ee} = 0$, so U_{eff} varies at the same rate as U_x. The kink is absent in Figure 12(b) because the effective potential in that case remains below E_F throughout the QPC; nevertheless, U_{ee} is still smaller near the center of the QPC relative to points that are farther away.

Figure 12(c) and (d) display the same plots at $T = 1\,K$. The kink in the effective potential is less pronounced at the higher temperature because the electron concentration does not vanish anymore when $U_{eff} > E_F$, and as a result the electron-electron interaction terms contribute to the effective potential, especially when U_{eff} is still very close to E_F. At points far from the center of the QPC, the effects of temperature on the effective potential are negligible.

In Figure 13(a) we plot the maximum effective potential $U_{eff,max}$ as a function of the maximum barrier height U_0 for different temperatures. At zero temperature, when the QPC is "pinched off," i.e. when U_0 lies above the Fermi energy and there are no electrons in the 1D channel, $U_{eff,max}$ decreases linearly with U_0. Then, when $U_{eff,max}$ drops below the Fermi level at $U_0 \approx 6\,meV$, it stays "pinned" (practically constant) over a 0.25 meV-range of U_0 values while the channel opens. For $U_0 \lesssim 5.75\,meV$, it decreases at a slow rate as the QPC channel becomes more populated with electrons. The pinning effect results from the compressibility peak in the 1D electron gas, in agreement with previous works.[46-48] This effect is also

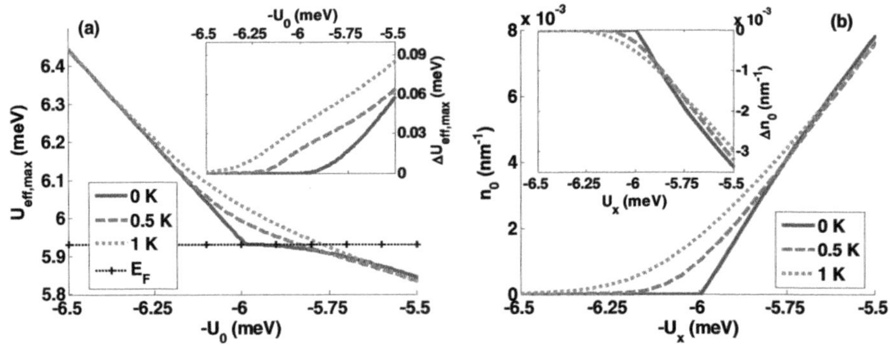

Figure 13. (a) Maximum effective potential vs. U_0, including exchange effects, for various temperatures. Inset: $\Delta U_{eff,max} \equiv U_{eff,max}(\text{w/o exchange}) - U_{eff,max}(\text{w/exchange})$ vs. U_0. (b) Electron density vs. U_x, including exchange effects, for various temperatures. Inset: $\Delta n \equiv n(\text{w/o exchange}) - n(\text{w/exchange})$. All parameters (apart from temperature) as in Figure 12.

present when the effective potential becomes smaller than E_F on both sides of the QPC while the 1D channel is still closed, and lead to the "kinks" observed in thc barrier profile (Figure 12(a)). At non-zero temperatures, when the electron concentration in the QPC is not zero for energies above E_F, the pinning is less effective.

The dependence of the concentration n on U_x is shown in Figure 13(b). When U_x decreases past the critical point at $U_x \approx 6\, meV$, n starts to increase linearly with $-U_x$, as can be inferred from Eq. (40) in the low-concentration limit. Meanwhile, when the temperature rises, n increases noticeably near the effective potential crossover but remains mostly unaffected well below pinch-off.

The insets of Figure 13(a) and (b) illustrate the changes in $U_{eff,max}$ and n if exchange effects are neglected. By neglecting the (negative) exchange term, the effective potential increases; consequently, the carrier concentration in the absence of exchange effects is lower relative to the case with exchange. Additionally, $U_{eff,max}$ becomes less sensitive to changes in U_0, which indicates that the absence of the exchange interaction enhances the pinning of the effective potential. When the temperature goes up, the increase in $U_{eff,max}$ becomes even more pronounced.

4.4. *The 0.3G_0 conductance kink; effects of temperature, QPC length and potential barrier shape on the conductance*

The conductance through the quantum point contact results from quantum-mechanical transmission through the effective potential barrier $U_{eff}(k_x, x)$, and is obtained using the Landauer formula:[49]

$$G = \frac{e^2}{h}\sum_{\sigma}\int dE\left(-\frac{\partial f_T}{\partial E}\right)\mathrm{T}_\sigma(E) \tag{43}$$

We calculate the transmission coefficient $\mathrm{T}_\sigma(E) = \mathrm{T}(E(k_x,\sigma))$ with the transfer matrix method,[30] which requires finding the effective potential at sample points along the x-axis.

A plot of the transmission coefficient $\mathrm{T}(E = E_F)$ as a function of U_0 (Figure 14(a)) reveals that, at $T=0$, there is an anomaly or kink near 0.35 when U_0 reaches the critical point of $\sim 6\, meV$ at which $U_{eff,max}$ crosses E_F. This kink results from the pinning of the effective potential just below E_F and is more pronounced when the exchange interaction is neglected (dashed curve), since in that case the effective potential is enhanced. By $T = 1\,K$, though, the anomaly is largely smoothed out, as displayed in Figure 14(b). This is a consequence of the thermal smearing of the effective potential pinning illustrated in Figure 13. At zero temperature, the height $U_{eff,max}$ of the effective potential barrier increases abruptly

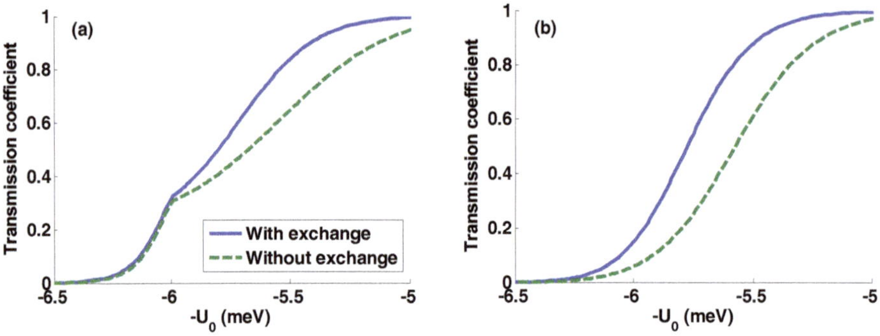

Figure 14. Transmission coefficient vs. maximum barrier height U_0 for (a) $T = 0$ K; (b) $T = 1$ K, showing the curves when the exchange interaction is present (solid line) or absent (dashed line). All other parameters as in Figure 12.

for a sufficiently-negative gate voltage, and therefore the transmission coefficient drops rapidly, leading to the anomaly in the transmission coefficient curve. However, due to the modulation of the potential barrier with increasing temperature, the changes in $U_{eff,max}$ are smoother and the transmission coefficient varies less suddenly, thus softening the kink. When the exchange interaction is neglected, the effective potential barrier is taller, thereby resulting in a lower transmission coefficient (dashed lines in Figure 14).

In Figure 15(a) we show the QPC conductance as a function of $-U_0$ for different temperatures, including exchange effects on the barrier. At $T = 0\,K$, the conductance is proportional to the transmission coefficient $\mathrm{T}(E = E_F)$, so the kink at $G \sim 0.35 G_0$ persists. However, as temperature is increased even by a few tenths of a kelvin, the anomaly is smeared out. This thermal smearing is confirmed in

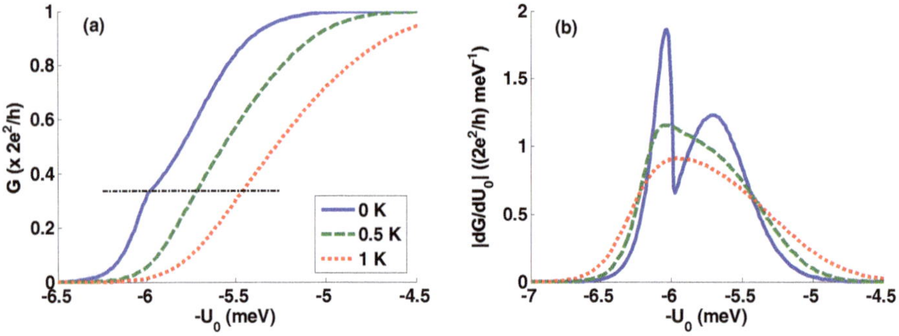

Figure 15. (a) Conductance vs. U_0 for different temperatures. For clarity, each successive curve is shifted to the right by 0.25 meV. The dot-dashed line specifies the location of the conductance kink at pinch-off. (b) Slope of the conductance, $|dG/dU_0|$, vs. U_0. Exchange effects are included. All parameters as in Figure 12.

Figure 15(b), where we plot the slope $|dG/dU_0|$ of the conductance. At zero temperature, the conductance slope exhibits a pair of peaks separated by an abrupt dip (signaling the kink in the conductance) at $U_0 \sim 6\ meV$; a sharp maximum before $U_0 = 6\ meV$ is followed by a broader and lower maximum, which signifies that the conductance anomaly is not merely a slope change, but is rather due to the onset of the 1D compressibility peak of the electron gas. However, as temperature increases the dip in the doubly-peaked conductance slope is smoothed out, leaving only a single and broad maximum.

Figure 16(a) displays the sensitivity of the zero-temperature conductance to variations in the length of the QPC. This is accomplished by varying the characteristic length x_0 of the potential barrier (defined as $U_x = U_0 \operatorname{sech}^2(x/x_0)$). A decrease in x_0, corresponding to a shorter QPC, causes the conductance anomaly to soften, broaden and shift upwards, towards $G \sim 0.4G_0$. Simultaneously, the conductance is spread out over a wider range of energies U_0 and its slope decreases. The upward displacement of the kink is a result of the narrower barrier, for which the transmission is enhanced at low energy compared to the conductance in longer QPCs.

A more significant effect on the conductance is observed if the profile of the U_x potential barrier changes. In Figure 16(b), we calculate the QPC conductance at $T = 0$ K when the potential barrier has the form $U_x = U_0\left[1+|x/x_0|^N\right]^{-1}$, where $N > 1$. Compared to the $\operatorname{sech}^2(x/x_0)$ potential shape, this function has a longer tail for $x > x_0$ and a steeper drop at $x=0$, with the effect being amplified as the value of the exponent N decreases. As can be seen, when N is reduced the anomaly is displaced upwards, from $G \sim 0.4G_0$ for $N = 2$ to $G \sim 0.6G_0$ when $N = 1.25$.

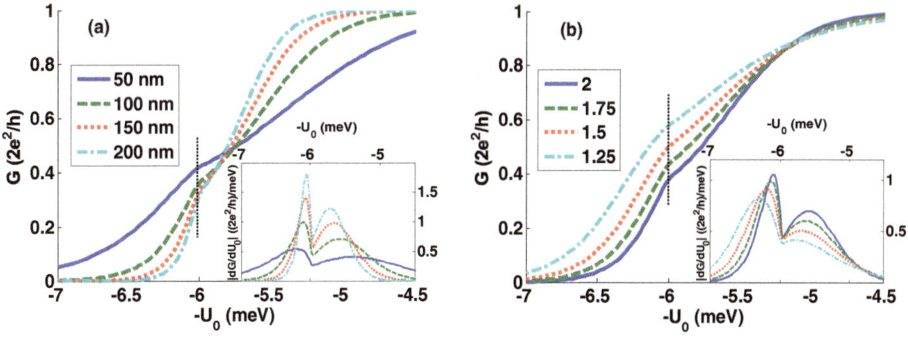

Figure 16. Conductance at $T = 0$ vs. U_0 for: (a) $U_x = U_0 \operatorname{sech}^2(x/x_0)$, for different x_0; (b) $U_x = U_0\,[1 + (x/x_0)^N]^{-1}$, for different N. Exchange effects are included. All parameters (except for x_0 in (a)) are the same as in Figure 12. The thin vertical dotted line shows the location of the conductance kink at pinch-off. Insets: slope of the conductance vs. U_0.

Additionally, the kink broadens and shifts to lower U_0 potential values. In general, the transmission is higher for lower-N potentials; the exception is when U_0 is well below the value at which the anomaly occurs, although at that point the conductance is almost equal to G_0 regardless of the value of N.

5. Charge conservation, spin polarization and the emergence of the 0.7 structure in a QPC

A comprehensive understanding of the "0.7 structure" or "anomaly" should account for its notable experimental features: (a) its universality, as it is observed in a wide variety of QPC configurations;[18-20] (b) its dependence on temperature T, as the shoulder widens into a plateau as temperature is increased above 0 K and becomes most evident at $T \sim 1-2\,K$, although the conductance and gate voltage values at the onset of the anomaly are both independent of temperature;[18,50,51] (c) its variability when an in-plane magnetic field is present, suggesting the existence of spin polarization even when the magnetic field is removed;[18,40,51-55] (d) its dependence on the source-drain bias V_{sd}, for which the anomaly shifts from $\sim 0.7G_0$ at low V_{sd} to $\sim 0.85G_0$ at high V_{sd};[56] and (e) its sensitivity to the length of the QPC as the shoulder drops to $\sim 0.5G_0$ for long constrictions.[57,58]

Although it is generally agreed that the 0.7 structure results from many-body effects, there continues to be a significant debate as to the precise cause of this phenomenon.[2,27] A variety of studies have attributed the anomaly to spin correlations in the carrier transport, either in the form of a Kondo-like effect caused by a quasi-bound state in the QPC,[20,59-64] or of spontaneous spin polarization in the constriction, [18,40,51-55] or both.[65] A spin-polarized ground state is forbidden in strictly 1D systems by the Lieb-Mattis theorem,[66] but could be achieved in realistic QPCs (and QWRs) since these devices have two- and three-dimensional structures; this possibility has generated additional interest in fully-electric manipulation of spin properties in quantum wires and point contacts.[67,68] However, none of the proposed theoretical explanations has fully clarified the origin of the 0.7 anomaly and simultaneously described all of the features (a)-(e) listed above.[2,3]

5.1. *Coexistence of spin-polarized channels under charge conservation*

We now consider the effects of spin polarization (as described in Section 4 for a quantum wire) on the electronic and transport properties of the quantum point contact at various temperatures. In a realistic QPC, the profile of the total carrier concentration $n = n_\uparrow + n_\downarrow$ is largely temperature-independent for fixed gate bias, because the split-gate system is only geometry- and material-dependent. Therefore, for convenience, the total concentration can be related to the zero-temperature Fermi energy $E_{F,0}$ above pinch-off. This observation is crucial as it guarantees the persistence of a conductance anomaly at $\sim 0.6-0.7G_0$, independently of temperature, whereas the spin-dependent concentrations n_σ are still functions of a temperature-dependent Fermi energy $E_F(T) \neq E_{F,0}$ (Eq. (19)).

We showed in Section 3.3 that, below a threshold value n^{onset} (corresponding to a threshold Fermi level $E_{F,0}^{onset}$), the only solution for $n_\sigma(T)$ and $E_F(T)$ is spin-unpolarized ($P=0$); we refer to it as the "0" configuration. For concentrations above n^{onset}, two additional spin-polarized solutions are possible, either with $P>0$ or $P<0$. They are denoted as the "+" and "–" configurations, respectively.

These spin configurations (or P-channels) in QPCs are not independent, but instead coexist in the 1D constriction, leading to two important consequences. First, given that the total electron charge in the split-gate capacitor remains constant with temperature (as mentioned earlier), the sum of the per-channel concentrations $\hat{n}_\sigma{}^\nu$ over all P-channels $\nu = (+),(0),(-)$ satisfies the condition $\hat{n}_\sigma{}^{(+)} + \hat{n}_\sigma{}^{(0)} + \hat{n}_\sigma{}^{(-)} = n/2$. Second, since the concentration in each P-channel is less than $n/2$, these P-channels are not filled up to the actual gate-imposed Fermi energy $E_{F,0}$, but instead up to their corresponding "effective" Fermi energies $\hat{E}_F{}^{(\pm)}$ (identical for both spin-polarized channels) and $\hat{E}_F{}^{(0)}$ (for the unpolarized channel), both of which are lower than $E_{F,0}$. At $T=0$, the Fermi energy reduces to $\hat{E}_F{}^\nu(T=0) = E_\uparrow{}^\nu\left(\hat{k}_{F,\uparrow}^\nu; T=0\right) = E_\downarrow{}^\nu\left(\hat{k}_{F,\downarrow}^\nu; T=0\right)$, where $\hat{k}_{F,\sigma}^\nu = \pi \hat{n}_\sigma{}^\nu$.

Figure 17(a) displays the temperature-dependent Fermi energies $E_F(T)$ at the center of the QPC as a function of the concentration at different temperatures. We set $\hbar\omega_y = 0.1\, meV$ and $N_A = 10^{14}\, cm^{-3}$. The constant confinement energy $E_z + \hbar\omega_y/2 = 15.47\, meV$ is subtracted from all energy values. At very low carrier concentrations the Fermi energies are strongly dependent on temperature, with lower E_F values corresponding to higher temperatures. However, as the carrier concentration increases, the Fermi energies coalesce towards the zero-temperature value $E_{F,0}$, i.e. the Fermi energy is temperature-independent and determined by the total carrier concentration. With the materials and confinement parameters above, the onset of polarization occurs at $n^{onset} = 6.4 \times 10^4\, cm^{-1}$, corresponding to $E_F^{onset} = 0.23\, meV$, at which point the polarized configurations become accessible. Figure 17(b) and (c) show the effective Fermi energies for the unpolarized and spin-polarized channels, respectively. Above the concentration threshold, because of the charge (carrier) conservation in the 1D channel, there is an abrupt drop of the effective Fermi energies at the polarization onset, from $E_F = E_F^{onset} = 0.23\, meV$ to $0.03\, meV$ at $0\, K$ and $0.08\, meV$ at $1\, K$, when charge is redistributed between the different channels because of thermal excitations. For the unpolarized configuration, $\hat{E}_F^{(0)}$ continues to drop as the concentration increases, with the dropping

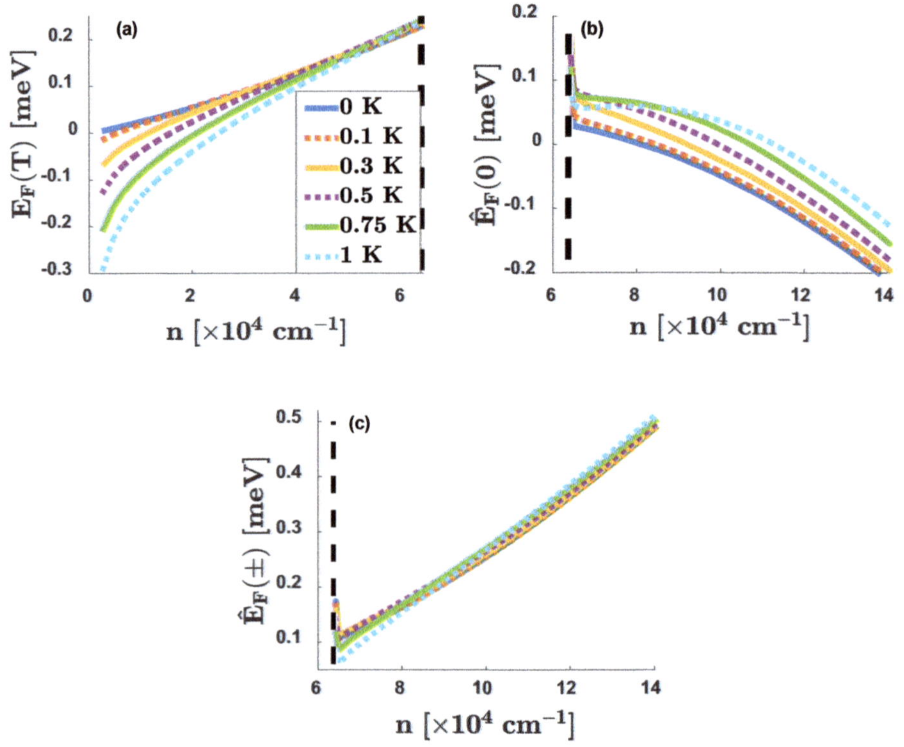

Figure 17. (a) Fermi energy $E_F(T)$ vs. total electron concentration *n* before the onset of spin polarization for different temperatures. (b) and (c) Effective Fermi energy $\hat{E}_F(T)$ vs. total electron concentration for the (b) unpolarized and (c) polarized channels, respectively. The vertical dashed lines indicate the concentration at the polarization onset. $\hbar\omega_y = 0.1\, meV$, $N_A = 10^{14}$ cm^{-3}.

rate being relatively temperature-independent except at the highest temperatures considered (0.75 K, 1 K). At $n \sim 1.4 \times 10^5\ cm^{-1}$, $\hat{E}_F^{(0)}$ decreases to $-0.2\, meV$ at $T = 0$. In general, the effective unpolarized Fermi energy is larger at higher temperatures, except for concentrations close to the polarization onset. Meanwhile, for the polarized channels, after the initial drop the effective Fermi energy $\hat{E}_F^{(\pm)}$ actually increases with increasing *n*, achieving a value of $0.5\, meV$ when $n = 1.4 \times 10^5\ cm^{-1}$. The initial drop is bigger for higher temperatures, but all curves tend to converge as *n* increases.

The carrier concentrations for the specific case of spin-up electrons in the different P-channels are shown in Figure 18(a-c) as functions of the total concentration. At the onset of polarization, all P-channels have the same concentration, $\hat{n}_\sigma^{(+)} = \hat{n}_\sigma^{(0)} = \hat{n}_\sigma^{(-)} = n/6$. For concentrations above the onset, the $P < 0$ and $P = 0$ channels depopulate. At zero temperature, this happens very rapidly, within

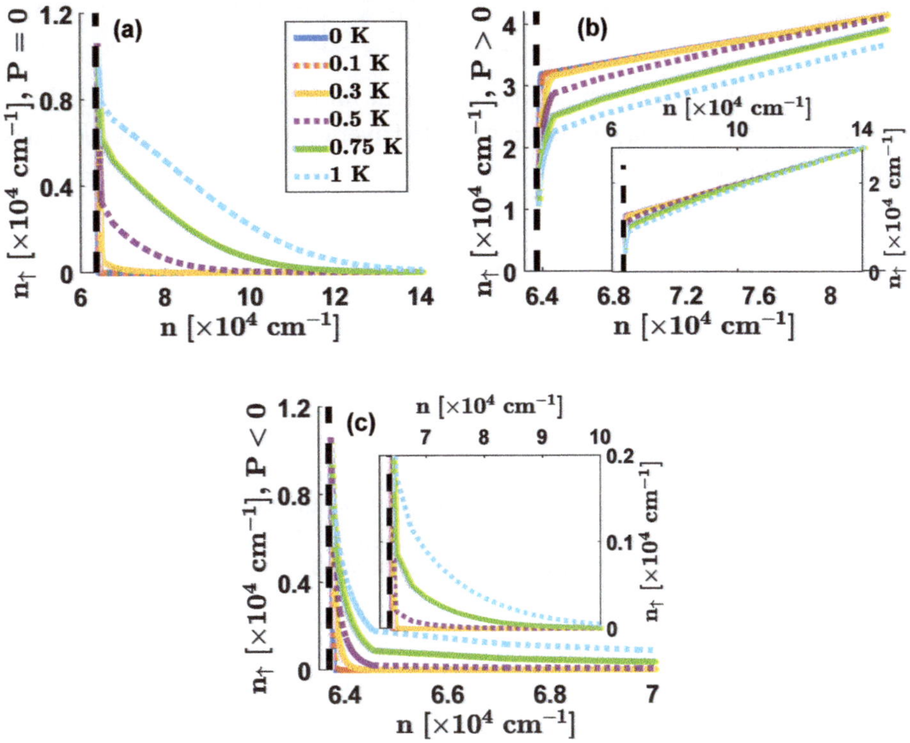

Figure 18. Per-channel concentrations of spin-up electrons vs. total concentration n for the three different channels: (A) unpolarized ($P = 0$); (B) net spin-up polarization ($P > 0$); (C) net spin-down polarization ($P < 0$). Insets: extended range. The vertical dashed lines indicate the concentration at the polarization onset.

a concentration range of $\sim 10^3\ cm^{-1}$ above n^{onset} (which corresponds to $\sim 1\ \mu eV$ above E_F^{onset}). Subsequently, all electrons are in the $P > 0$ channel. The depopulation rate decreases with increasing temperature, especially for the $P = 0$ channel. As a result, the carrier concentration in the $P > 0$ channel decreases too. These behaviors are a direct consequence of the variations of $\hat{E}_F^{(0)}$ and $\hat{E}_F^{(\pm)}$ relative to the minimum electron energy in each P-channel.

5.2. *The 0.7G_0 conductance plateau at the onset of polarization*

The conductance through the QPC for each P-channel is calculated with the Landauer formula (Eq. (43)), modified to account for the effective Fermi energies of each channel:

$$G_\sigma{}^\nu = \frac{e^2}{h}\int_{-\infty}^{+\infty} dE\left[-\frac{\partial f_T\left(E;\hat{E}_F{}^\nu\right)}{\partial E}\right]\mathrm{T}_\sigma{}^\nu\left(E\right) \tag{44}$$

Here, $\mathrm{T}_\sigma{}^\nu\left(E\right)$ is the transmission coefficient for electrons of spin σ in the P-channel ν. The total conductance is a weighted average of the per-channel values, summed over both spins:

$$G = \frac{1}{n}\sum_{\sigma,\nu}\hat{n}_\sigma{}^\nu G_\sigma{}^\nu \tag{45}$$

For our calculations we use the following transmission coefficient function:

$$\mathrm{T}_\sigma{}^\nu\left(E\right) = \left\{1+\gamma\exp\left[-\frac{E-U^\nu_{eff,\sigma}\left(E\right)}{\Gamma}\right]\right\}^{-1} \tag{46}$$

This expression is dictated by the results shown in Section 4.4, and its parameters account for variations in the shape and length of the barrier. The dimensionless parameter γ determines the transmission coefficient at pinch-off (i.e., when $E = U_{eff}$). In particular, in long QPCs the transmission coefficient drops to a lower value,[29] which leads to an increase in γ. Meanwhile, very smooth barriers (as in slower-rising, i.e., weaker, confinement potentials) experience a reduction of the transmission slope after the pinch-off,[29] which corresponds to a large energy parameter Γ modulating the transmission.

Figure 19(a)-(c) shows the conductance above pinch-off for spin-up electrons in each of the three configurations, with respect to the Fermi energy for different temperatures. Here, we use the Fermi energy $E_{F,0}$ at $T = 0\,K$ that characterizes the carrier concentration in the channel as a function of gate voltage, independently of temperature; since $E_{F,0} \propto n \propto V_g$, these plots are equivalent to plots of $G\left(V_g\right)$, the conductance as a function of gate bias. We set $\gamma = 7/3$ to obtain a realistic transmission coefficient $\mathrm{T}=0.3$ at channel pinch-off.[29] At the polarization threshold, all conductances are identical and significantly lower than just before the threshold as carriers are redistributed among the three channels. As $E_{F,0}$ increases and the $P=0$ and $P<0$ channels are depopulated, the corresponding conductances drop (as expected, because the potential barriers are much higher than the respective effective Fermi energies). The drop is less pronounced for higher temperatures, reflecting the fact that more electrons can now populate those two channels as the temperature rises. Meanwhile, in the $P>0$ channel, the conductance rises with the QPC Fermi energy $E_{F,0}$, but the rise is slower at higher temperatures. We find that $\Gamma = 0.0335\,meV$ yields a conductance of $0.7G_0$ at the onset of polarization.

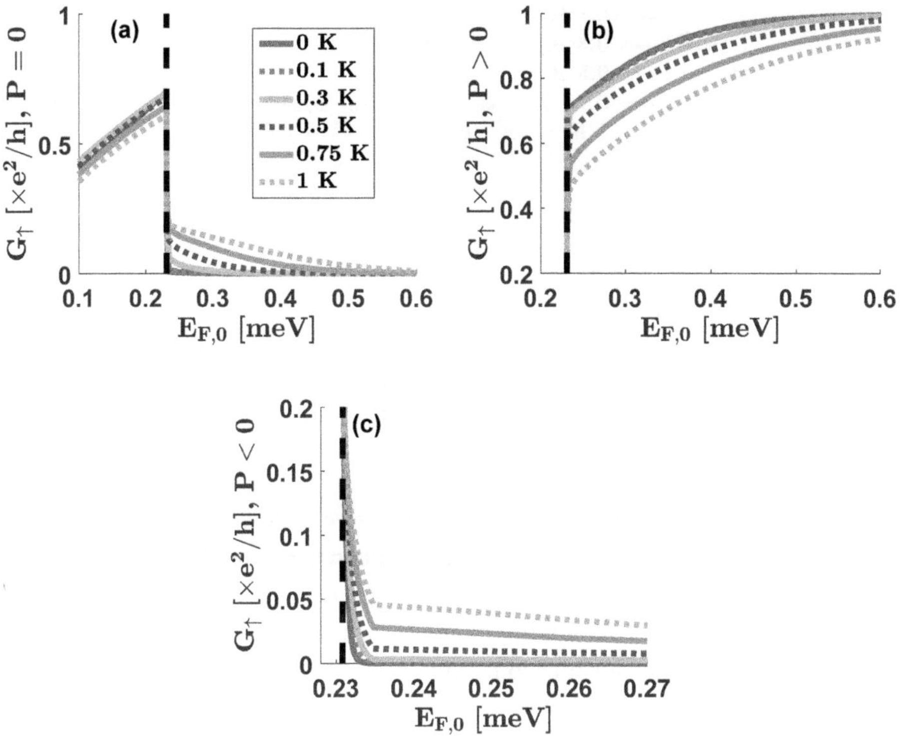

Figure 19. Per-channel conductance for spin-up electrons versus Fermi energy for different channels. (A) unpolarized ($P=0$); (B) net spin-up polarization ($P>0$); (C) net spin-down polarization ($P<0$). $\gamma=7/3$, $\Gamma=0.0335\,meV$. The vertical dashed line indicates the value of the Fermi energy at the onset of polarization.

In Figure 20, we plot the total conductance, averaged over all spin channels, as a function of the zero-temperature Fermi energy $E_{F,0}$ used to calculate the electron charge in the QPC induced by the gate voltage, independent of temperature. The zero of the Fermi energy $E_{F,0}$ refers to the channel pinch-off, where a small "kink" in the conductance at $G\sim 0.25-0.3G_0$ and $T=0\,K$ observed experimentally[26] is due to the singular nature of the 1D density of states when the channel emerges with increasing gate biases.[29,46]

This kink vanishes promptly once the temperature increases by a few hundreds of mK.[29] This happens because the singularity in the density of states is smeared out by the temperature-dependent Fermi function derivative in the Landauer formula (Eq. (44)), thereby causing the conductance at $T>0$ to differ from the zero-temperature value. In this situation, a vanishingly small carrier concentration may not be conserved for a particular Fermi level as the temperature rises, as seen in

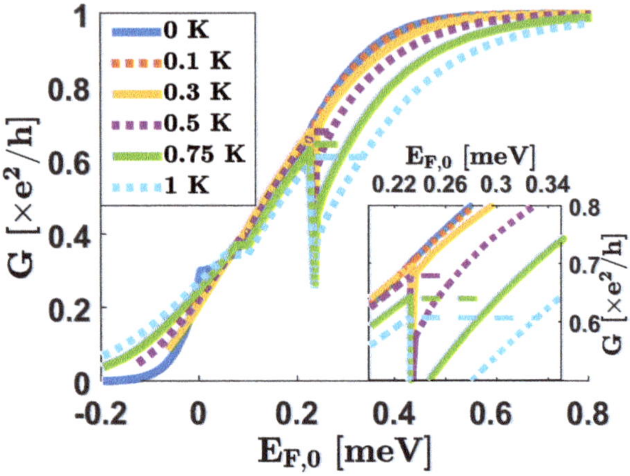

Figure 20. Total conductance versus Fermi energy for different temperatures. The dashed lines correspond to the conductance plateaus that replace the abrupt conductance drops at the onset of polarization. Inset: close-up of the conductance plateaus at the onset of polarization.

Figure 17(a). We call this regime "weakly capacitive". However, for $E_{F,0}$ continues to increase, gate capacitors overtakes the influence of the 1D density of states. For $E_{F,0} \geq 0.1\, meV$, the capacitive action of the gate dominates, and at a given $E_{F,0}$ the total electron concentration, uniquely defined by the Fermi level (Figure 17(a)), remains the same for all temperatures. On Figure 20, we have indicated with a shaded stripe the region where the conductance for the two capacitive regimes should match.

Above a fixed $E_{F,0} = E_F^{onset} = 0.2\, meV$, i.e. at a fixed threshold concentration n^{onset} above pinch-off (corresponding to the onset of polarization), the conductance drops, as electrons redistribute among the different channels with different barrier heights and the per-channel conductance (Figure 19) decreases. This sudden reduction occurs for the same conductance value $\sim 0.7G_0$ at all temperatures. After the drop, the conductance rises again as the gate bias increases, but the rise becomes slower as the temperature increases. A similar drop, although not as abrupt, in the conductance at the anomaly onset has been observed in earlier measurements in QPCs that were fabricated on undoped GaAs/AlGaAs heterostructures, either with split Schottky gates[19,69] or with a top gate between a pair of side gates.[70,71] The absence of a pronounced negative differential resistance (NDR) at $\sim 0.7G_0$, where a shoulder or a weak NDR are instead observed experimentally, results from the electrostatic nature of the confinement that controls the QPC barrier height relative to the Fermi level, the latter of which is fixed by its value in the 2D leads. Indeed, while increasing the bias of the confining gates lowers the barrier, the thermal carrier redistribution among the polarization channels with lower effective Fermi energy tends to increase it (see Figure 21), in order to recover the situation with only one unpolarized channel that existed at concentrations under the threshold. The competition between these two effects softens the NDR and results in an experimental shoulder that widens with higher temperature as the population of the $P = 0$ channel, characterized by a lower conductance (Figure 19), increases with temperature, at the expense of the conductance of the polarized channels.

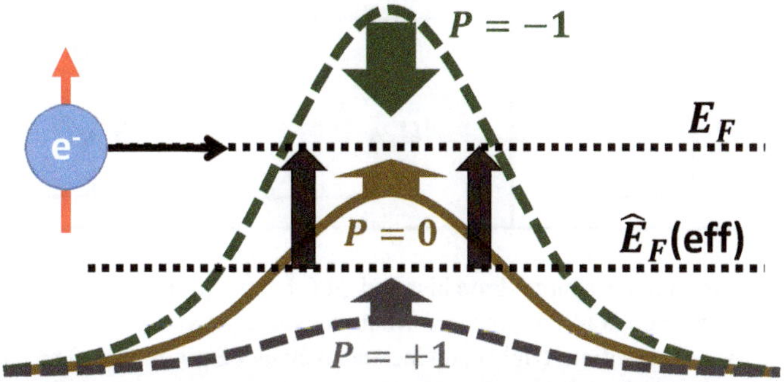

Figure 21. Diagram of the potential barrier for each spin-polarized configuration, as seen by an incident spin-up electron just after the onset of polarization. The thick vertical arrows show the changes in the barrier due to the capacitive nature of the split gates; the dotted lines correspond to the Fermi energy and the effective Fermi energy just after polarization.

5.3. *Discussion*

The widening of the conductance shoulder with temperature is consistent with all experimental observations.[18-20] Furthermore, our model is also consistent with the other experimental features (c-e) mentioned in Section 4.1. For instance, by grounding the source and applying a larger source-drain bias at constant gate bias, the bottom of the conduction band in the QPC will drop while the gate bias will maintain the same lateral potential, which will make the y-parabolic confinement potential steeper (see Figure 22(a)). As a result, the confinement energy $\hbar\omega_y$ will increase. Then, as illustrated in Section 3.4, the onset concentration for spin polarization will also increase[28], which consequently leads to a shift in the conductance shoulder towards higher values (feature (d)). Meanwhile, an in-plane magnetic field will cause one of the spin-polarized channels in our model to become depopulated at a faster rate, lowering the total conductance even more, so that the conductance anomaly will drop towards the spin-split result $0.5G_0$ (feature (c)). Finally, as shown in Section 4.5, the transmission value at pinch-off decreases in long QPC channels; or, in other words, if the length of the QPC is increased, the energy of the incident electron must increase too if the conductance is to remain the same, as illustrated in Figure 21(b). As the onset concentration n^{onset} is length-independent, the anomalous shoulder will shift to a lower conductance value (feature (e)) in agreement with the experimentally observed drop of the conductance anomaly from $0.7G_0$ to $0.5G_0$ for long QPCs.[57]

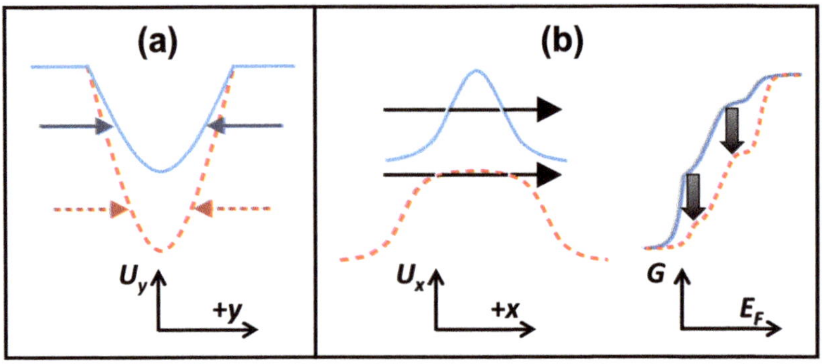

Figure 22. Effect of variations in source-drain bias and QPC length. (a) Lateral confinement potential U_y; the dashed line corresponds to the potential when the source-drain bias is increased. (b) (Left) Potential barrier U_x versus position; (right) conductance versus Fermi energy. The dashed line corresponds to a longer quantum point contact.

6. Acknowledgments

We thank Shinsei Ryu for useful discussions. This work was supported by a grant from the Research Board at the University of Illinois at Urbana-Champaign (UIUC). We also thank the Beckman Institute and the Department of Physics at UIUC for their continued support.

7. References

1. H. van Houten and C. Beenakker, Quantum point contacts, *Phys. Today*. **49** (7), 22–27, (1996).
2. A. P. Micolich, What lurks below the last plateau: experimental studies of the 0.7x2e2/h conductance anomaly in one-dimensional systems, *J. Phys.: Condens. Matter*. **23** (44), 443201, (2011).
3. A. Micolich, Quantum point contacts: Double or nothing?, *Nat. Phys.* **9** (9), 530–531 (Sept., 2013). ISSN 1745-2473.
4. J. P. Leburton, Size effects on polar optical phonon scattering of 1-d and 2-d electron gas in synthetic semiconductors, *Journal of Applied Physics*. **56** (10), 2850–2855, (1984).
5. S. Briggs and J. P. Leburton, Size effects in multisubband quantum wire structures, *Phys. Rev. B*. **38**, 8163–8170 (Oct, 1988).
6. S. Briggs, D. Jovanovic, and J. P. Leburton, Intersubband population inversion in quantum wire structures, *Applied Physics Letters*. **54** (20), 2012–2014, (1989).
7. D. Jovanovic, S. Briggs, and J. P. Leburton, Intersubband resonant effects of dissipative transport in quantum wires, *Phys. Rev. B*. **42**, 11108–11113 (Dec, 1990).

8. C. Beenakker and H. van Houten. Quantum transport in semiconductor nanostructures. In eds. H. Ehrenreich and D. Turnbull, *Semiconductor Heterostructures and Nanostructures*, vol. 44, *Solid State Physics*, pp. 1 – 228. Academic Press, Eindhoven, (1991).
9. S. A. Wolf, D. D. Awschalom, R. A. Buhrman, J. M. Daughton, S. von Molnár, M. L. Roukes, A. Y. Chtchelkanova, and D. M. Treger, Spintronics: A spin-based electronics vision for the future, *Science*. **294** (5546), 1488–1495, (2001).
10. K. A. Matveev, Conductance of a quantum wire in the wigner-crystal regime, *Phys. Rev. Lett.* **92** (10), 106801 (Mar, 2004).
11. J. S. Meyer and K. A. Matveev, Wigner crystal physics in quantum wires, *J. Phys.: Condens. Matter*. **21** (2), 023203, (2009).
12. S. Tarucha, T. Honda, and T. Saku, Reduction of quantized conductance at low temperatures observed in 2 to 10 um-long quantum wires, *Solid State Commun.* **94** (6), 413 – 418, (1995). ISSN 0038-1098.
13. A. Yacoby, H. L. Stormer, N. S. Wingreen, L. N. Pfeiffer, K. W. Baldwin, and K. W. West, Nonuniversal conductance quantization in quantum wires, *Phys. Rev. Lett.* **77**, 4612–4615 (Nov, 1996).
14. O. M. Auslaender, H. Steinberg, A. Yacoby, Y. Tserkovnyak, B. I. Halperin, K. W. Baldwin, L. N. Pfeiffer, and K. W. West, Spin-charge separation and localization in one dimension, *Science*. **308** (5718), 88–92, (2005).
15. Y. Jompol, C. J. B. Ford, J. P. Griffiths, I. Farrer, G. A. C. Jones, D. Anderson, D. A. Ritchie, T. W. Silk, and A. J. Schofield, Probing spin-charge separation in a tomonaga-luttinger liquid, *Science*. **325** (5940), 597–601, (2009).
16. D. A. Wharam, T. J. Thornton, R. Newbury, M. Pepper, H. Ahmed, J. E. F. Frost, D. G. Hasko, D. C. Peacock, D. A. Ritchie, and G. A. C. Jones, One-dimensional transport and the quantisation of the ballistic resistance, *J. Phys. C*. **21** (8), L209, (1988).
17. B. J. van Wees, H. van Houten, C. W. J. Beenakker, J. G. Williamson, L. P. Kouwenhoven, D. van der Marel, and C. T. Foxon, Quantized conductance of point contacts in a two-dimensional electron gas, *Phys. Rev. Lett.* **60** (9), 848–850 (Feb, 1988).
18. K. J. Thomas, J. T. Nicholls, M. Y. Simmons, M. Pepper, D. R. Mace, and D. A. Ritchie, Possible spin polarization in a one-dimensional electron gas, *Phys. Rev. Lett.* **77** (1), 135–138 (Jul, 1996).
19. S. Nuttinck, K. Hashimoto, S. Miyashita, T. Saku, Y. Yamamoto, and Y. Hirayama, Quantum point contacts in a density-tunable two-dimensional electron gas, *Jpn. J. Appl. Phys.* **39** (Part 2, No. 7A), L655–L657, (2000).
20. S. M. Cronenwett, H. J. Lynch, D. Goldhaber-Gordon, L. P. Kouwenhoven, C. M. Marcus, K. Hirose, N. S. Wingreen, and V. Umansky, Low-temperature fate of the 0.7 structure in a point contact: A kondo-like correlated state in an open system, *Phys. Rev. Lett.* **88** (22), 226805 (may, 2002).
21. L. I. Glazman, G. B. Lesovik, D. E. Khmel'nitskii, and R. I. Shekhter, Reflectionless quantum transport and fundamental ballistic-resistance steps in microscopic constrictions, *JETP Lett.* **48**, 238, (1988).
22. M. Büttiker, Quantized transmission of a saddle-point constriction, *Phys. Rev. B*. **41** (11), 7906–7909 (Apr, 1990).
23. R. Fitzgerald, Quantum point contact mysteries reexamined, *Phys. Today*. **55** (5), 21–23, (2002).

24. R. de Picciotto, L. N. Pfeiffer, K. W. Baldwin, and K. W. West, Nonlinear response of a clean one-dimensional wire, *Phys. Rev. Lett.* **92**, 036805 (Jan, 2004).
25. H. Kothari, A. Ramamoorthy, R. Akis, S. M. Goodnick, D. K. Ferry, J. L. Reno, and J. P. Bird, Linear and nonlinear conductance of ballistic quantum wires with hybrid confinement, *J. Appl. Phys.* 103 (1): 013701, (2008).
26. T.-M. Chen, A. C. Graham, M. Pepper, I. Farrer, and D. A. Ritchie, Bias-controlled spin polarization in quantum wires, *Appl. Phys. Lett.* 93 (3): 032102, (2008).
27. K.-F. Berggren and M. Pepper, Electrons in one dimension, *Phil. Trans. R. Soc. A.* **368** (1914), 1141–1162, (2010).
28. A. Sánchez and J. Leburton. Tunable magnetic phases in quasi-one-dimensional systems, (2015).
29. A. X. Sánchez and J.-P. Leburton, Temperature modulation of the transmission barrier in quantum point contacts, *Phys. Rev. B.* **88**, 075305 (Aug, 2013).
30. B. Jonsson and S. Eng, Solving the schrodinger equation in arbitrary quantum-well potential profiles using the transfer matrix method, *IEEE J. Quantum Electron.* **26** (11), 2025 –2035 (nov, 1990). ISSN 0018-9197.
31. R. Winkler, *Spin-Orbit Coupling Effects in Two-Dimensional Electron and Hole Systems*. vol. 191, *Springer Tracts in Modern Physics*, (Springer, Berlin, 2003).
32. J. P. Perdew and Y. Wang, Accurate and simple analytic representation of the electron-gas correlation energy, *Phys. Rev. B.* **45**, 13244–13249 (Jun, 1992).
33. P. Matagne. *Quantum Modeling of Single-Electron Charging in Vertical Quantum Dots*. PhD thesis, Université de Liège, (2001).
34. A. D. Güçlü, C. J. Umrigar, H. Jiang, and H. U. Baranger, Localization in an inhomogeneous quantum wire, *Phys. Rev. B.* **80**, 201302 (Nov, 2009).
35. L. Shulenburger, M. Casula, G. Senatore, and R. M. Martin, Correlation effects in quasi-one-dimensional quantum wires, *Phys. Rev. B.* **78**, 165303 (Oct, 2008).
36. T. J. Thornton, M. Pepper, H. Ahmed, D. Andrews, and G. J. Davies, One-dimensional conduction in the 2d electron gas of a gaas-algaas heterojunction, *Phys. Rev. Lett.* **56**, 1198–1201 (Mar, 1986).
37. K. J. Thomas, M. Y. Simmons, J. T. Nicholls, D. R. Mace, M. Pepper, and D. A. Ritchie, Ballistic transport in one-dimensional constrictions formed in deep two-dimensional electron gases, *Appl. Phys. Lett.* **67** (1), 109–111, (1995).
38. S. M. Reimann and M. Manninen, Electronic structure of quantum dots, *Rev. Mod. Phys.* **74**, 1283–1342 (Nov, 2002).
39. K. Hess, *Advanced Theory of Semiconductor Devices*. (Wiley-IEEE Press, New York, 2000).
40. L. P. Rokhinson, L. N. Pfeiffer, and K. W. West, Spontaneous spin polarization in quantum point contacts, *Phys. Rev. Lett.* **96** (15), 156602 (Apr, 2006).
41. A. C. Graham, D. L. Sawkey, M. Pepper, M. Y. Simmons, and D. A. Ritchie, Energy-level pinning and the 0.7 spin state in one dimension: Gaas quantum wires studied using finite-bias spectroscopy, *Phys. Rev. B.* **75**, 035331 (Jan, 2007).
42. K.-F. Berggren, P. Jaksch, and I. Yakimenko, Effects of electron interactions at crossings of zeeman-split subbands in quantum wires, *Phys. Rev. B.* **71**, 115303 (Mar, 2005).
43. A. Lassl, P. Schlagheck, and K. Richter, Effects of short-range interactions on transport through quantum point contacts: A numerical approach, *Phys. Rev. B.* **75**, 045346 (Jan, 2007).
44. E. Lieb and D. Mattis, Theory of ferromagnetism and the ordering of electronic energy levels, *Phys. Rev.* **125**, 164–172 (Jan, 1962).

45. T. Ando, A. B. Fowler, and F. Stern, Electronic properties of two-dimensional systems, *Rev. Mod. Phys.* **54** (2), 437–672 (Apr, 1982).
46. S. Lüscher, L. S. Moore, T. Rejec, Y. Meir, H. Shtrikman, and D. Goldhaber-Gordon, Charge rearrangement and screening in a quantum point contact, *Phys. Rev. Lett.* **98** (19), 196805 (May, 2007).
47. K. Hirose, S.-S. Li, and N. S. Wingreen, Mechanisms for extra conductance plateaus in quantum wires, *Phys. Rev. B.* **63** (3), 033315 (Jan, 2001).
48. S. Ihnatsenka and I. V. Zozoulenko, Origin of the "0.25 anomaly" in the nonlinear conductance of a quantum point contact, *Phys. Rev. B.* **79**, 235313 (Jun, 2009).
49. S. Datta, *Electronic Transport in Mesoscopic Systems*. Cambridge Studies in Semiconductor Physics and Microelectronic Engineering, (Cambridge University Press, 1997). ISBN 9780521599436.
50. K. J. Thomas, J. T. Nicholls, M. Pepper, W. R. Tribe, M. Y. Simmons, and D. A. Ritchie, Spin properties of low-density one-dimensional wires, *Phys. Rev. B.* **61** (20), R13365–R13368 (May, 2000).
51. K. J. Thomas, J. T. Nicholls, N. J. Appleyard, M. Y. Simmons, M. Pepper, D. R. Mace, W. R. Tribe, and D. A. Ritchie, Interaction effects in a one-dimensional constriction, *Phys. Rev. B.* **58**, 4846–4852 (Aug, 1998).
52. D. J. Reilly, T. M. Buehler, J. L. O'Brien, A. R. Hamilton, A. S. Dzurak, R. G. Clark, B. E. Kane, L. N. Pfeiffer, and K. W. West, Density-dependent spin polarization in ultra-low-disorder quantum wires, *Phys. Rev. Lett.* **89** (24), 246801 (Nov, 2002).
53. A. A. Starikov, I. I. Yakimenko, and K.-F. Berggren, Scenario for the 0.7-conductance anomaly in quantum point contacts, *Phys. Rev. B.* **67** (23), 235319 (Jun, 2003).
54. P. Havu, M. J. Puska, R. M. Nieminen, and V. Havu, Electron transport through quantum wires and point contacts, *Phys. Rev. B.* **70**, 233308 (Dec, 2004).
55. K.-F. Berggren and I. I. Yakimenko, Nature of electron states and symmetry breaking in quantum point contacts according to the local spin density approximation, *J. Phys.: Condens. Matter.* **20** (16), 164203, (2008).
56. A. Kristensen, H. Bruus, A. E. Hansen, J. B. Jensen, P. E. Lindelof, C. J. Marckmann, J. Nygård, C. B. Sørensen, F. Beuscher, A. Forchel, and M. Michel, Bias and temperature dependence of the 0.7 conductance anomaly in quantum point contacts, *Phys. Rev. B.* **62** (16), 10950–10957 (Oct, 2000).
57. D. J. Reilly, G. R. Facer, A. S. Dzurak, B. E. Kane, R. G. Clark, P. J. Stiles, R. G. Clark, A. R. Hamilton, J. L. O'Brien, N. E. Lumpkin, L. N. Pfeiffer, and K. W. West, Many-body spin-related phenomena in ultra low-disorder quantum wires, *Phys. Rev. B.* **63**, 121311 (Mar, 2001).
58. O. P. Sushkov, Conductance anomalies in a one-dimensional quantum contact, *Phys. Rev. B.* **64**, 155319 (Sep, 2001).
59. Y. Meir, K. Hirose, and N. S. Wingreen, Kondo model for the "0.7 anomaly" in transport through a quantum point contact, *Phys. Rev. Lett.* **89** (19), 196802 (Oct, 2002).
60. K. Hirose, Y. Meir, and N. S. Wingreen, Local moment formation in quantum point contacts, *Phys. Rev. Lett.* **90** (2), 026804 (Jan, 2003).
61. T. Rejec and Y. Meir, Magnetic impurity formation in quantum point contacts, *Nature.* **442** (7105), 900–903 (aug, 2006). ISSN 0028-0836.
62. S. Ihnatsenka and I. V. Zozoulenko, Conductance of a quantum point contact based on spin-density-functional theory, *Phys. Rev. B.* **76**, 045338 (Jul, 2007).

63. Y. Meir, The theory of the '0.7 anomaly' in quantum point contacts, *J. Phys.: Condens. Matter.* **20** (16), 164208, (2008).
64. M. J. Iqbal, J. P. de. Jong, D. Reuter, A. D. Wieck, and C. H. van der Wal, Split-gate quantum point contacts with tunable channel length, *J. Appl. Phys.* 113 (2): 024507, (2013).
65. T. Song and K.-H. Ahn, Ferromagnetically coupled magnetic impurities in a quantum point contact, *Phys. Rev. Lett.* **106**, 057203 (Feb, 2011).
66. E. Lieb and D. Mattis, Ordering energy levels of interacting spin systems, *J. Math. Phys.* **3** (4), 749–751, (1962).
67. P. Debray, S. M. S. Rahman, J. Wan, R. S. Newrock, M. Cahay, A. T. Ngo, S. E. Ulloa, S. T. Herbert, M. Muhammad, and M. Johnson, All-electric quantum point contact spin-polarizer, *Nat. Nanotechnol.* **4**, 759–764, (2009).
68. J. Wan, M. Cahay, P. Debray, and R. S. Newrock, Spin texture in quantum point contacts in the presence of lateral spin orbit coupling, *J. Nanoelectron. Optoelectron.* **6** (2), 95–101, (2011).
69. Y. Hirayama, K. Muraki, and T. Saku, Two-dimensional electron gas formed in a back-gated undoped heterostructure, *Appl. Phys. Lett.* **72** (14), 1745–1747, (1998). ISSN 00036951.
70. B. E. Kane, G. R. Facer, A. S. Dzurak, N. E. Lumpkin, R. G. Clark, L. N. Pfeiffer, and K. W. West, Quantized conductance in quantum wires with gate-controlled width and electron density, *Appl. Phys. Lett.* **72** (26), 3506–3508, (1998).
71. D. J. Reilly, Phenomenological model for the 0.7 conductance feature in quantum wires, *Phys. Rev. B.* **72** (3), 033309 (Jul, 2005).

Chapter 4

Coherent Spin Transport in Inorganic Semiconductor Quantum Wires

Md. Iftekhar Hossain*, Saumil Bandyopadhyay*,†, Jayasimha Atulasimha‡ and Supriyo Bandyopadhyay*

**Department of Electrical and Computer Engineering, Virginia Commonwealth University, Richmond, Virginia 23284, USA*

†Department of Electrical Engineering and Computer Science, Massachusetts Institute of Technology, Cambridge, Massachusetts 02139, USA

‡Department of Mechanical and Nuclear Engineering, Virginia Commonwealth University, Richmond, Virginia 23284, USA
sbandy@vcu.edu

Spin transport in compound semiconductor quantum wires, in which only the lowest conduction subband is occupied by electrons, is unique because the D'yakonov-Perel' (DP) spin relaxation mechanism (the dominant spin relaxation channel in semiconductors) can be nearly eliminated if the axis of the wire coincides with certain crystallographic directions, or if the wire is polycrystalline. This leads to long spin diffusion lengths, making single-subband quantum wires the ideal system to study coherent spin transport. A large number of experiments have been performed in quantum wire spin valve structures with a view to studying coherent injection, transport and detection of spins, both at low temperatures and room temperature. Here, we review some of these experiments and the lessons learned. These systems have potential applications in high-performance devices – both electronic and optical – and they are discussed briefly.

1. Introduction

Spintronic devices usually rely on long spin relaxation times and/or lengths for optimum performance. Therefore, suppression or elimination of spin relaxing mechanisms through careful material choice and structural engineering is an important consideration. Conduction electrons in *compound* semiconductors usually do not experience intrinsic spin orbit coupling ($\vec{L} \cdot \vec{S}$ coupling) because the conduction band minimum in such semiconductors is at the Γ-point (Brillouin zone center) where the orbital wavefunctions are s-type with orbital quantum number $\vec{L} = 0$. However, they exhibit bulk (crystallographic) inversion asymmetry that causes Dresselhaus spin-orbit interaction [1] which relaxes spin via the D'yakonov-Perel' (DP) [2] and the Elliott-Yafet (EY) [3] mechanisms. On the other hand, *elemental* semiconductors like Si or Ge are centrosymmetric crystals and hence experience no Dresselhaus interaction, but their conduction band valleys are at the Brillouin zone edge (not center). Consequently, the orbital wave functions of electrons in such semiconductors are not s-type and the electrons do experience intrinsic ($\vec{L} \cdot \vec{S}$) spin-orbit coupling which relaxes spin. Si will experience weaker intrinsic spin-orbit coupling than Ge since the coupling strength varies as the fourth power of the atomic weight and Si is a lighter element than Ge. Thus, there are complex considerations in determining the optimal material systems for spin transport.

Many narrow gap compound semiconductors like InSb and InAs have considerably higher electron mobilities than elemental semiconductors (primarily because of the low effective mass of electrons in these materials), which make them attractive for device applications. However, being non-centrosymmetric, they suffer from the Dresselhaus interaction. More importantly, any structural inversion asymmetry in these semiconductors caused by internal or external electric fields will generate strong Rashba-type spin-orbit interaction [4] because the Rashba interaction is stronger in narrower gap semiconductors than in wider gap semiconductors. This makes narrow gap semiconductors particularly vulnerable to the DP mechanism. When one adds to that the fact that the DP relaxation rate is inversely proportional to carrier mobility, it appears that narrow gap semiconductors, with high electron mobility and strong Rashba interaction, will experience rapid spin relaxation due to the DP mechanism and hence should be *avoided* for spintronic applications.

All that changes if one can eliminate the DP relaxation in narrow gap compound semiconductors via innovative structural engineering. Such systems will still have strong Rashba and Dresselhaus spin-orbit interactions, but no DP relaxation, which will make them extremely *desirable*. The strong Rashba interaction is a sought after property in many spintronic device applications, e.g. the Datta-Das spin field effect transistor [5], which, paradoxically, cannot tolerate spin relaxation. Similarly, strong Dresselhaus interaction is a cherished property for a spintronic transistor based on modulating the Dresselhaus interaction with an external potential [6]. This transistor too cannot tolerate spin relaxation. Obviously, such devices can benefit enormously from narrow gap semiconductor structures where electrons will experience strong Rashba and Dresselhaus spin orbit interaction and yet no DP spin relaxation. The DP relaxation is the dominant spin relaxation mechanism in most compound semiconductors [7; 8; 9; 10; 11; 12; 13], particularly at room temperature, and hence its elimination (or strong suppression) can increase the spin relaxation time (and length) considerably.

The all-important question then becomes how does one suppress or eliminate the DP relaxation in a semiconductor structure? One way to accomplish that is to utilize a quantum wire and ensure that all the electrons occupy a single conduction subband. If the wire is oriented along certain principal crystallographic directions such as [100], or if the wire is polycrystalline, then the DP relaxation is quenched despite strong Rashba and Dresselhaus interactions [14; 15; 16; 17; 18; 19]. This happens solely owing to single subband occupancy. Fortunately, narrow gap semiconductors have small electron effective mass which increases the energy separation between the subbands in a quantum wire and makes single subband occupancy easier to realize, even at room temperature. We have focused on studying spin transport in 50-nm diameter InSb quantum wires where 96% of the electrons occupy the lowest conduction subband at equilibrium and at room temperature. These structures exhibit a 10-fold increase in the spin relaxation time compared to bulk or quantum wells of the same material, suggesting strong suppression of the DP mechanism due to single subband occupancy [20]. Experiments with these systems are described in Sections 3 and 4. We then discuss spin transport studies in a few other semiconductor systems and conclude with a forecast of device prospects.

2. The "spin valve" effect and Hanle conductance oscillation

Before we discuss coherent spin transport in quantum wire systems, we briefly describe two effects that are routinely used to probe coherent spin injection into a semiconductor, coherent spin transport through the semiconductor, and coherent detection of spins that have traveled through the semiconductor. These effects are observed in a so-called "spin valve" structure that consists of the semiconductor and two ferromagnetic contacts. In our case, the semiconductor is a quantum wire. The contacts are made shape-anisotropic with easy axis of magnetization along the axis of the quantum wire. We will assume that they have the same sign of tunneling spin polarization. When a strong magnetic field is applied in one direction along the wire axis, both contacts easily magnetize in the direction of the field and their magnetizations become mutually parallel. Spins injected by one contact into the quantum wire then transmit easily through the other contact (provided the spins have not relaxed in transit), resulting in low resistance of the spin valve structure.

When the magnetic field is subsequently reversed and scanned in the opposite direction, the magnetization of the contact with the lower coercivity flips first (at the coercive field of that contact) placing the two contacts temporarily in a mutually anti-parallel configuration. At this point, the spin valve's resistance increases because spins injected by one contact are blocked by the other (again, assuming that the spins do not relax during transit). Finally when the reversed magnetic field's strength exceeds the coercive field of the other contact as well, the latter reverses magnetization as well and the two contacts once again have parallel magnetization. At that point, the resistance drops. Thus, one should observe a resistance "peak" between the coercive fields of the two magnetic contacts during the reverse scan of the magnetic field. The same peak will be observed again if the field is brought to zero and then forward scanned. These peaks are called *spin valve signals.* A schematic representation of this feature in the longitudinal (magnetic field collinear with quantum wire axis) magnetoresistance trace is shown in Fig. 1(a).

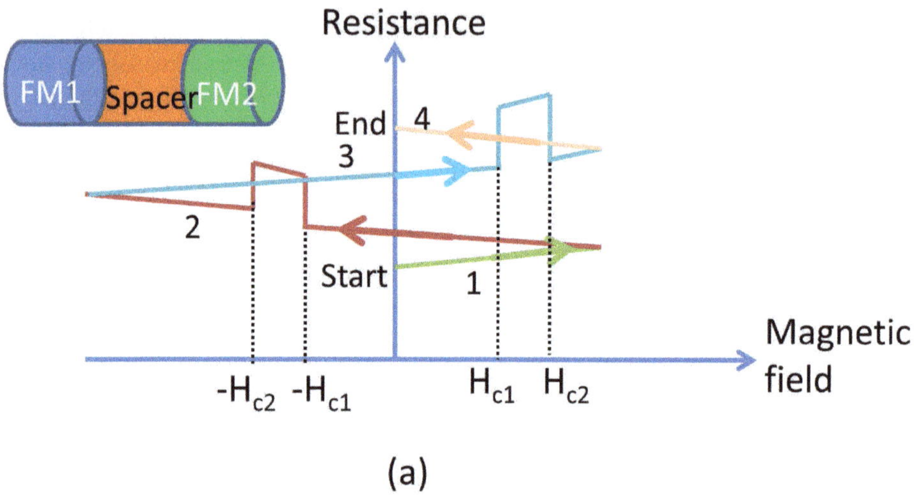

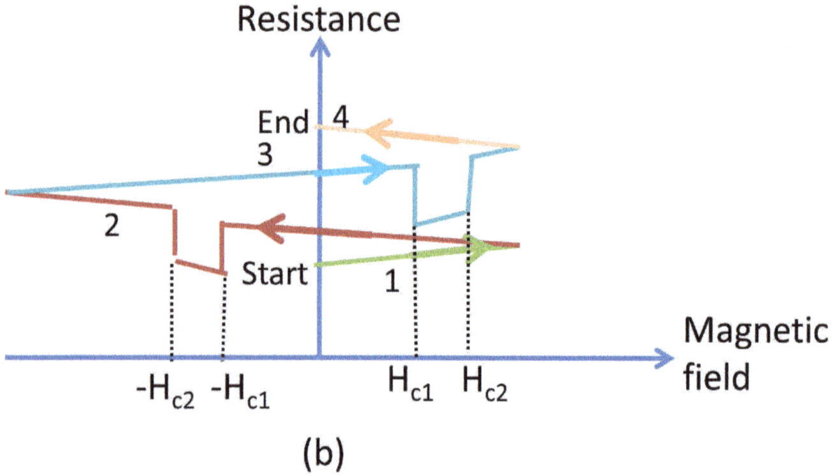

Fig. 1. Schematic representation of spin valve peaks and troughs in longitudinal magnetoresistance scans. The magnetic field is always along the axis of the spin valve nanowire (FM = ferromagnet): (a) longitudinal magnetoresistance trace showing the spin valve peaks. A small slope is artificially imposed on the trace for clarity. The start and end points of the magnetic field scan are indicated. Trace '1' is the first forward scan starting from zero field, trace '2' is the first reverse scan, trace '3' is the second forward scan, and trace '4' is the final reverse scan that brings the field back to zero at the end point. The coercive fields of the two ferromagnets are H_{c1} and H_{c2}. (b) Longitudinal magnetoresistance trace showing inverse spin valve troughs.

An interesting feature emerges if electrons in the semiconductor quantum wire spacer flow between the two contacts by tunneling through defect sites in the quantum wire. If the electron energy happens to be resonant with the quantized energy of a defect site, then electrons resonantly tunnel through that site. If the defect is close enough to one contact, then the spin polarization of that contact becomes effectively reversed, meaning that when the two contacts have parallel magnetization, they behave as if the magnetizations are anti-parallel, and vice versa [21]. In this case, the spin valve resistance peak becomes a trough and the effect is referred to as the inverse spin valve signal. It is shown in Fig. 1(b). Since defects are random, one expects to see the normal spin valve effect in some samples and the inverse version in others. Both types of behavior have been observed in nominally identical samples produced in the same run [22].

The observation of the spin valve peak (or the inverse spin valve trough) in the longitudinal (field parallel to wire) magnetoresistance of a spin valve structure, between the coercive fields of the two contacts, is evidence of coherent spin injection, transport and detection. Clearly, these features would not be observed if: (1) the injecting contact did not preferentially inject spins of one polarization, (2) the spins were randomized in the wire, and/or (3) the detecting contact did not preferentially transmit spins of one polarization.

In addition to the spin valve effect, there is another effect that suggests coherent spin injection, transport and detection. It is the Hanle effect which makes the conductance of a spin valve oscillate in a transverse magnetic field applied perpendicular to the direction of current flow (i.e. perpendicular to the nanowire axis). These oscillations are called *Hanle oscillations* and arise from coherent spin precession of injected spins about the magnetic field as they traverse the spacer layer between the two ferromagnetic contacts [23; 24]. The spin valve's contacts are initally magnetized along the wire axis with a strong longitudinal magnetic field. Next, a magnetic field is applied transverse to the wire axis and gradually scanned up to a field strength that is considerably lower than the coercive field of either contact so that neither contact's magnetization is affected by this field. Here, the injecting contact injects spins polarized in the direction of the wire axis. The transverse magnetic field is perpendicular to the spin orientation and hence makes the spins precess about itself (Larmor precession). This is shown in Fig. 2. The rate of spin precession (Larmor frequency) can be changed by varying the transverse field strength, so that the angle Φ through which the spins precess as they traverse the distance between the two ferromagnetic contacts can be altered with the field. The quanity Φ is related to the flux density B of the transverse magnetic field according to

$$\Phi = \frac{g\mu_B B}{\hbar} \langle \tau_t \rangle , \qquad (1)$$

where g is the Landé g-factor in the spacer material, μ_B is the Bohr magneton and $\langle \tau_t \rangle$ is the average transit time of spin carriers through the spacer layer.

If Φ is an odd multiple of π, then the resistance of the structure will be high since the spins that arrive at the detecting contact are anti-parallel to that contact's magnetization and hence will be blocked. On the other hand, if Φ is an even multiple of π, then the resistance will be low since the arriving spins are parallel to the magnetization of the detecting contact. Clearly, this will happen only if the magnetizations of the two contacts are mutually parallel, i.e. one observes

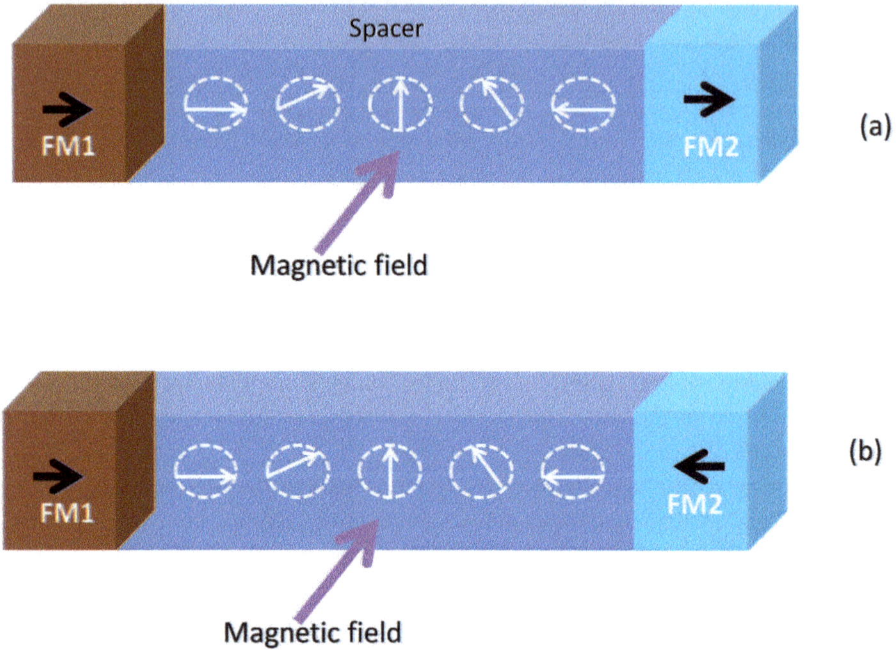

Fig. 2. Explanation of the Hanle oscillations in transverse magnetoresistance. (a) The contacts are magnetized parallel to each other at zero magnetic field and such a configuration will show the normal spin valve effect in the longitudinal magnetoresistance. The left contact injects spins which precess about the transverse magnetic field. When the spins arrive at the right contact, their transmission probability depends on the angle through which they have precessed in transit. If the angle is an odd multiple of π, the transmission probability (and hence the conductance) will be minimum, while if the angle is an even multiple of π, the transmission probability (and conductance) will be maximum. (b) The contacts are magnetized anti-parallel to each other at zero magnetic field. In this case, one would have observed the inverse spin valve effect in the longitudinal magnetoresistance. The transmission probability and the conductance would be maximum if the spins precess by an odd multiple of π and minimum if the spins precess by an even multiple of π.

the normal spin valve signal in the structure during measurement of longitudinal magnetoresitance. The situation would be reversed if one observes the inverse spin valve signal in the longitudinal magnetoresistance since that would imply that the two contacts have anti-parallel magnetizations.

The conductance G of the structure will vary with the magnetic flux density as

$$G = G_0 cos^2\left(\frac{\Phi}{2}\right) \qquad \texttt{normal spin valve}$$
$$G = G_0 sin^2\left(\frac{\Phi}{2}\right) \qquad \texttt{inverse spin valve,} \qquad (2)$$

where G_0 is the maximum conductance of the structure and Φ is given by Equation (1). A diagrammatic explanation of the Hanle effect is shown in Fig. 2.

In the event that magnetic field is not exactly perpendicular to the wire axis, there can be a phase shift (ϕ) in the oscillations, resulting in

$$G = G_0 cos^2\left(\frac{\Phi+\phi}{2}\right) \qquad \texttt{normal spin valve}$$

$$G = G_0 sin^2\left(\frac{\Phi+\phi}{2}\right) \qquad \texttt{inverse spin valve,} \qquad (3)$$

which means that the conductance is not maximum (or minimum) at zero magnetic field strength. This is sometimes referred to as the *oblique Hanle effect* [25].

Obviously, the Hanle oscillations could not be observed unless the injecting contact injected spins coherently, the spins then traversed the wire without complete randomization, and the detecting contact transmitted the arriving spins coherently depending on their polarizations. Therefore, the simultaneous observation of both the spin valve (or inverse spin valve) signal in the longitudinal magnetoresistance and the Hanle conductance oscillation in the transverse magnetoresistance is strong evidence of spin coherence. More importantly, these experiments reveal a wealth of information in one-dimensional structures, such as the spin diffusion length, spin relaxation time, and even non-spin-related transport parameters such as carrier mobility, carrier density, etc. We exemplify this in the next section.

3. Spin transport in InSb nanowires

InSb is a narrow gap semiconductor and therefore has strong Rashba spin-orbit coupling and also a large gyromagnetic ratio (g-factor). Consequently, it is a preferred material for Datta-Das type spin field effect transistors (SPINFET) where the current flowing between the source and the drain contacts of the transistor is modulated with a gate potential that changes the Rashba spin-orbit interaction strength in the transistor's channel. Both the source and drain in the Datta-Das transistor are ferromagnetic and act as spin polarizers and analyzers. The device is essentially a spin valve structure with one difference: the gate which controls the spin-orbit interaction strength in the channel. The spin-orbit interaction acts as an effective magnetic field [16] and changing that with the gate potential changes the spatial rate of spin precession in the transistor's channel, which leads to modulation of the source-to-drain current. The source-to-drain conductance therefore oscillates as the gate potential is varied. The reader will understand that this is essentially the Hanle conductance oscillation, except that instead of a real transverse magnetic field, here we have an effective magnetic field due to spin-orbit interaction. The effective magnetic field is varied by varying the gate potential.

In the case of the usual Hanle oscillation, one did not need any spin-orbit interaction. In fact, spin-orbit interaction would be undesirable since it would relax spin. However, the Datta-Das transistor cannot do without spin-orbit interaction since that is the source of the effective magnetic field which causes the Hanle-like oscillation in the source-to-drain conductance as the gate potential is varied. The

Datta-Das transistor would prefer strong spin-orbit coupling, but that also causes rapid spin relaxation because of the DP relaxation. If the spins relax too much, the gate voltage will no longer be able to modulate the source-to-drain current. Thus, we have a conflicting requirement: we desire strong spin orbit interaction so that we can modulate the source-to-drain current with a reasonably small gate potential, and yet we do not wish to see the strong spin relaxation that comes with strong spin-orbit interaction strength. The dichotomy is resolved if we are able to suppress the DP relaxation by fashioning the transistor channel out of a quantum wire and confining all the channel electrons to the lowest conduction subband. In that case, we can have strong spin-orbit interaction and yet not experience strong spin relaxation! The ideal channel prototype is therefore an InSb quantum wire. It has strong Rashba interaction, large g-factor and small effective mass that makes the energy separation between subbands large and allows single subband occupancy in wires of reasonable diameter at room temperature. We now describe how such quantum wires are self-assembled.

Arrays of parallel InSb nanowire spin valve structures, each of ~50 nm diameter and consisting of Co, InSb and Ni layers, were fabricated by sequentially electrodepositing Co, InSb and Ni in 50-nm diameter pores of nanoporous anodic alumina films [20; 26]. A representative diagram is shown in Fig. 3. Each nanowire is a "spin valve"; one ferromagnetic contact injects spin-polarized electrons via tunneling through the Schottky barrier formed at the ferromagnet/semiconductor interface, and the other ferromagnetic contact transmits the electrons to varying degrees depending on their spin polarization. Thus, the injecting contact acts as a spin polarizer and the detecting contact acts as a spin analyzer. In each nanowire, ~96% of the electrons occupy the lowest subband at room temperature (see section 3.3) so that there is essentially single subband occupancy. Approximately 10^8 parallel wires were electrically contacted from the top and bottom, forming an assembly of 10^8 parallel resistors. Energy dispersive analysis of x-ray (EDAX) spectra of the samples showed peaks corresponding to Co, Ni, In and Sb which are the constituents of the spin valves [20].

3.1. *Spin valve effect*

Fig. 4 shows the longitudinal magnetoresistance traces of a sample with $\sim 10^8$ nanowire spin valves in parallel measured at room temperature. One can observe the spin valve troughs at ~-1800 Oe during reverse scan and ~+1900 Oe during forward scan. These troughs are not due to the anisotropic magnetoresistance (AMR) effect since that effect is associated with the resistances of the ferromagnetic contacts alone which are negligible compared to the resistance of the InSb nanowire. The spin valve resistance consists of three resistances in series: the injecting ferromagnetic contact, the InSb nanowire spacer and the detecting ferromagnetic contact. Since the ferromagnets are metals, the InSb wire resistance dominates and we do not expect to see any AMR effect. The fractional trough depths $\Delta R/R$ are about

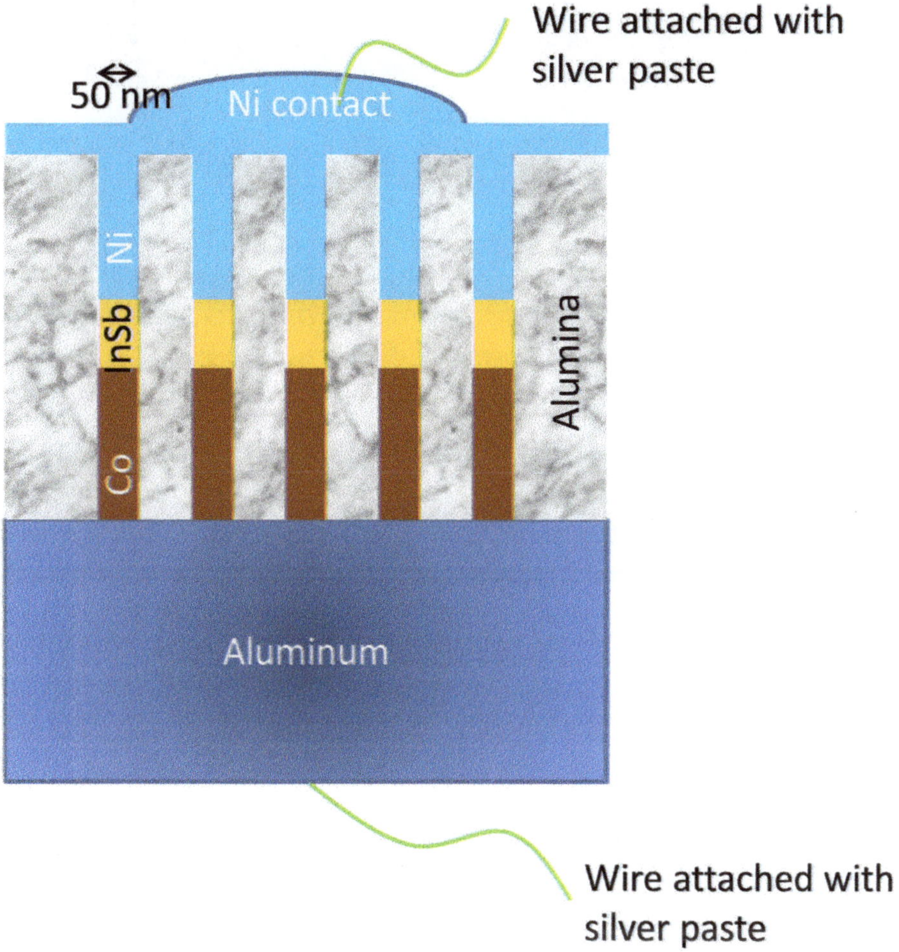

Fig. 3. Schematic of the self-assembled nanowire spin valves produced by sequentially electrodepositing Co, InSb and Ni in 50 nm diameter nanopores in anodic alumina films.

4.9×10^{-4}.

There is an abundance of formulas (and underlying theory) that relates $\Delta R/R$ to the spin diffusion length L_{InSb} in the InSb nanowire [27; 28; 29; 30; 31; 32; 33]. We will use the simple formula from [27]:

$$\frac{\Delta R}{R} \approx \frac{2P_{Co}P_{Ni}\exp\left(-L/L_{InSb}\right)}{1 - P_{Co}P_{Ni}\exp\left(-L/L_{InSb}\right)}, \tag{4}$$

where L_{InSb} is the spin diffusion length in the spacer, L is the length of the spcaer layer, and $P_{Co(Ni)}$ is the spin polarization in Co(Ni). If the spin polarizations and

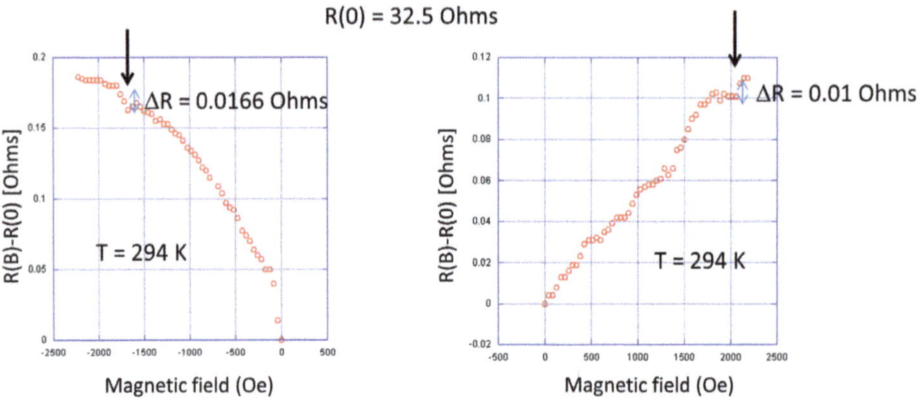

Magnetic field is parallel to axis of spin valve

$$\frac{\Delta R}{R(0)} = \frac{2P_1P_2e^{-L/L_s}}{1-P_1P_2e^{-L/L_s}} = 4.9\times 10^{-4}$$

Fig. 4. Room-temperature longitudinal magnetoresistance of a sample consisting of more than 100 million Co-InSb-Ni nanowire spin valves in parallel. The spin valve troughs are identified with vertical arrows. Reproduced from ref. [20] with permission of Wiley-VCH.

the length of the spacer layer are known, the above relation yields the value of the spin diffusion length.

The spin polarizations in bulk nickel and cobalt at 0 K temperature have been reported as 0.33 and 0.42, respectively [34], but at room temperature, the value reported in 50 nm thin films of cobalt is only 0.1 [35]. There is one report that the value increased to 0.16 when the resistance of the junction between cobalt and a non-magnetic spacer increased [36], but this is not relevant to our work. Furthermore, interface effects always reduce spin polarization [37], and we have a very large interface-to-volume ratio in our cobalt nanocontacts. Therefore, it is reasonable to expect that the spin polarization in cobalt nanocontacts is no more than 0.1. There is no equivalent study for nickel, but it is reasonable to assume a similar 4-fold reduction in the spin polarization and assume the spin polarization to be no more than 0.075 in nickel nanocontacts. Based on these spin polarization values and the measured resistance ratio $\frac{\Delta R}{R}$ in Fig. 2, the spin diffusion length in the InSb spacer is ~3.4 times smaller than the spacer layer length. The latter (estimated from transmission electron micrograph of individual spin valves) is about 25 nm, which makes the spin diffusion length in the InSb nanowire 7.3 nm at room temperature.

3.2. *Hanle effect*

Figure 5 shows the transverse magnetoresistance of the sample at room temperature. There is a periodic oscillation in the resistance or conductance (equally spaced minima and equally spaced maxima) with a period of ~315 Oe. This is the signature of the Hanle effect. The oscillations are very distorted and non-sinusoidal because the sample is composed of $\sim 10^8$ nanowire spin valves in parallel and different nanowires produce slightly different periods (owing to different transit times through the spacer layer), and ensemble averaging over the periods distorts the shape of the oscillation, making it appear non-sinusoidal.

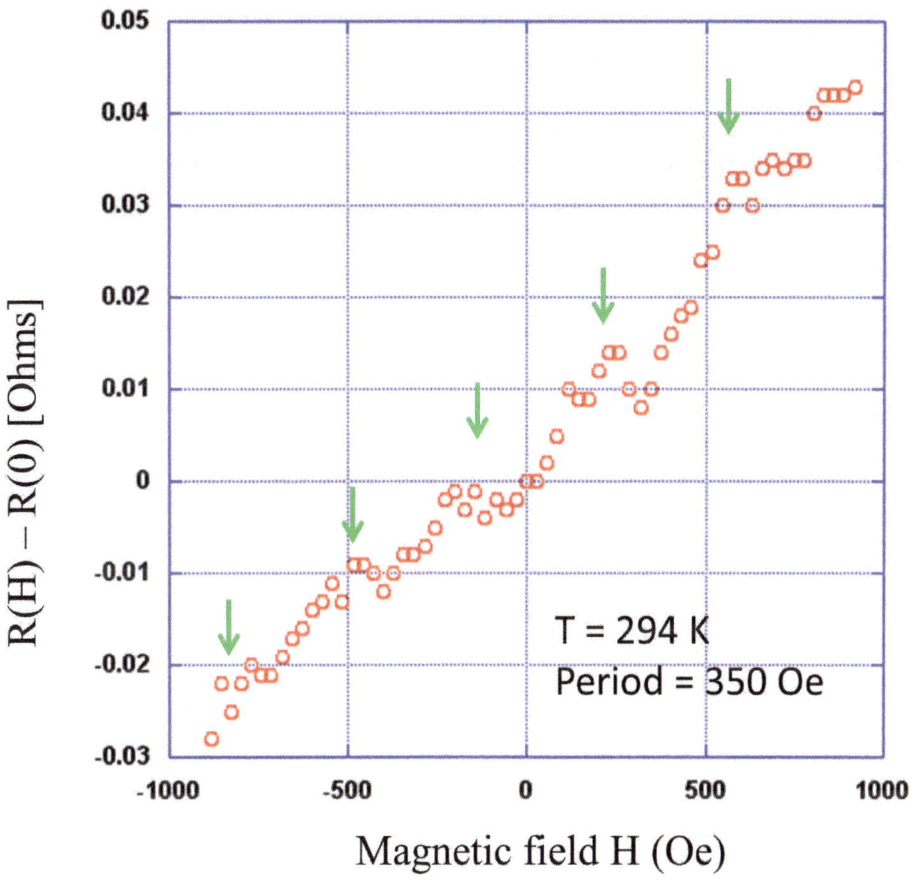

Fig. 5. Room temperature (295 K) transverse magnetoresistance plot showing Hanle oscillations in spin valve samples consisting of $\sim 10^8$ Co-InSb-Ni nanowire spin valves. Reproduced from ref. [20] with permission of Wiley-VCH.

There are two sources that introduce a spread in the transit time τ_t and hence a spread in the period of the Hanle oscillation. The first is inhomogeneous broadening due to a spread in the spacer layer width L among the different nanowires, and the second is homogeneous broadening due to a spread in the electron velocity even within the same nanowire. As long as the distribution in the spacer layer width and carrier velocity (and hence the distribution in the period) is somewhat peaked, we will be able to observe a periodic behavior (with the period corresponding to the peak in the distribution), but the oscillation will be non-sinusoidal. In other words, successive minima and maxima will be equally spaced in the magnetic field, but the minima may not occur exactly midway between two successive maxima and vice versa. This is precisely what we see in Fig. 5. The homogeneous broadening is fortunately small and the spread in velocity is much less than the thermal spread since electrons resonantly tunnel through one or more defect sites (which is why we observe the inverse spin valve effect) and resonant tunneling acts as a velocity filter. Only electrons whose energies are very close to the resonant level can tunnel through, which reduces the spread in kinetic energy and electron velocity. Therefore, most of the spread in the oscillation period accrues from the inhomogeneous broadening. The variation in the spacer layer width (ascertained from transmission electron microscopy) is 25% - 35%, which should result in a similar variation in the period. This variation distorts the oscillation, making it non-sinusoidal, but still very much discernible.

Notice that the Hanle oscillations are not symmetric about zero transverse magnetic field. This usually happens when the transverse magnetic field is slightly misaligned and not exactly perpendicular to the axis of the nanowire spin valve [38]. The magnetic field can never be exactly perpendicular to every wire's axis since the wires themselves are not exactly mutually parallel. Hence, we will always observe some symmetry-breaking. Note also that , the zero-field resistance is not a maximum and that there is a shift. This happens when the magnetizations of the two contacts are not exactly collinear.

The period of the Hanle oscillation should however still scale approximately as $1/sin\Theta$, where Θ is the angle between the wire axis and the magnetic field. This was tested and found to hold.

There is a background magnetoresistance which is due to thermal drift. The resistance increases with time possibly due to sample heating by the current.

From the observed average period B_{period} in the magnetic flux density, one can calculate the ensemble averaged transit time of electrons through the InSb spacer layer according to the relation

$$\langle \tau_t \rangle = \frac{h}{|g|\mu_B B_{period}}, \tag{5}$$

where h is the Planck constant.

The g-factors of materials are known to change in nanostructures from the bulk value, but in InSb quantum dots, the g-factor has been reported to be -52 in the lowest subband [39], which is close to the bulk value of −51. Based on this, it is found that $\langle \tau_t \rangle = 44$ ps. In one dimensional nanowires with a single occupied subband, the spin relaxation time bears the same ratio to the transit time as the spin diffusion length to the spacer layer length. The spin valve experiment showed this ratio is ~3.4. Hence the spin relaxation time in the InSb spacer is 13 ps. Note that these are true *transport* spin lifetimes, since they are measured in a transport experiment. Previous measurements of spin lifetimes have been carried out with optical techniques and not transport methods [40; 41; 42], which may not yield the true transport lifetime.

The room temperature spin relaxation time in *epilayers* of InSb has been reported as low as 2.5 ps [40] and as high as 300 ps [41], but in quantum wells, it drops to 1.2 - 4.2 ps [42]. It has been suggested theoretically that intrinsic inversion symmetry breaking at the interfaces of a quantum well can decrease spin relaxation time by over an order of magnitude [43]. The calculated spin relaxation time in InAs/GaSb quantum wells is only 0.9 ps [43]. In InSb nanowires of 50 nm diameter, we will expect the spin relaxation time to be even shorter than in quantum wells because of the larger interface-to-volume ratio, and hence shorter than 1 ps at room temperature. Therefore, the measured relaxation time of 13 ps is roughly an order of magnitude *longer* than what is expected.

The drift velocity v_d of electrons in the InSb spacer layer is determined from the quantity $L/ < \tau_t >$ and turns out to be 25 nm/44 ps = 568 m/sec. The average electric field E in the spacer layer is estimated from V/L where V is the voltage dropped over the sample during the magnetoresistance measurements. The voltage drop was about 30 mV, which yields that the electric field is 30 mV/25 nm = 1.2 MV/m. Therefore, the effective drift mobility μ is $v_d/E = 4.7$ cm^2/V-sec. This is four orders of magnitude lower than the value reported in bulk or quantum wells, presumably owing to severe interface roughness scattering brought about by the nanowires having a large interface-to-volume ratio.

The mobility degradation should increase the Elliott-Yafet spin relaxation rate by four orders of magnitude compared to bulk or quantum wells since this rate is inversely proportional to mobility [3]. Yet, the spin relaxation rate had *decreased* by nearly one order of magnitude. This can only happen if Elliott-Yafet is not the dominant spin relaxation mechanism but D'yakonov-Perel' is (this was theoretically predicted in ref. [11]), and the D'yakonov-Perel' mechanism has been suppressed or eliminated in the nanowires. The suppression can happen either because of the formation of persistent spin helix state [44] that is not relavant in our case, or because of strictly one dimensional confinement of carrier motion resulting from single subband occupancy [19; 14]. The next subsection presents theoretical calculations to show that indeed 96% of the electrons reside in the lowest subband so that these nanowires experience single subband transport.

3.3. *Calculation of relative subband occupancy in the InSb spacer layers*

The nanowires are cylindrical in shape, and if we assume that their axes are along the z-direction, then the wavefunctions and energies of electrons in different subbands of the InSb segment are given by [45]

$$\psi_{n,m}(k_z) = \frac{1}{N} J_n(k_{n,m} r) \exp(in\theta) \exp(ik_z z)$$
$$E_{n,m} = \frac{\hbar^2}{2m^*}\left(k_{n,m}^2 + k_z^2\right), \tag{6}$$

where J_n is the Bessel function of n-th order, N is a normalization constant, k_z is the electron's wavevector component along the wire axis (z-axis), m^* is the electron's effective mass, and r, θ are the radial and angular coordinates on the cylinder's circular cross-section.

In order to find $k_{n,m}$, we apply the boundary condition that the wavefunction vanishes at the interface between InSb spacer and the surrounding alumina, i.e. when $r = R$, where R (= 25 nm) is the nanowire's radius. Therefore, we first find the zeros of the Bessel function for $n = 0$, which yields $k_{0,1} = 2.4048/(25\text{ nm}) = 9.6\times10^7\text{ m}^{-1}$, $k_{0,2} = 5.5201/(25\text{ nm}) = 2.2\times10^8\text{ m}^{-1}$, $k_{0,3} = 8.6537/(25\text{ nm}) = 3.5\times10^8\text{ m}^{-1}$ and $k_{0,4} = 11.7915/(25\text{ nm}) = 4.7\times10^8\text{ m}^{-1}$. From these values, we compute that the energies of the lowest four subbands are 0.9 kT, 4.8 kT, 12.1 kT and 21.8 kT at room temperature, assuming that the effective mass of an electron in InSb is 0.0145 times the free electron mass [46].

The occupancy of the subbands at equilibrium is determined by the Fermi-Dirac probability. This depends on the placement of the Fermi level with respect to the subband energies. One can find the placement from the carrier concentration in the nanowires, which is related to the Fermi energy E_F. The conductances of the nanowires are measured, which yield the conductivity σ of the spacer layer. This is related to the carrier concentration n and mobility μ according to $\sigma = qn\mu$, where q is the electron charge. From the value of n thus obtained, one determines the Fermi energy and hence the occupation probability for each subband at any temperature. This exercise yields that at room temperature, ~96% of the electrons occupy the lowest subband in the InSb spacer. Thus, the samples tested indeed have nearly single subband occupancy.

4. Modulating the DP relaxation rate by controlling relative subband occupation with an external agent

Since the DP relaxation is the dominant spin relaxation mechanism in compound semiconductors like InSb [7; 8; 9; 10; 11; 12] at room temperature, its elimination can increase the spin relaxation time and spin relaxation length in these semiconductors considerably. That is an important result since spin-relaxation is the spoiler in most spin-based devices and applications. More importantly, controlled modulation of the spin relaxation rate by an external agent that can alter relative subband occupancy, and thus change the DP relaxation rate in a semiconductor spin valve, can open up new possibilities and lead to novel device paradigms. One possibility is a spintronic infrared detector [47]. Let us assume that in the dark, only the lowest conduction subband in a semiconductor quantum wire is occupied by electrons and the DP relaxation is quenched. Under illumination, infrared photons can excite electrons from the lowest subband to higher subbands and cause muti-subband transport which revives the DP relaxation. This can alter spin relaxation rate in the quantum wire spacer of a spin valve and change its resistance. The change in resistance can be utilized to detect the infrared light, just as in a normal photodetector [47].

Consider a tri-layered nanowire spin valve as shown in Fig. 6 whose two ferromagnetic contacts have opposite signs of tunneling spin-polarization. The injecting contact mostly injects its majority spins, and if they do not flip during transit through the spacer layer, then they are blocked by the detecting contact because the latter's majority spins are anti-parallel to the arriving spins. In that case, the resistance of the spin valve is high. However, if the injected spins flip (relax) in transit, then they will become parallel to the majority spins at the right contact and transmit easily. If the spin relaxation rate can be increased with IR light, then that will increase the likelihood of the injected spins flipping and thereafter being transmitted easily by the right contact. This will cause an increase in the current through the nanowire under IR illumination, which is the physical basis of the photodetection mechanism.

One can estimate the light-to-dark current contrast ratio of the photodetector as follows: Assume that IR light induces only intra-band transitions in the conduction band of the spin valve's spacer layer and not inter-band transitions from the valence to the conduction band which would have changed the carrier concentration. This assumption would be particularly true for wide-gap semiconductor spacers where the bandgap vastly exceeds the IR photon energy. Also assume that the current in the spacer is mostly due to drift of injected electrons and not diffusion. In that case, it is given by

$$I = \frac{qv_d n_l}{4}\left[1 + \zeta_1\zeta_2 - 2\zeta_1\zeta_2 e^{-\frac{L}{L_s}}\right], \tag{7}$$

where q is the electron charge, v_d is the drift velocity of electrons, n_l is the linear electron concentration in the spacer layer, L is the length of the spacer layer, ζ_1 is the spin injection efficiency at the injecting (left) contact, ζ_2 is the spin detection efficiency at the detecting (right) contact, and L_s is the spin relaxation length in the spacer layer, ensemble averaged over electron velocity.

In deriving the above relation, two assumptions were made. First, even in single subband transport (when DP relaxation is absent), any spin-orbit interaction will make the spins precess as they travel, but the angle δ by which they precess in traversing the distance L is the same for all electrons (coherent rotation) [16]. This precession can be neglected since δ is very small when L is small. Second, in multi-

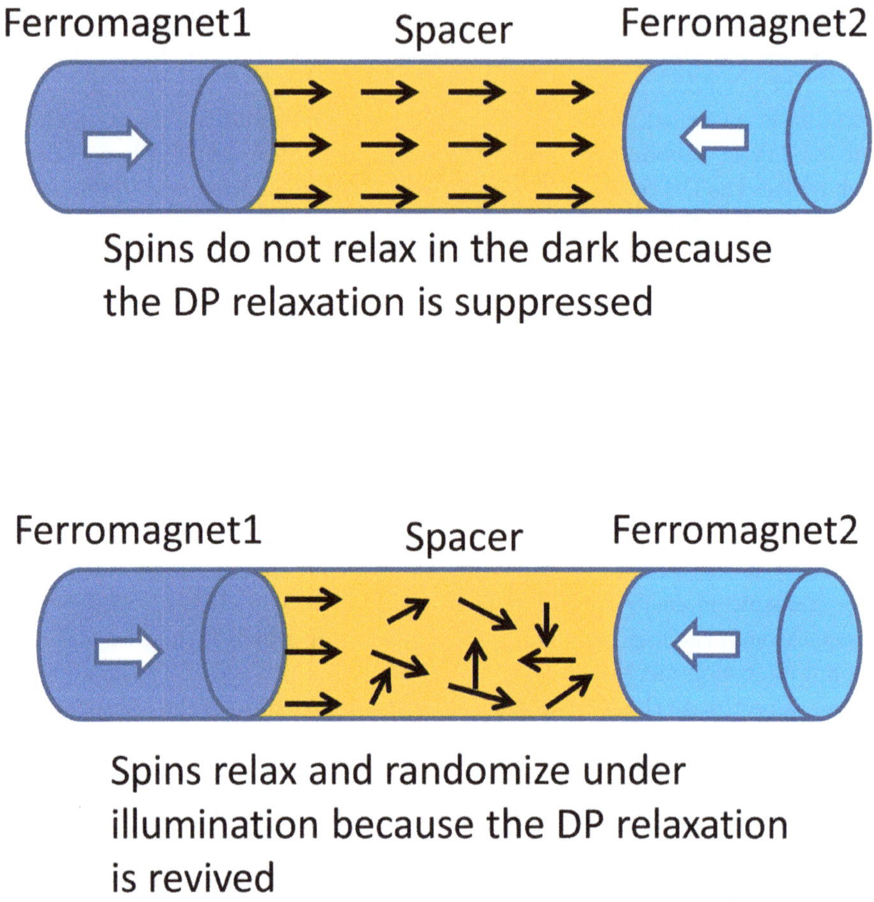

Fig. 6. A nanowire spin valve whose two ferromagnetic contacts have opposite signs of tunneling spin polarization. (a) If there is no spin relaxation in the spacer layer, spins injected by the injecting contact are blocked by the detecting contact and the spin valve's resistance is high. (b) Infrared light induces spin relaxation and randomizes the spins. The flipped spins transmit through the detecting contact and the resistance drops.

subband transport when DP relaxation is present, different electrons will precess by different amounts in traversing the distance L (incoherent rotation) which causes spin relaxation in space and is the basis of DP relaxation. It is tacitly assumed that this relaxation is captured by an exponential decay of spin polarization with distance. That, however, may be only approximately true [48].

Assuming that IR light does not change ζ_1 or ζ_2 appreciably, the light-to-dark contrast ratio will be

$$\texttt{Contrast ratio} = \frac{I_{light}}{I_{dark}} = \frac{1+\zeta_1\zeta_2 - 2\zeta_1\zeta_2 e^{-\frac{L}{L_l}}}{1+\zeta_1\zeta_2 - 2\zeta_1\zeta_2 e^{-\frac{L}{L_d}}}, \tag{8}$$

where L_l and L_d are the spin relaxation lengths under illumination and in the dark, respectively. In the event $L_l, L_d \gg L$, the above expression simplifies to

$$\begin{aligned}\texttt{Contrast ratio} &\approx \frac{1-\zeta_1\zeta_2 + 2\zeta_1\zeta_2 (L/L_l)}{1-\zeta_1\zeta_2 + 2\zeta_1\zeta_2 (L/L_d)} \\ &\approx \frac{L_d}{L_l}\,(\texttt{if}\ \zeta_1 \approx \zeta_2 \approx 1)\,. \end{aligned}\tag{9}$$

Ideally $L_d \to \infty$, which will make the contrast ratio diverge. This will also make the dark current zero and eliminate any standby power dissipation. Of course, making this happen is a very tall order since making $\zeta_1 = \zeta_2 = 1$ is an immense challenge.

To demonstrate the detector principle, nanowire spin valves were electrochemically self-assembled as described before, but with one important difference. The Co electrodeposition was carried out for a very small duration and the Ni deposition for a much longer duration.

Because of the longer deposition time, the Ni nanocontacts have cylindrical shapes conforming to the pores in which they are produced and their axes are collinear with the nanowires' common axis. Therefore, their easy axes of magnetization is always along the wire axis. However, since the Co deposition is carried out for a much shorter duration, the Co layers in the spin valves are very short and therefore do not necessarily have cylindrical shapes with their easy axes of magnetization along the nanowire axis. In fact, they are more like thin discs as shown in Fig. 7(a). Consequently, their magnetizations may not be collinear with the nanowires' common axis when no magnetic field is present, and maybe, in fact, perpendicular to the nanowire axis. The spins injected by the Co nanocontact partially transmit through the Ni nanocontact since the magnetizations of the two contacts are mutually perpendicular as opposed to parallel or anti-parallel. As a result, the spin valve resistance is intermediate.

If we apply a high enough magnetic field directed along the wire axis, the magnetizations of the short Co contacts will ultimately rotate to align along the field and hence along the nanowire axis. For each Co nanocontact, this will happen abruptly when the magnetostatic energy due to the magnetic field just overcomes the shape anisotropy energy barrier. If the shape anisotropy energy barrier does not vary too much from one Co nanocontact to another among the millions of nanowire spin valves that are simultaneously probed by the large area contacts, then all nanocontacts undergo the rotation at nearly the *same* field. At that threshold field, the majority spins in the Co and Ni contacts should suddenly become parallel, and the (spin-dependent) transmission probability of the electrons should jump, causing the resistance of the entire sample to drop *abruptly* at that field. On the other hand, if the shape anisotropy energy barriers vary significantly from one Co nanocontact to another among the many millions that are probed, then different nanocontacts will rotate at different fields and the resistance of the sample should drop *gradually* with increasing field and not abruptly. The behavior just described applies to the normal spin valve effect observed in the longitudinal magnetoresistance. If the sample exhibits the inverse spin valve effect, then the resistance of the spin valve will jump up abruptly at the threshold field instead of dropping.

In Fig 7(b), we show the room-temperature magnetoresistance of one sample measured in the dark and under illumination by an infrared (IR) lamp radiating in the wavelength range 2-5 μm. The magnetoresistance was measured with the magnetic field directed parallel to the axes of the wires. The lamp was kept far enough away from the sample to avoid heating effects. The resistance was monitored as a function of time during illumination to see if it drifted with time. Since no drift was observed, there are no discernible thermal effects on resistance due to the IR lamp.

The data taken in the dark (left panel of Fig. 7(b)) show a reproducible abrupt jump in the resistance by 2.5% when the magnetic field strength exceeds $\pm$650 Oe*. Clearly, this is the threshold field at which the magnetizations of the Co nanocontacts overcome their shape anisotropy energy barriers and rotate to align along the nanowire axis. The fact that the resistance rise is abrupt indicates that the variation in the shape anisotropy energy barrier among the different Co nanocontacts in this sample is small.

Note that beyond the threshold field, the resistance actually drops continuously. This is probably due to the fact that the magnetic field increases the energy separation between subbands and decreases inter-subband scattering in the InSb spacer layers, thereby slightly decreasing the resistance. This is unrelated to any spin-sensitive effect. However, the continuous decrease in magnetoresistance beyond the threshold field gives additional support to the conclusion that the sudden step increase at the threshold field is due to spin-polarized transport since it bucks the trend of the resistance decreasing with increasing magnetic field.

When the magnetoresistance is measured under constant IR illumination (right panel of Fig. 7(b)), we observe the same resistance jump at around $\pm$650 Oe magnetic field, but this time the resistance change is only 0.4%, which is considerably less than that in the dark. The more than 6-fold decrease in the step size can only happen if spin-polarized transport has been weakened by IR light. The IR light causes multi-subband transport in the InSb spacer layers of the spin valves by exciting electrons to the higher subbands and therefore shortens the spin relaxation length by reviving DP relaxation. The shorter spin relaxation length decreases spin polarization of the current and hence suppresses the step increase in the magnetoresistance at the threshold field.

*This indicates that the sample would have exhibited the inverse spin valve effect had the Co contact been similar to the Ni contact.

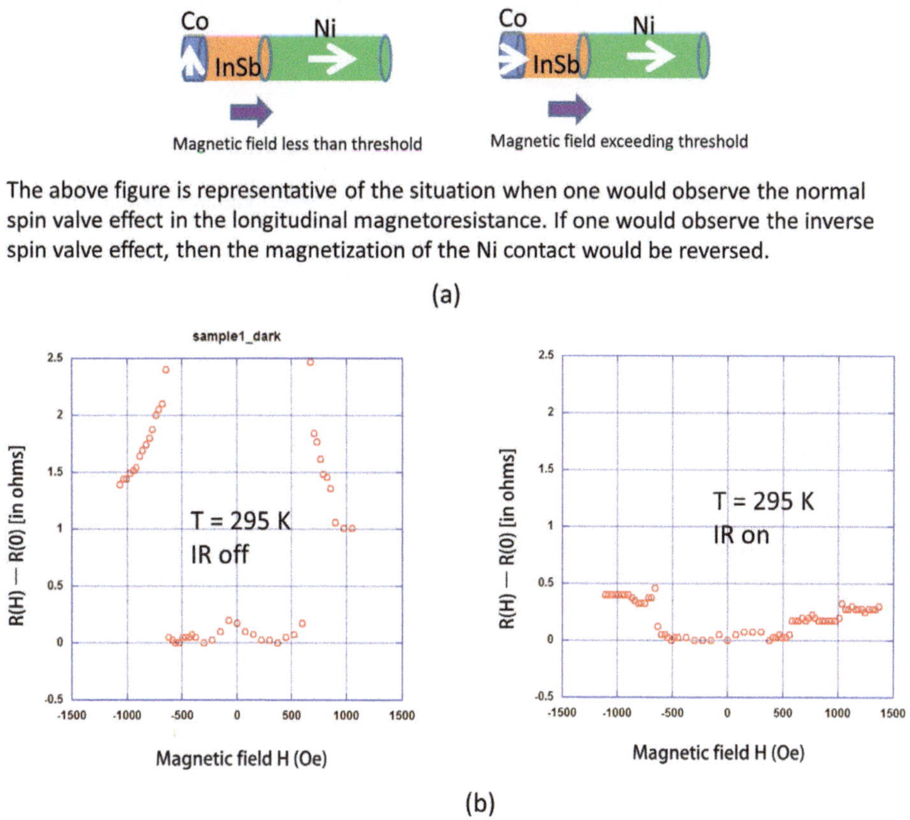

Fig. 7. (a) A nanowire spin valve where one contact is much smaller than the other and its shape anisotropy causes the easy axis of magnetization to point perpendicular to the wire axis. The other contact's easy axis of magnetization is along the wire axis. A strong magnetic field along the wire axis can overcome the shape anisotropy of the small contact and make its magnetization rotate to align in the direction of the field and the nanowire axis. (b) Room-temperature magnetoresistance of a Co-InSb-Ni nanowire spin valve sample in the dark (left panel) and under illumination by an infrared lamp radiating in the wavelength range 2-5 μm (right panel). The zero-field dark resistance was 4.2 ohms. Reproduced from ref. [47] with permission from the Institute of Physics.

We can obtain at least an order estimate for the average spin relaxation length in the InSb spacer layers – both in the dark and under IR illumination – from the modified Jullieré formula for drift transport [27]:

$$\frac{\Delta R}{R} = -\frac{2P_1P_2e^{-L/L_s}}{1 - P_1P_2e^{-L/L_s}}, \tag{10}$$

where $\Delta R/R$ is, again, the relative change in resistance at the step, P_1, P_2 are the spin polarizations of the two contacts, L is the average width of the InSb layer and L_s is the average spin relaxation length. Transmission electron microscopy of released nanowire spin valves showed $L \simeq$40 nm. Because the Co nanocontacts

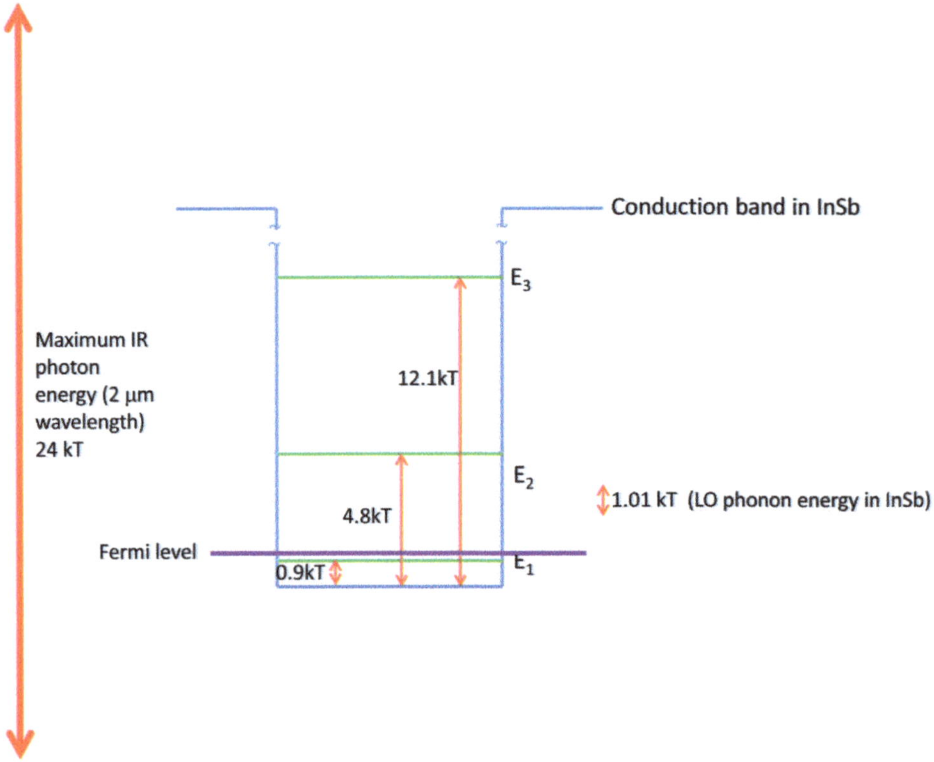

Fig. 8. Energy states in the conduction band of the InSb nanowire used in the spin valve experiment.

are small, but the Ni contact is large, it was assumed that $P_1 = 0.1$ (for Co) [20; 36] and $P_2 = 0.33$ (for Ni) [34]. Therefore, from the measured $\Delta R/R$, $L_s \approx 40$ nm in this sample at room temperature in the dark. Under IR illumination, the quantity $\Delta R/R$ drops to 0.4% from 2.5%, indicating that the average spin relaxation length L_s has decreased three-fold to ~14 nm. Therefore, $L_d = 40$ nm and $L_l = 14$ nm.

It is reasonable to assume that $P_1 \approx \zeta_1$ and $P_2 \approx \zeta_2$. If one uses the values $\zeta_1 = 0.1$ (ultrasmall Co contact), $\zeta_2 = 0.33$ (much larger Ni contact), $L = 40$ nm, $L_d = 40$ nm and $L_l = 14$ nm in Equation (8) to find the relative change in resistance between dark and illuminated conditions, we find that the quantity $(R_d - R_l)/R_l = 2\zeta_1\zeta_2\left(e^{-L/L_d} - e^{-L/L_l}\right)/\left(1 + \zeta_1\zeta_2 - 2\zeta_1\zeta_2 e^{-L/L_d}\right) = 2\%$, where R_d is the dark resistance and R_l is the resistance under illumination. Experimentally, this quantity was measured as 2% (see Fig 7), which indicates excellent agreement between theory and experiment.

The IR source used in the experiment was a broadband lamp that radiated in the wavelength range between 2 and 5 μm corresponding to photon energies between 9.4 kT and 24 kT at room temperature. The energy diagram within the conduction band is shown in Fig. 8. Clearly, the IR light can induce transitions between the subbands and cause multi-subband transport. The phonon energy in InSb is however too small to cause inter-subband transitions. That is why a single subband can be occupied in the dark and multiple subbands under illumination. Thus, light can modulate the D'yakonov-Perel' spin relaxation rate in the InSb nanowire.

Since the photon energy exceeds the bandgap energy of InSb (6.5 kT), IR illumination will also induce inter-band transition (valence-to-conduction band) in the InSb spacer layer. The resistance of the sample at zero magnetic field (when significant spin-polarized transport does not occur) however decreased by only 0.15% under illumination indicating that the inter-band transition (which will increase the electron and hole population in the InSb spacer and therefore decrease the resistance) is not significant. This is because the light intensity from the IR lamp is too weak for significant interband transitions to occur. It is therefore also too weak to cause significant intraband (or inter-subband) transitions, and yet the spin relaxation length decreased by a factor of three under illumination. This is due to the fact that even a slight departure from single subband transport can increase the DP relaxation rate considerably and quench spin-polarized transport. It is an encouraging observation since it portends high detectivity and small noise equivalent power for photodetectors predicated on this effect.

5. Room-temperature spin transport in a single GaN nanowire

GaN is a wide bandgap semiconductor used in high power electronics and solid state lighting. Because of the wide bandgap, the Rashba spin-orbit coupling constant in this compound semiconductor material is small. However, the non-centrosymmetricity of the GaN crystal, which crystallizes in the wurtzite or zincblende form, results in Dresselhaus interaction. Overall, however, spin-orbit coupling in GaN is quite weak, which is conducive to spin-coherent transport.

In addition to experiencing relatively weak spin-orbit interaction, electrons in GaN also move relatively slowly in an electric field owing to the small mobility accruing from a heavy effective mass. Therefore, the D'yakonov-Perel' relaxation rate, which is directly proportional to electron mobility (with the proportionality constant increasing with spin-orbit coupling strength) will be small in GaN leading to long spin relaxation time and length. It has been predicted that the spin relaxation time in GaN should be about three orders of magnitude longer than in GaAs at all temperatures [49].

The spin relaxation time and length in a *single* GaN nanowire has been measured at room temperature and reported as 100 ps and 260 nm, respectively [50]. The nanowires used in the experiment were epitaxially grown on a silicon substrate using plasma-assisted molecular beam epitaxy and released into isopropyl alcohol. The suspension, containing a relativley low density of nanowires, is drop cast on to a silicon substrate covered with a SiO_2 layer and the alcohol is allowed to evaporate. This leaves behind isolated GaN nanowires on the SiO_2 surface. Contact regions are defined on an isolated nanowire using electron beam lithography and contacts are formed by e-beam evaporation of 1-nm thick MgO and 60-nm thick FeCo [50]. The magnetic contacts inject and extract electrons into and from the GaN nanowire by tunneling though the thin MgO layer. That alleviates the resistivity mismatch problem between metallic FeCo and semiconducting GaN [51] and results in efficient spin injection and detection.

Because of the heavy effective mass, multiple subbands are occupied in a GaN nanowire at room temperature, resulting in some D'yakonov-Perel' (DP) spin relaxation. However, because the D'yakonov-Perel' mechanism is rather weak in GaN, as stated before, both the spin valve effect and the Hanle conductance oscillations could be observed at room temperature [50]. By fitting the Hanle oscillation characteristic to an expression, the room-temperature spin relaxation time and length were estimated to be 100 ps and 260 nm, respectively. These numbers are about an order of magnitude larger than that reported in InSb single-subband nanowires [20; 47] even though the DP relaxation would have been much more suppressed in the InSb nanowires. One possible explanation for this observation is that the electron mobility in the anodic-alumina-template-grown InSb nanowires is much poorer than in the molecular-beam-epitaxy-grown GaN nanowires, and the spin-orbit coupling strength is also much higher in InSb. Hence, the Elliott-Yafet spin relaxation is much stronger in the InSb nanowires. The sum of the Elliott-Yafet and D'yakonov-Perel' relaxation rates in the InSb nanowires exceeds that in the GaN nanowires, which is why the InSb nanowires have the smaller spin relaxation length and time.

6. Spin transport in Ge nanowires

Ge is a centrosymmetric material because it is an elemental semiconductor. Therefore, an electron in Ge should experience no Dresselhaus interaction if the electron current flows along a principal crystallographic direction such as the [100] direction. However, Ge is also an indirect bandgap semiconductor in which the lowest conduction band valley is at the Brillouin zone edge [L-valley]. The orbital nature of electrons in the L-valley are not s-type and hence the $\vec{L} \cdot \vec{S}$ coupling is non-zero. Therefore, electrons in Ge can experience the atomic $\vec{L} \cdot \vec{S}$ coupling, in addition to any Rashba spin-orbit coupling.

A number of spin transport experiments in Ge have been reported in the literature [52; 54] at low temperatures. One of us and coworkers fabricated Ge nanowires with the anodic alumina template method and carried out two-terminal (local) measurements of spin transport. The normal spin valve effect was observed, but complicated by the simultaneous presence of the anisotropic magnetoresistance (AMR). The spin valve effect resulted in magnetoresistance peaks while the AMR effect caused magnetoresistance troughs. Cross-section transmission electron micrographs of Co-Ge-Ni nanowire spin valve structures, produced by sequential electrodeposition of Co, Ge and Ni inside 50-nm diameter pores in anodic alumina films are shown in Fig. 9(a).

Figure 9(b) shows the longitudinal magnetoresistance traces at 1.9 K and 100 K temperatures. In the low temperature trace, one observes sharp peaks due to the normal spin valve effect superimposed on a background magnetoresistance that arises from a mixture of anisotropic magnetoresistance effect and Hall effect. Since we are not in the Coulomb blockade regime, we do not expect to see any magneto-Coulomb effects [53]. From the spin valve peaks, the spin relaxation length was estimated to be 400 nm at 1.9 K and 180 nm at 100 K [52]. The spin relaxation time was estimated as 4 ns at 1.9 K and 810 ps at 100 K [52].

Spin transport in Ge nanowires has also been studied by others using spin injection from a metallic ferromagnet into a Ge nanowire via tunneling through a MgO layer [54]. This involved measurement of non-local spin valve signals in four-terminal structures where the current is fed into and extracted from two ferromagnetic terminals, and the voltage is measured between two other ferromagnetic terminals as shown in Fig. 10. The non-local resistance is the ratio of the voltage to current and exhibits spin valve troughs in a varying magnetic field directed along the wire axis (between the coercive fields of the two current-injecting contacts), from which the spin relaxation length can be extracted. The advanatge of the non-local measurement is that it eliminates spurious effects such as anisotropic magnetoresistance. The authors of ref. [54] suggested that in order to fit their measured spin valve signal data, the spin relaxation length in their Ge nanowire had to be at least 100 μm at 4.2 K. This is nearly three orders of magnitude larger than what was reported in ref. [52]. Curiously, it is also almost three orders of magnitude larger than what has been reported in bulk Ge at 4.2 K [55]. The bulk Ge value is $\sim$580 nm at 4.2 K [55] which is, interestingly, close to the value (400 nm) reported in Ge nanowires at 1. 9 K in ref. [52]. It is possible however that the spin relaxation length in Ge nanowires can exceed that in bulk Ge due to suppression of the Elliott-Yafet mechanism. The suppression occurs because the mobility in nanowires can be larger (despite increased interface roughness scattering due to larger surface-to-volume ratio) because of the quasi one-dimensional constriction of phase space for scattering. Ref. [56] reported an electron spin relaxation length of 2.5 μm and spin relaxation time of 7.2 ns at 45 K temperature in n-type Ge nanowires.

The giant spin relaxation length reported in ref. [54], if true, would be a remarkable result. However, the spin relaxation length in silicon nanowires has been reported as a mere 6 μm at 4 K [57] and silicon should exhibit a longer spin relaxation length than Ge because it is a lighter element and hence should subject electrons to weaker spin-orbit interaction. Consequently, there is some skepticism about the 100 μm estimate [57] and it would be important to measure the spin relaxation length by different techniques such as Hanle oscillations or optical methods to either confirm or dispute the $\sim$100 μm estimate.

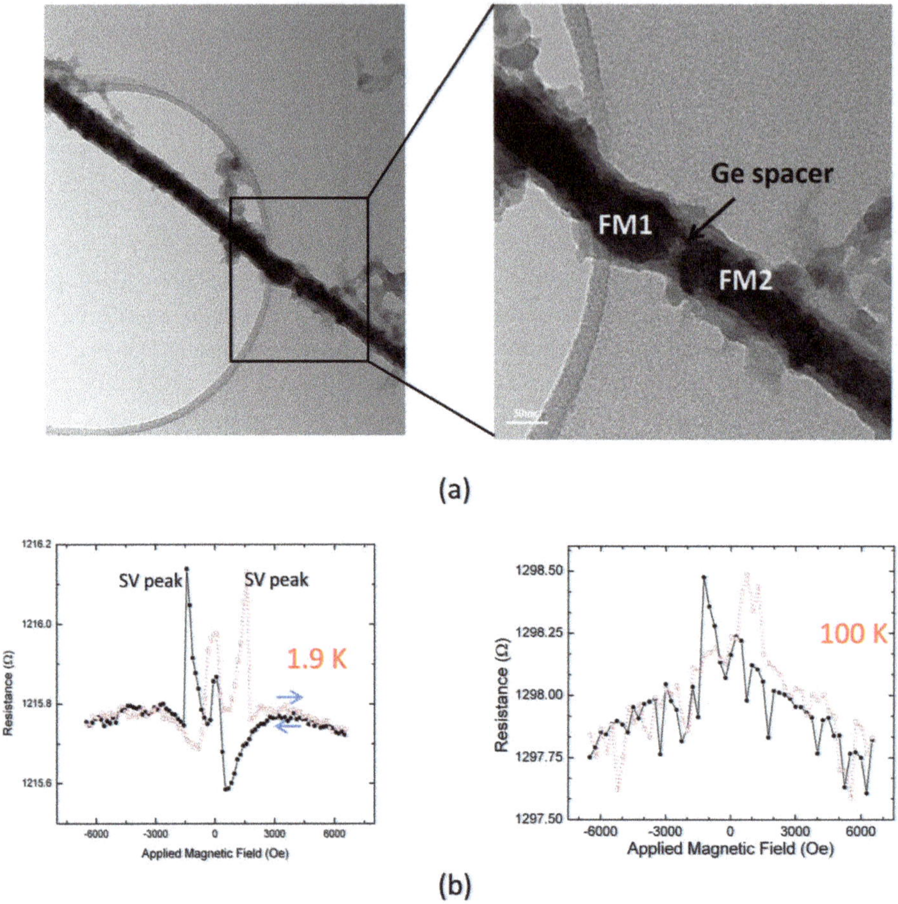

Fig. 9. (a) Cross-section transmission electron micrograph of a nanowire spin valve consisting of Co(FM1)-Ge-Ni(FM2). The scale bar in the micrograph is 50 nm. The nanowires are released from their porous alumina host by dissolving the alumina in hot chromic/phosphoric acid and captured on TEM grids for imaging. The capture is accomplished by soaking the TEM grid in the acid suspension containing the nanowires. (b) Measured longitudinal magnetoresistance of Co-Ge-Ni nanowire spin valves at two different temperatures – 1.9 K and 100 K. The sweeps from high positive to high negative field is shown in black while the reverse sweep is shown in pink. In the low temperature trace, the peaks are due to the normal spin valve (SV) effect and there is a complicated background that shows hint of an anisotropic magnetoresistance effect. There is no discernable SV peak or AMR trough at 100 K. This figure is reproduced from ref. [52] with permission from the American Institute of Physics.

7. Spin transport in silicon nanowires

Silicon is the workhorse material for electronics. It is a centrosymmetric crystal (ideally no Dresselhaus interaction) albeit electrons will experience the atomic $\vec{L} \cdot \vec{S}$ coupling because silicon is an indirect bandgap semiconductor with the lowest conduction band valley at the Brillouin zone edge (X-valley). The $\vec{L} \cdot \vec{S}$ interaction

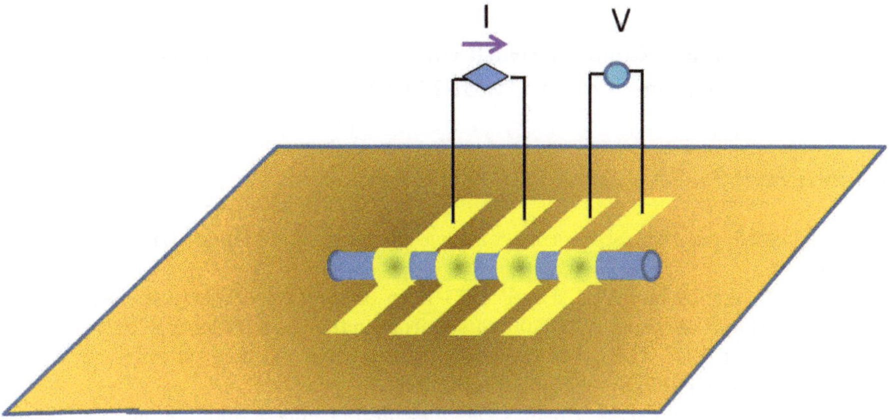

Fig. 10. Four-terminal, non-local spin valve measurement.

strength is proportional to the fourth power of the atomic number of the element [16] and hence Si has an advantage over Ge since it is a lighter element.

Spin transport in Si has been studied quite exhaustively [25] and there are reports of successful spin injection and detection in silicon nanowires using oxide tunnel barriers integrated with ferromagnetic contacts. Non-local spin transport experiments were carried out in four-terminal spin valve samples and from the observed troughs in the longitudinal magnetoresistance, the spin relaxation length was estimated as $\sim$6 μm at 4 K. Hanle oscillations have also been observed in Si nanowires where spin injection occurs by tunneling through graphene tunnel barriers [58] and the spin realxation length estimated from those measurements is 300-350 nm at 10 K. The large spread in the reported values of the spin relaxation length suggests that further experiments are needed (as in the case of Ge nanowires) to understand spin transport in Si nanowires.

8. Conclusion

A large number of experiments have been reported in a variety of semiconductor nanowires to probe coherent spin transport. Generally, compound semiconductor materials, with direct bandgap, exhibit shorter spin relaxation lengths (InN: 130 nm at 4 K [59], InAs: 200 nm at 10 K [60]) while elemental semiconductors, with indirect bandgap, exhibit longer spin relaxation lengths despite the $\vec{L} \cdot \vec{S}$ coupling. Since some semiconductor nanowires exhibit reasonably long spin coherence lengths at room temperature, they may have a role as "spin-coherent communicators" to transfer spin coded information between closely spaced spintronic devices and this, if realized, will be a major advance in both spin-based classical information processing systems and quantum computing and communication systems.

Acknowledgement

The work at Virginia Commonwealth University is supported by the US National Science Foundation under grant CMMI-1301013.

References

[1] G. Dresselhaus 1955 Spin-orbit coupling effects in zinc-blende structures *Phys. Rev.* **100** 580.
[2] M. I. D'yakonov and V. I. Perel' 1971 Spin orientation of electrons associated with the interband absorption of light in semiconductors *Sov. Phys. JETP* **33** 1053.
[3] R. J. Elliott 1954 Theory of the effect of spin-orbit coupling on magnetic resonance in some semiconductors *Phys. Rev.* **96** 266.
[4] Y. A. Bychkov and E. I. Rashba 1984 Oscillatory effects and the magnetic susceptibility of carriers in inversion layers *J. Phys. C: Solid State Phys.* **17** 6039.
[5] S. Datta and B. Das 1990 Electronic analog of the electro-optic modulator *Appl. Phys. Lett.* **56** 665.
[6] S. Bandyopadhyay and M. Cahay 2004 Alternate spintronic analog of the electro-optic modulator *Appl. Phys. Lett.* **85** 1814.
[7] M. I. D'yakonov and V. I. Perel' 1972 Spin relaxation of conduction electrons in non-centrosymmetric semiconductors *Sov. Phys. Solid State* **13** 3023.
[8] N. S. Averkiev and L. E. Golub 1999 Giant spin relaxation anisotropy in zinc-blende heterostructures *Phys. Rev. B* **60** 15582.
[9] N. S. Averkiev, L. E. Golub and M. Willander 2002 Spin realxation anisotropy in two-dimensional semiconductor systems *J. Phys: Condens. Matt.* **14** R271.
[10] J. Kainz, U. Rössler and R. Winkler 2003 Anisotropic spin splitting and spin relaxation in asymmetric zinc-blende semiconductor quantum structures *Phys. Rev. B* **68** 075322.
[11] P. H. Song and K. W. Kim 2002 Spin relaxation of conduction electrons in bulk III-V semiconductors *Phys. Rev. B* **66** 035207.
[12] S. Patibandla, G. M. Atkinson, S. Bandyopadhyay and G. C. Tepper 2010 Competing D'yakonov-Perel' and Elliott-Yafet spin relaxation in germanium *Physica E* **42** 1721.
[13] A. Bournel, P. Dollfus, P. Bruno and P. Hesto 1998 Gate-induced spin precession in an $In_{0.53}Ga_{0.47}As$ two-dimensional electron gas *Eur. Phys. J. Appl. Phys.* **4** 1.
[14] S. Pramanik, S. Bandyopadhyay and M. Cahay 2005 Spin relaxation in the channel of a spin field-effect transistor *IEEE Trans. Nanotechnol.* **4** 2.
[15] T. P. Pareek and P. Bruno 2002 Spin coherence in a two-dimensional electron gas with Rashba spin-orbit interaction *Phys. Rev. B* **65**, 241305(R).
[16] S. Bandyopadhyay and M. Cahay 2015 *Introduction to Spintronics*, 2nd. edition (CRC Press, Boca Raton).
[17] A. A. Kiselev and K. W. Kim 2000 Progressive suppression of spin relaxation in two-dimensional channels of finite width *Phys. Rev. B* **61** 13115.
[18] A. G. Mal'shukov and K. A. Chao 2000 Waveguide diffusion modes and slowdown of D'yakonov-Perel' spin relaxation in narrow two-dimensional semiconductor channels *Phys. Rev. B* **2000**, *61*, R2413.
[19] A. W. Holleitner, V. Sih, R. C. Myers, A. C. Gossard and D. D. Awschalom 2006 Suppression of spin relaxation in submicron InGaAs wires *Phys. Rev. Lett.* **97** 036805.
[20] S. Bandyopadhyay, Md. I. Hossain, H. Ahmad, J. Atulasimha and S. Bandyopadhyay 2014 Coherent spin transport and suppression of spin relaxation in InSb nanowires

with single subband occupancy at room temperature *Small* **10** 4379. Please see the supplementary material accompanying this paper for exhaustive thereotical analyses and characterization of the InSb self-assembled nanowires.
[21] E. Y. Tsymbal, A. Sokolov, I. F. Sabirianov and B. Doudin 2003 Resonant inversion of tunneling magnetoresistance *Phys. Rev. Lett.* **90** 186602.
[22] S. Pramanik, S. Bandyopadhyay, K. Garre and M. Cahay 2006 Normal and inverse spin valve effect in organic semiconductor nanowires and the background monotonic magnetoresistance *Phys. Rev. B* **74** 235329.
[23] Y. Takamura and S. Sugahara 2012 Analysis and control of the Hanle effect in metal-oxide-semiconductor inversion channels *J. Appl. Phys.* **111**, 07C323.
[24] Y. Fukuma, L. Wang, H. Idzuchi, S. Takahashi, S. Maekawa and Y. C. Otani 2011 Giant enhancement of spin accumulation and long distance spin precession in metallic lateral spin valves. *Nature Mater.* **10** 527.
[25] I. Appelbaum, B. Huang and D. J. Monsma 2007 Electronic measurement and control of spin transport in silicon *Nature* **447** 295
[26] X. Zhang, Y. Hou, G. Meng and L. Zhang 2005 Fabrication of highly ordered InSb nanowire arrays by electrodeposition in porous anodic alumina membranes *J. Electrochem. Soc.* **152** C664.
[27] Z. H. Xiong, D. Wu, Z. V. Vardeny and J. Shi 2004 Giant magnetoresistance in organic spin valves *Nature* **427** 821.
[28] F. J. Jedema, A. T. Filip and B. J. van Wees 2001 Electrical spin injection and accumulation at room temperature in an all-metal mesoscopic spin valve *Nature* **410**, 345.
[29] A. Fert and H. Jaffrés Conditions for efficient spin injection from a ferromagnetic metal into a semiconductor *Phys. Rev. B* **64** 184420
[30] M. Tran, H. Jaffrés, C. Deranlot, J,-M. George, A. Fert, A. Miard and A. Memaître 2009 Enhancement of the spin accumulation at the interface between a spin-polarized tunnel junction and a semiconductor *Phys. Rev. Lett.* **102** 036601.
[31] A. N. Chantis and D. L. Smith 2008 Theory of electrical spin-detection at a ferromagnet/semiconductor interface *Phys. Rev. B* **78** 235317.
[32] T. Valet and A. Fert 1993 Theory of perpendicular magnetoresistance in magnetic multilayers *Phys. Rev. B* **48** 7099.
[33] B. Dlubak, M. B. Martin, C. Deranlot, B. Servet, S. Xavier, R. Mattana, M. Sprinkle, C. Berger, W. A. De Heer and F. Petroff 2012 Highly efficient spin transport in epitaxial graphene on SiC *Nature Phys.* **8** 557.
[34] E. Y. Tsymbal, O. N. Mryasov and P. R. Leclair 2003 Spin dependent tunneling in magnetic tunnel junctions *J. Phys: Condens. Matt.* **15** R109.
[35] F. J. Jedema, M. V. Costache, H. B. Heersche, J. J. A. Baselmans and B. J. van Wees 2002 Electrical detection of spin accumulation and spin precession at room temperature in metallic spin valves. *Appl. Phys. Lett.* **81** 5162.
[36] S. O. Valenzuela and M. Tinkham 2004 Spin-polarized tunneling in room-temperature mesoscopic spin valves *Appl. Phys. Lett.* **85** 5914.
[37] P. A. Dowben and R. Skomski 2004 Are half-metallic ferromagnets half metals? *J. Appl. Phys.* **95** 7453.
[38] J. Li, B. Huang, I. Appelbaum 2008 Oblique Hanle effect in semiconductor spin transport devices *Appl. Phys. Lett.* **92** 142507.
[39] H. A. Nilsson, P. Caroff, C. Thelander, Thelander, M. Larsson, J. B. Wagner, L. E. Warnersson, L. Samuelson and H. Q. Xu 2009 Giant, level-dependent g-factors in InSb nanowire quantum dots *Nano Lett.* **9** 3151.

[40] K. L. Litvinenko, L. Nikzad, B. N. Murdin, C. R. Pidgeon, J. J. Harris, T. Zhang and L. F. Cohen 2007 Spin lifetime in high quality InSb epitaxial layers grown on GaAs *J. Appl. Phys.* **101** 083105.
[41] P. Murzyn, et al. 2004 Electron spin lifetimes in $Hg_{0.78}Cd_{0.22}Te$ and InSb. *Physica E* **20** 220.
[42] M. Bhowmick, R. N. Kini, K. Nontapot, N. Goel, S. J. Chung, T. D. Mishima, M. B. Santos and G. A. Khodaparast 2010 Probe of interband relaxations of photo-excited carriers and spins in InSb based quantum wells *Phys. Procedia* **3** 1161.
[43] J. T. Olesberg, W. H. Lau, M. E. Flatté, C. Yu, E. Altunkaya, E. M. Shaw, T. C. Hasenberg and T. F. Boggess 2001 Interface contributions to spin relaxation in a short-period InAs/GaSb superlattice *Phys. Rev. B.* **64** 201301(R).
[44] Y. Kunihashi, M. Kohda and J. Nitta 2009 Enhancement of spin lifetime in gate-fitted InGaAs narrow wires *Phys. Rev. Lett.* **102** 226601 and references therein.
[45] P. C. Sercel and K. J. Vahala 1991 Polarization dependence of optical emission and absorption in quantum wires *Phys. Rev. B.* **44**, 5681.
[46] S. M. Sze 1981 *Physics of Semiconductor Devices*, 2nd. edition (Wiley-Interscience, New York).
[47] Md. I. Hossain, S. Bandyopadhyay, J. Atulasimha and S. Bandyopadhyay 2015 Modulating spin relaxation in nanowires with infrared light at room temperature *Nanotechnology* **26** 281001.
[48] S. Pramanik, S. Bandyopadhyay and M. Cahay 2004 Decay of spin-polarized hot carrier current in a quasi one-dimensional spin valve structure *Appl. Phys. Lett.* **84** 266.
[49] S. Krishnamurthy, M. von Schilfgaarde and N. Newman 2003 Spin lifetimes of electrons injected into GaAs and GaN *Appl. Phys. Lett.* **83** 1761.
[50] H. Kum, J. Heo, S. Jahangir, A. Banerjee, W. Guo and P. Bhattacharya 2012 Room-temperature single GaN nanowire spin valves with FeCo/MgO tunnel contacts *Appl. Phys. Lett.* **100**, 182407.
[51] E. I. Rashba 2000 Theory of electrical spin injection: Tunnel contacts as a solution of the conductivity mismatch problem *Phys. Rev. B* **62** 16267(R).
[52] S. Patibandla, S. Pramanik, S. Bandyopadhyay and G. C. Tepper 2006 Spin relaxation in a germanium nanowire *J. Appl. Phys.* **100** 044303.
[53] H. Shimada, K. Ono and Y. Ootuka 1998 Magneto-Coulomb oscillations in ferromagnetic single electron transistors *J. Phys. Soc. Jpn.* **67** 1359.
[54] E-S Liu, J. Nah, K. M. Varahramyan and E. Tutuc 2010 Lateral spin injection in germanium nanowires *Nano Lett.* **10** 3297.
[55] Y. Zhou, W. Han, L-T Chang, F. Xiu, M. Wang, M. Oehme, I. A. Fischer, J. Schulze, R. K. Kawakami and K. L. Wang 2011 Electrical spin injection and transport in germanium *Phys. Rev. B* **84** 125323.
[56] J. Tang, T. Nie and K. L. Wang 2014 Spin transport in Ge nanowires for diluted magnetic semiconductor-based nonvolatile transpinor *ECS Trans.* **64** 613.
[57] S. Zhang, S. A. Dayeh, Y. Li, S. A. Crooker, D. L. Smith and S. T. Picraux 2013 Electrical spin injection and detection in silicon nanowires through oxide tunnel barriers *Nano Lett.* **13** 430.
[58] O. M. J. van't Erve, A. L. Friedman, C. H. Li, J. T. Robinson, J. Connell, L. J. Lauhon and B. T. Jonker 2015 *Nature Communications* **6** 7541.
[59] S. Heedt, C. Morgan, K. Weis, D. E. Bürgler, R. Calarco, H. Hardtdegen, D. Grützmacher and Th. Schäpers 2012 Electrical spin injection into InN semiconductor nanowires *Nano Lett.* **12** 4437.
[60] A. E. Hansen, M. T. Björk, C. Fasth, C. Thelander and L. Samuelson 2005 *Phys. Rev. B* **71** 205328.

Chapter 5

Signatures of Spin Polarization In Four-Gate Quantum Point Contact Structures

A. Jones[*], *M. Cahay*[*,‡], *I. I. Yakimenko*[†] *and K.-F. Berggren*[†]

[*]Department of Electrical Engineering and Computing Systems,
University of Cincinnati, Cincinnati, Ohio 45221, USA
[†]Department of Physics, Chemistry, and Biology, Linköping University,
S-58183 Linköping, Sweden
[‡]*marc.cahay@uc.edu*

Using a Non-equilibrium Green's function (NEGF) approach, a set of experiments is suggested which can provide indirect evidence of the fine and non-local electrostatic tuning of the onset of spin polarization in two closely spaced quantum point contacts (QPCs) with two sets of in-plane side gates (SGs) in the presence of lateral spin-orbit coupling (LSOC). The conductance of the two closely spaced QPCs or four-gate QPC is studied for different biasing conditions applied to two leftmost and rightmost SGs. When calculated as a function of the common sweep voltage V_{sweep} applied to two of the SGs, the conductance plots show several conductance anomalies, i.e., below $G_0 = 2e^2/h$, characterized by intrinsic bistability, i.e., hysteresis loops due to a difference in the conductance curves for up and down sweeps of the common gate voltage. The hysteresis loops are related to the co-existence of multistable spin textures in the narrow channel of the four-gate QPC and are very sensitive to the biasing conditions on the four SGs. The shape of the conductance anomalies and size of hysteresis loops are different when the biasing conditions on the leftmost and rightmost SGs are swapped. This rectifying behavior is an additional indirect evidence for the onset of spontaneous spin polarization in nanoscale devices made of QPCs. These results show that the onset and fine tuning of conductance anomalies in QPC structures are highly sensitive to the non-local action of closely spaced SGs. This effect must therefore be taken into account in the design of all electrical spin valves making use of middle gates to fine tune the spin precession between QPC based spin injector and detector contacts.

1. Introduction

To realize the full potential of spin-based devices, ways must be found to inject, manipulate, and detect the spin of the electron by *purely electrical* means. This remains the central challenge in spintronics [1,2]. It would provide a breakthrough in applications of new quantum devices and circuits, with application ranging from quantum information processing to quantum computing[3]. In recent years, spin orbit coupling (SOC) has been envisioned as a possible tool for all-electrical spin control and generation of spin-polarized currents. SOC has been used to modulate spin polarized currents by taking advantage of symmetry-breaking factors such as interfaces, electric fields, strain, and crystalline directions[4,5]. There have been several recent experimental reports that an asymmetry in the potential applied between the two side gates (SGs) of a QPC can be used to create strongly spin polarized currents using both top and in-plane SGs, different materials (GaAs and InAs based heterostructures), two-dimensional electron gas (2 DEG) with different electron mobility, and a wide variety of heterostructure design and QPC dimensions[6-10].

Following the work of Thomas et al.[11], there have been many experimental reports of anomalies in the conductance of QPCs appearing at non-integer multiples of G_0, including $0.25G_0$, $0.5G_0$, and $0.7G_0$ [12-16]. Thomas et al. suggested that the $0.7G_0$ conductance anomaly was an indirect evidence of a spontaneous ferromagnetic spin polarization in the narrow portion of the QPC. Since then, there has been mounting evidence that the number and location of these conductance anomalies can be tuned by deliberately introducing a broken symmetry in the QPC's electrostatic confining potential[17-20].

Over the last six years, starting with the work of Debray et al.[7], Cahay et al. have shown that LSOC, resulting from the lateral in-plane electric field of the confining potential of InAs- and GaAs-based QPCs with in-plane SGs, can be used to create a strongly spin-polarized current by purely electrical means, i.e., *in the absence of any applied magnetic field*[8-10]. This is realized by tuning the electron confinement potential of the channel using asymmetric bias voltages on the SGs. In both InAs and GaAs QPCs, a plateau at $G \cong 0.5G_0$ has been observed in the ballistic conductance of the QPC *in the absence of* a magnetic field – *an indirect signature of complete spin polarization.* Das et al. have observed several other anomalous (i.e., $G < G_0$) conductance plateaus in asymmetrically biased $In_{0.52}Al_{0.48}As$/InAs QPCs with in-plane SGs in the presence of LSOC[10]. It was found that the number and location of the anomalous conductance plateaus strongly depend on the polarity of the offset bias. The anomalous plateaus appear only over an intermediate range of offset bias of several volts. They are quite robust, being observed over a maximum range of nearly 1V for the common sweep voltage applied to the two SGs. In the past, Wan et al. have used a NEGF approach to calculate the conductance of a small QPC with LSOC and found three *sufficient* conditions for the generation of a strongly spin polarized current[21,22]: (a) an asymmetric lateral confinement, (b) a LSOC induced by the lateral confining potential of the QPC, and (c) a strong electron–electron (e-e) interaction. *NEGF simulations have shown that LSOC is instrumental only in triggering the initial spin imbalance; it is the strong e-e interaction that enhances it and results in the strong spin polarization*[21,22]. The NEGF approach was used to study in detail the ballistic conductance of asymmetrically biased QPCs in the presence of LSOC for a wide range of QPC dimensions and gate bias voltage. Various conductance anomalies were predicted below the first quantized conductance plateau (G_0); these were shown to occur due to spontaneous spin polarization in the narrowest portion of the QPC. It was found that the number of conductance anomalies increases with the aspect ratio (length/width) of the QPC constriction[22]. Based in the NEGF approach, the occurrence of other conductance anomalies besides the 0.5 G_0 was interpreted as evidence for the sensitivity of the QPC spin polarization to defects (surface roughness, impurity and dangling bond scattering) generated during the etching process that forms the QPC side walls[10]. All conductance anomalies were shown to be fingerprints of spin textures in the narrow portion of the QPC[22].

More recently, hysteresis was observed between the forward and reverse sweeps of a common mode bias applied to the two in-plane SGs of asymmetrically biased InAs and GaAs QPCs[23,24]. The size of the hysteresis loop was found to increase with the amount of bias asymmetry ΔV_G between the two SGs and depends on the polarity of ΔV_G. These results were found to be in qualitative agreement with NEGF simulations which show that the conductance plots versus common gate voltage applied to the two SGs can have single or multiple hysteresis loops depending on the QPC dimensions and biasing conditions. Numerical simulations have shown that (a) the hysteresis appears only for sufficiently long QPCs if the e-e interaction is strong enough, (b) the shape of the hysteresis loops depends on the polarity and magnitude of ΔV_G and the strength of the e-e interaction, and (c) the hysteresis loops are affected by the presence of dangling bonds on the sidewalls[24]. The rich plethora of hysteresis loops is intimately related to a wide variety of metastable spin textures and linked to the onset of a net spin polarization in the narrow portion of the QPC for biasing conditions leading to conductance anomalies, that is, less than G_0. Therefore, the observation of hysteresis loops constitutes another *indirect proof of spontaneous spin polarization* in the narrow portion of the QPC.

A new device consisting of a QPC with two closely spaced in-plane SGs was proposed to create a tunable all electrical spin polarizer[25] (see Fig.1). The first set of SGs near the source contact is asymmetrically biased to create spin polarization in the channel of the QPC. The same bias is applied on the SGs near the drain and varied to maximize the amount of spin conductance polarization of the QPC. The range of common mode bias on the first set of SGs over which a maximum spin conductance polarization can be achieved was shown to be much broader for the four-gate structure compared the case of a QPC with a single pair of in-plane SGs[25].

Hereafter, a more thorough analysis of the appearance and control of the conductance anomalies in four-gate QPCs is performed, including a study of the hysteresis loops appearing due to a difference between the conductance characteristics of the device for forward and reverse sweeps of the common gate voltage applied to some of the SGs. As will be shown in the numerical examples below, conductance anomalies in four-gate QPCs can appear even in the case of symmetric biasing conditions on the four SGs. In this case, it will be shown that it is the asymmetry in the confining potential along the direction of current flow which creates a small numerical imbalance between the spin density associated to spin-up and spin-down electrons. The latter is amplified in the presence of strong e-e interaction for specific range of (symmetric biases) on the SGs. The spatial asymmetry along the direction of current flow is also responsible for a rectification effect on the conductance curves of the four-gate QPCs which is especially strong in the hysteresis region near the conductance anomalies.

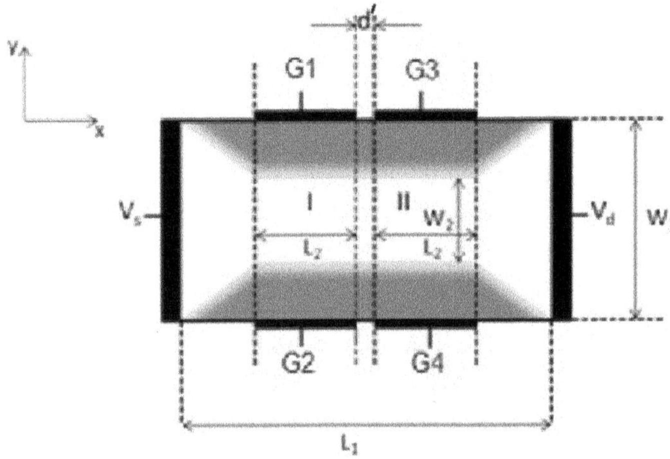

Figure 1: Schematic of four-gate QPC[25]*. Different biasing conditions are applied to the four gates as listed in Tables A, B, and C. The current flows in the x-direction. In all simulations, the dimensions of the device are set equal to:* $W_1 = 48nm$, $W_2 = 16nm$, $L_1 = 80nm$, $L_2 = 22nm$, *and* $d' = 4nm$.

2. Model and NEGF simulations

Unless otherwise specified, NEGF simulations were performed for the four-gate InAs QPC depicted in Fig. 1 using the following parameters: $W_1 = 48$ nm, $L_1 = 80$ nm, $W_2 = 16$ nm, $L_2 = 22$ nm, $d' = 4$ nm. A summary of the NEGF approach is given in the Appendix. Further details are given in refs. [21,22]. The effective mass in the InAs channel was set equal to $m^* = 0.023m_0$, where m_0 is the free electron mass. All calculations were performed at a temperature $T = 4.2\ K$. The potential at the source, V_s, was set equal to 0V and the one at the drain, V_d, to 0.3 mV in all simulations. The conductance of the constriction was studied as a function of the sweeping (common-mode) potential V_{sweep}. The Fermi energy was equal to 106.3 meV in the source contact and 106 meV in the drain contact, ensuring single-mode transport through the QPC. At the interface between the narrow portion of the QPC and vacuum, the conduction band discontinuities at the bottom and the top interface were modeled, respectively, as

$$\Delta E_c(y) = \frac{\Delta E_c}{2}\left[1 + \cos\frac{\pi}{d}\left(y - \frac{w_1 - w_2}{2}\right)\right], \tag{1}$$

and

$$\Delta E_c(y) = \frac{\Delta E_c}{2}\left[1 + \cos\frac{\pi}{d}\left(\frac{w_1 + w_2}{2} - y\right)\right], \tag{2}$$

to achieve a smooth conductance band change, where d was selected equal to 1.6nm to represent a gradual variation of the conduction band profile from the inside of the quantum wire to the vacuum region. A similar grading was also used along the walls going from the wider part of the channel to the central constriction of the QPC. This gradual change in $\Delta E_c(y)$ is responsible for the LSOC that triggers the spin polarization of the QPC in the presence of an asymmetry in V_{sg1} and V_{sg2}. ΔE_c in Eqns. (1) and (2) was set equal to 4.5 eV. The device and material parameters are selected such that NEGF simulations of a QPC with only two SGs of length L_2 (See Fig.1) selected small enough (less than 26 nm, according to the simulations of ref. [22]) so that no net spin polarization can be created by varying the common sweep voltage on the two gates after a bias asymmetry is applied between them. In this case, no conductance anomaly around 0.5 G_0 was observed as a function of V_{sweep} in the NEGF simulations[22]. Following Lassl et al.[26], we used a Hartree-Fock approximation to include the effects of e-e interactions in the QPC. More specifically, the e-e interaction was taken into account by considering a repulsive Coulomb contact potential, $V_{int}(x,y;x',y') = \gamma\,\delta(x-x')\,\delta(y-y')$, where γ indicates the e-e interaction strength[27]. Unless otherwise specified, the strength of the parameter γ in the interaction self-energy characterizing the strength of the electron-electron interaction was set equal to $3.7\hbar^2/2m^*$ and the parameter β characterizing the strength of the LSOC was set equal 200 Å^2 [21,22].

3. Numerical Results

Using the physical dimensions and physical parameters listed above, a first series of conductance calculations of the four-gate QPC were carried for the biasing conditions listed in Table A. Cases 4 through 6 correspond to an interchange of the biasing conditions on the four gates considered in cases 1 through 3, respectively. Hereafter, we also show plots of the spin conductance polarization $\alpha = (G_{up} - G_{down})/(G_{up} + G_{down})$ versus V_{sweep}, where G_{up} and G_{down} are the contributions to the total conductance from the spin-up and spin-down electrons, respectively. These represent the contributions to the total conductance due to the electrons injected from the contacts with a spin component S_z, equal to +1/2 and -1/2, respectively.

Table A: Different sets of biasing conditions on the SGs of the four-gate of the QPC device shown in Fig.1. For cases 4, 5, and 6, the biasing conditions on gates 1 and 2 and gates 3 and 4 are flipped compared to those for cases 1, 2, and 3, respectively. The geometrical dimensions of the four-gate QPC are given in the caption of Fig.1.

Case	$V_{sg1}(V)$	$V_{sg2}(V)$	$V_{sg3}(V)$	$V_{sg4}(V)$
#1	$-0.2 + V_{sweep}$	$0.2 + V_{sweep}$	0.0	0.0
#2	$-0.2 + V_{sweep}$	$0.2 + V_{sweep}$	-0.1	-0.1
#3	$-0.2 + V_{sweep}$	$0.2 + V_{sweep}$	-0.2	-0.2
#4	0.0	0.0	$-0.2 + V_{sweep}$	$0.2 + V_{sweep}$
#5	-0.1	-0.1	$-0.2 + V_{sweep}$	$0.2 + V_{sweep}$
#6	-0.2	-0.2	$-0.2 + V_{sweep}$	$0.2 + V_{sweep}$

The conductance plots were calculated for both the up and down sweeps of V_{sweep} to assess the strong dependence of the hysteresis loops in the conductance curves on the specific biasing conditions on the four side gates. As will be shown hereafter, the onset of hysteresis is linked to the presence of multistable spin density configurations in the narrow portion of the dual QPC as illustrated via two-dimensional contour plots of spin texture profile $n_\uparrow$ (x,y) - $n_\downarrow$ (x,y) as a function of V_{sweep}.

First, the potentials of the leftmost side gates in Fig.1 were set equal to $V_{sg1} = -0.2 + V_{sweep}$ and $V_{sg2} = 0.2 + V_{sweep}$, where V_{sweep} is the common sweep voltage. The potential on the other two SGs were set to three different sets of values: (case 1, Table A): (V_{sg3}, V_{sg4}) = (0.0,0.0) V, (case 2 Table A): (V_{sg3}, V_{sg4}) = (-0.1,-0.1) V, and (case 3, Table A) (V_{sg3}, V_{sg4}) = (-0.2,-0.2) V. Plots of the up and down sweeps of the conductance vs V_{sweep} for case 1 are shown in Fig.2 and the corresponding plot of spin conductance polarization, α, are shown in Fig.3.

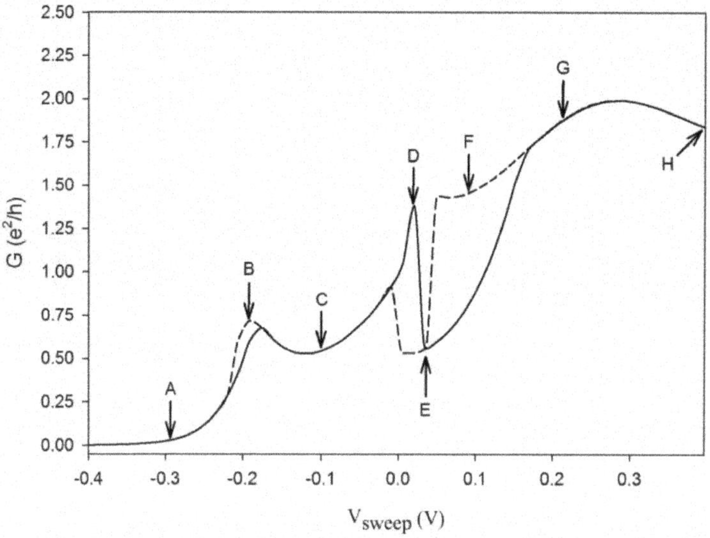

Figure 2: Conductance versus V_{sweep} of a four-gate QPC with the dimensions listed in Fig.1. The biases on the different gates are set to the conditions from Table A, Case 1. The full and dash curves are the conductance versus V_{sweep} for the up and down sweeps of the common gate voltage V_{sweep} on gates 1 and 2, respectively. Two-dimensional contour plots of the spin texture profile $n_{\uparrow}(x,y) - n_{\downarrow}(x,y)$ for the up and down sweeps of V_{sweep} corresponding to points A through H are shown in Figures 4 and 5.

Figure 2 shows the presence of two conductance anomalies, i.e., occurring at values below G_0. For the up sweep, Fig. 2 shows that there are sharp drops (i.e., conductance anomalies) linked to the onset of a negative differential region (NDR) in the conductance curve appearing above -190 mV and 20 mV.

Hysteresis loops are due to a difference between the up and down sweeps in the conductance curves appear in the regions around the conductance anomalies. The range of the hysteresis loops is better resolved by looking at a plot of the spin conductance polarization α for the up and down sweeps, as shown in Fig.3.

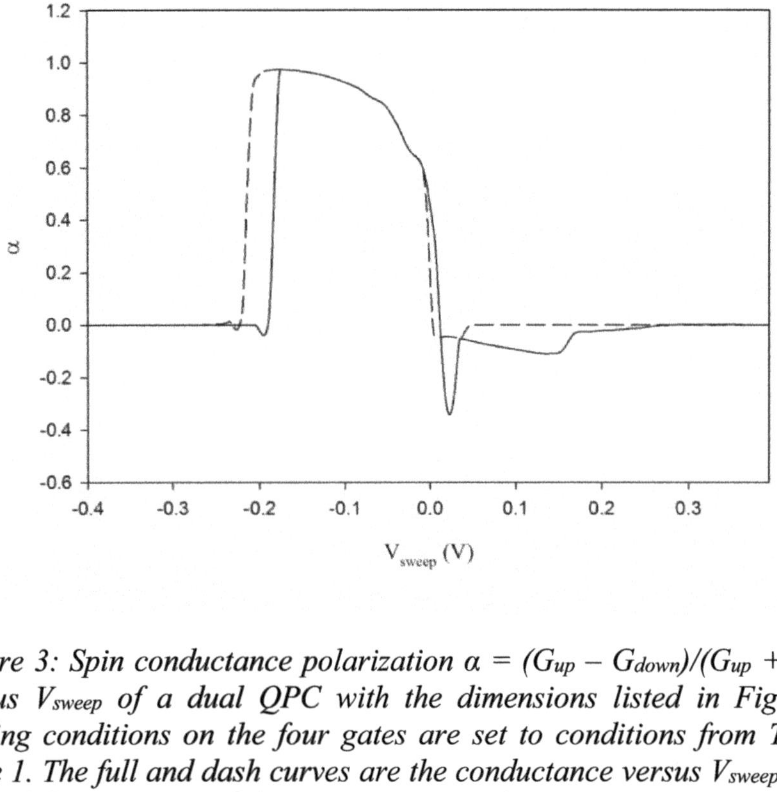

Figure 3: Spin conductance polarization $\alpha = (G_{up} - G_{down})/(G_{up} + G_{down})$ *versus* V_{sweep} *of a dual QPC with the dimensions listed in Fig.1. The biasing conditions on the four gates are set to conditions from Table A Case 1. The full and dash curves are the conductance versus* V_{sweep} *for the up and down sweeps of the common gate voltage* V_{sweep} *on gates 1 and 2, respectively.*

Near the leftmost conductance anomaly in Fig.2, α for the up and down sweeps are different over a range of V_{sweep} extending from -250 mV to -190 mV. The size of the hysteresis loop around the second anomaly is much broader and appears over a V_{sweep} range extending from -20 mV to 320mV. To better understand the origin of the hysteresis loops, Figures 4 and 5 show two-dimensional contour plots of spin texture profile $n_{\uparrow}(x,y) - n_{\downarrow}(x,y)$ for the specific values of V_{sweep} corresponding to points A, B, C, D, E, F, G, and H in Figure 2.

For the up sweep, the spin texture changes abruptly from what we refer to as a spin Hall configuration (corresponding to opposite spin polarities on both sides of the narrow channel of the four-gate QPC device) to a single lobe associated to accumulation of spin-up electrons over the entire narrow portion of the four-gate QPC past the threshold for the first NDR. This single lobe configuration survives until the onset of the second NDR region at point D. Past that point, the spin texture is composed of two lobes with opposite polarities, with a majority spin-up and spin-down configurations between the two leftmost and rightmost SGs, respectively. This two-lobe configuration evolves gradually towards a three-lobe configuration past 170 mV (after which conductance curves for the up and down sweeps are nearly equal) with two lobes of opposite spin polarities between the two leftmost SGs and a single lobe between the two rightmost SGs. This three-lobe configuration gradually switches to a spin density wave which smoothly evolves into a spin Hall configuration as the conductance value approaches G_0, as shown in Fig.6. For the down sweep, the three-lobe configuration is not observed and the spin Hall configuration survives until it abruptly switches to a two-lobe configuration at point E and then a single lobe configuration below point D. This single lobe configuration persists until V_{sweep} falls below the leftmost point in the first hysteresis loop (around -250 mV) at which point the spin texture regain a spin Hall nature.

Figure 7 shows the conductance curves for the up and down sweeps for the biasing conditions corresponding to cases 2 and 3 in Table A. Comparing to the results of case 1 depicted in Fig.2, the threshold for the rise of the conductance curves shifts to larger values of V_{sweep}, from about -0.3, to -0.2 and -0.1V, for the biasing conditions corresponding to cases 1, 2, and 3, respectively. This results from the fact that when the biases on the gates G_3 and G_4 is more negative, there is an increased potential barrier the electrons must surmount in the narrow portion of the QPC between the rightmost SGs. In the up sweep, the first peak (conductance anomaly) in the conductance curve shifts to a lower value as the biasing conditions on the rightmost gates are changed from case 1, to cases 2 and 3, respectively. This indicates that the location of the conductance anomaly can be adjusted by varying the potential on the rightmost SGs, a feature already pointed out in ref. [25]. The size of the leftmost hysteresis loop is only slightly affected by the biasing conditions on the rightmost SGs.

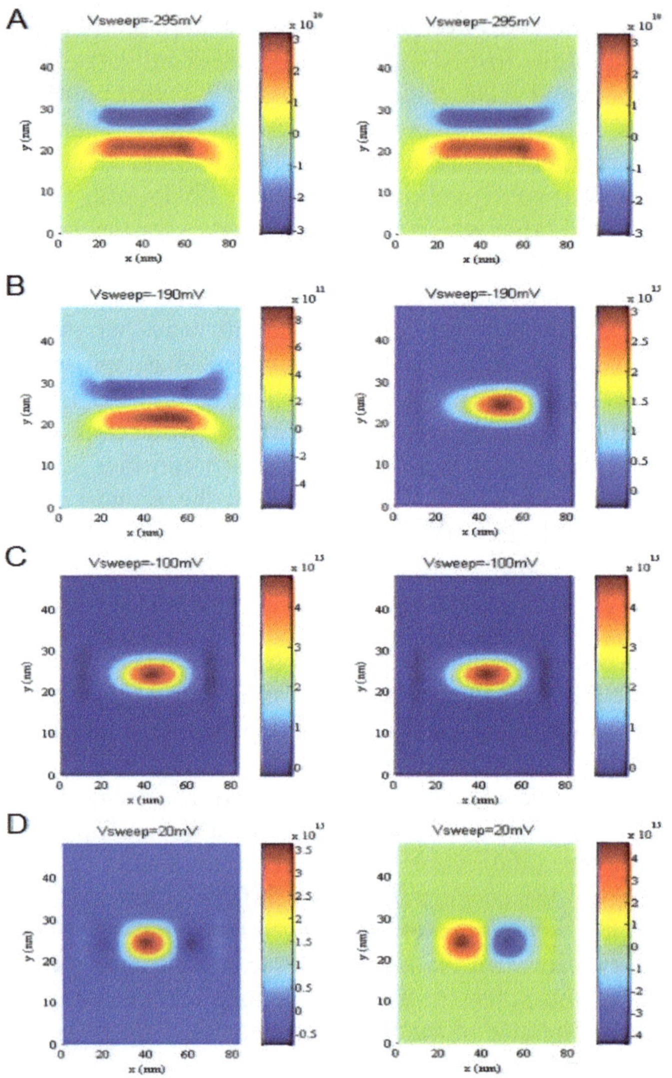

Figure 4: Two-dimensional contour plots of spin texture profile $n_\uparrow (x,y)$ - $n_\downarrow (x,y)$ for values of V_{sweep} corresponding to points A, B, C, and D in Fig.2. It uses the dimensions listed in Fig.1. These correspond to values of V_{sweep} equal to -295mV, -190mV, -100mV, and 20mV, respectively. In each frame, the left and right figures are the spin texture profile through the four-gate QPC device for the up and down voltage sweeps, respectively.

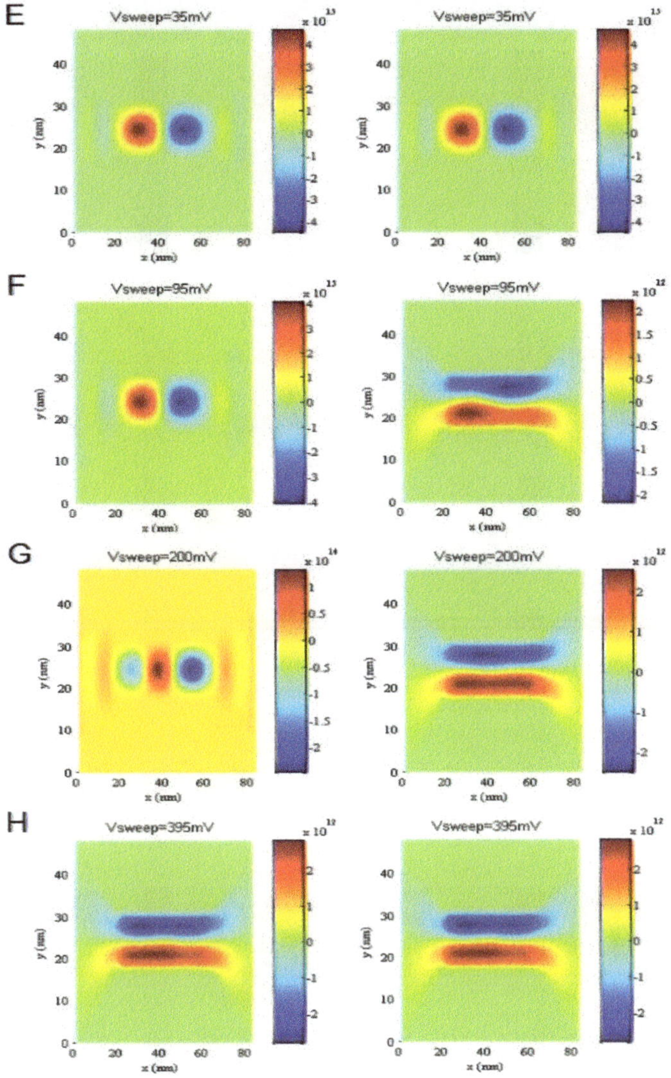

Figure 5: Two-dimensional contour plots of spin texture profile $n_\uparrow (x,y) - n_\downarrow (x,y)$ for values of V_{sweep} corresponding to points E, F, G, and H in Fig.2. It uses the dimensions listed in Fig.1. These correspond to values of V_{sweep} equal to 35mV, 95mV, 200mV, and 395mV, respectively. In each frame, the left and right figures are the spin texture profile through the four-gate QPC device for the up and down voltage sweeps, respectively.

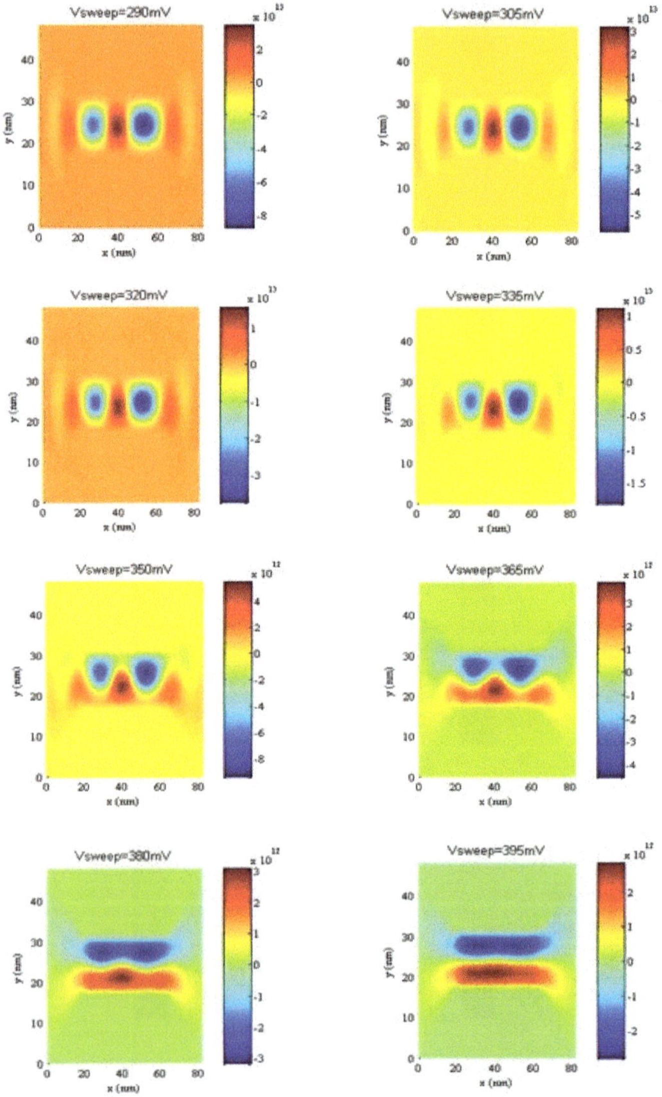

Figure 6: Two-dimensional contour plots of spin texture profile $n_{\uparrow}(x,y) - n_{\downarrow}(x,y)$ for values of V_{sweep} from 290 to 395 mV associated to the conductance curve for case1 of Table A. The spin texture evolves smoothly from a three-lobe, to a spin density wave, and finally to a spin Hall configuration as the conductance approaches the normal conductance plateau G_0.

The latter have a more drastic influence on the location of the second conductance peak (which is located around 1.4, 1.0, and 0.8 G_0 for cases 1, 2, and 3, respectively) and the overall shape of the second hysteresis loop. Comparing the results of Figures 2 and 7, the conductance curves past the second peak is split into two regions. Point E in Fig.2, where the conductance plots for the up and down sweeps are equal, evolves in a broader region as the potential on the rightmost SGs is made more negative. The size of the hysteresis loop near the second conductance peak in Fig. 7 decreases drastically and the hysteresis loop just below the conductance plateau G_0 occurs over a smaller region of V_{sweep}. Even though not shown here, the spin textures associated to the regions where the two conductance curves for the up and down sweeps coincide past the second peak were found to be identical. In agreement with the spin texture results shown in Figures 4 and 5 for case 1, the spin textures were found to be different at any value of V_{sweep} for which the conductance curves for the up and down sweeps differ in Fig.7. These results agree with those reported in ref. [25] for a QPC with only two gates where the appearance of hysteresis loops were shown to be intimately related to the existence of anomalous conductance plateaus and linked to the formation of multistable spin textures in these regions[28].

Because of their sensitivity to the biasing conditions on the gates of the four-gate QPC, the occurrence and evolution of hysteresis loops in the conductance plots could be used as a testbed to study the tunability of the spin polarization in narrow multi-gate QPC devices. In Figure 7, the conductance curves were obtained by varying V_{sweep} up to a maximum value of 0.6V. Originally, the maximum value of V_{sweep} was set at 0.4 V in calculating the conductance curves in Fig.7, as in Fig.2. In that case, the upper hysteresis loop was not observed in Fig.7. This can be understood from the results of Fig.7 which shows the V_{sweep} value for which the two conductance curves are equal for the up and down sweeps is past 0.4 V for both cases 2 and 3. Similarly to the study to the magnetization density (M) versus magnetic field H of a ferromagnet in which the hysteresis is not observed unless H is swept until M reaches its saturation value, hysteresis in the conductance curve of a four-gate QPC is not observed at large value of V_{sweep} unless the conductance plateau G_0 has been reached.

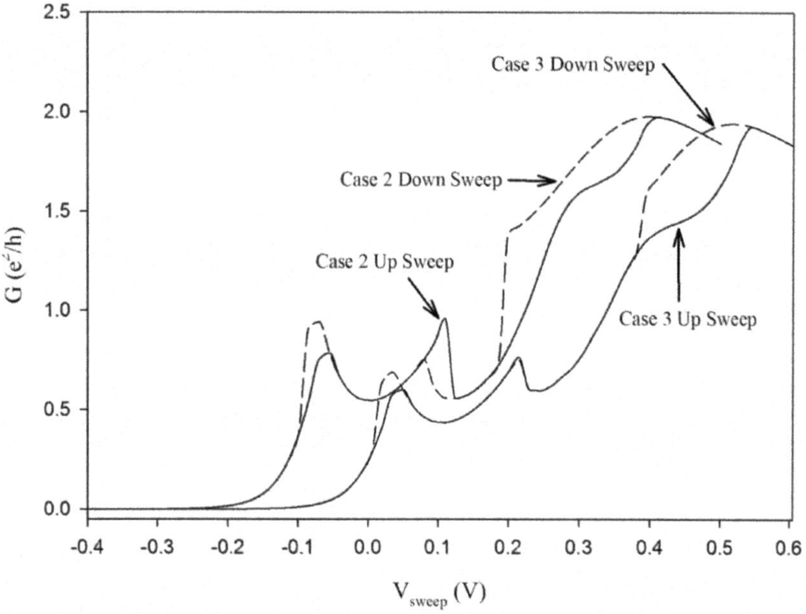

Figure 7: Conductance versus V_{sweep} of a four-gate QPC with the dimensions listed in Fig.1 for the biasing conditions corresponding to cases 2 and 3 from Table A. The full and dash curves are the conductance versus V_{sweep} for the up and down sweeps of the common gate voltage V_{sweep} on gates 1 and 2, respectively.

In analogy to the case of a ferromagnet where hysteresis in the M-H curve is linked to different microscopic configurations of the magnetic domains within the material[29], the hysteresis in the conductance curves of the four-gate QPC is linked to different spin textures within the narrow portion of the four-gate QPC, as illustrated in Figures 4 and 5. Figure 8 is a plot of the spin conductance polarization α versus V_{sweep} of a dual QPC for the biasing conditions corresponding to cases 2 and 3 from Table A. This figure clearly shows the three different regions of V_{sweep} corresponding to the hysteresis loops shown in Fig.7. Interestingly, for case 2, the sign of α is opposite for the up and down sweeps in the region of V_{sweep} between the two conductance peaks. This means that spin polarization of a four-gate QPC could be toggled from spin-up to spin-down by simply adjusting the potential of the rightmost gates and sweeping V_{sweep} back and forth between -0.1 and 0.15 V (Case 2 of Table A). This extreme sensitivity of α is to be expected because of the quadrupole nature of the electric field in the narrow portion of the QPC and its drastic influence on the intricate relationship between the Hartree and exchange-correlation potentials. This intricate interplay is interesting from a pure physics standpoint. It should have an even more drastic influence on the onset of spin polarization in more complicated QPC based structures but may be much harder to fully control in practical implementations of spin polarizers and detectors in a large scale QPC based circuits.

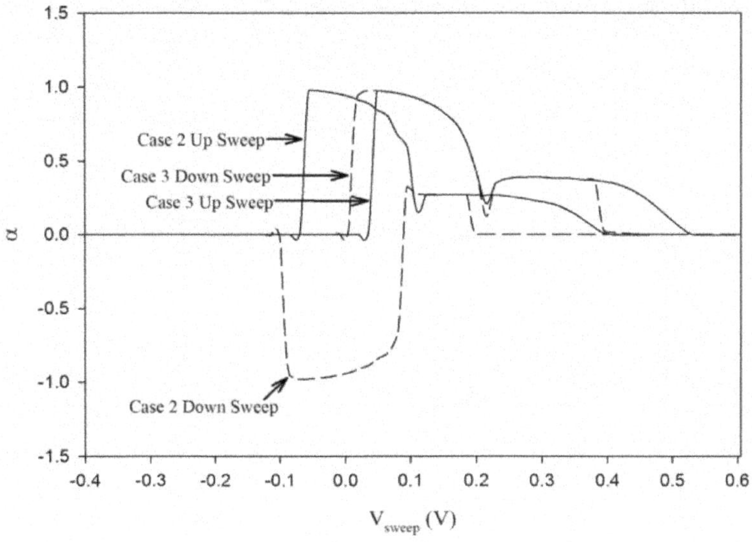

Figure 8: Spin conductance polarization $\alpha = (G_{up} - G_{down})/(G_{up} + G_{down})$ *versus* V_{sweep} *of a dual QPC with the dimensions listed in Fig.1 for the biasing conditions corresponding to cases 2 and 3 from Table A. The full and dash curves are the conductance versus* V_{sweep} *for the up and down sweeps of the common gate voltage* V_{sweep} *on gates 1 and 2, respectively.*

Next, the conductance and spin polarizations curves vs V_{sweep} were calculated for cases 4, 5, and 6 associated to a switching of the biasing conditions on the leftmost and rightmost SGs compared to cases 1, 2, and 3, respectively. Figures 9(a), 9(b), and 9(c) show a comparison of the up and down sweep conductance calculations for cases 1 and 4, cases 2 and 5, and cases 3 and 6, respectively. Figure 9 shows that the locations of the peaks in the conductance curves, their locations along the V_{sweep} axis, and the sizes of the hysteresis loops are affected by the switching of the biasing conditions of the leftmost and rightmost SGs. This rectifying behavior could easily be tested experimentally by reversing the role of the source and drain contacts. This would provide a further indirect evidence of the sensitivity of the spin texture in the narrow portion of the QPC on the electrostatics within the device and its control of the interplay of the Hartree and exchange-correlations potentials.

Next we performed NEGF simulations of the same four-gate QPC but for the three biasing conditions listed in Table B.

Table B: A different set of biasing conditions on the SGs of the four-gate of the QPC device shown in Fig.1. In this case, the biases on the two leftmost gates are identical and the potential difference $V_{sg4} - V_{sg3}$ is set equal to 0.2, 0.1, and 0.0V, for cases 1, 2, and 3, respectively. The geometrical dimensions of the four-gate QPC are given in the caption of Fig.1.

Case	$V_{sg1}(V)$	$V_{sg2}(V)$	$V_{sg3}(V)$	$V_{sg4}(V)$
#1	V_{sweep}	V_{sweep}	-0.2	0.0
#2	V_{sweep}	V_{sweep}	-0.15	-0.05
#3	V_{sweep}	V_{sweep}	-0.1	-0.1

In this case, the two leftmost SGs are kept at the same common gate voltage V_{sweep}. Because of the symmetric bias in that portion of the QPC, no spin polarization is expected to be triggered in that region. However, with the use of an asymmetric bias on the two rightmost SGs, it is expected that some level of asymmetry in the potential energy profile will be inferred in the leftmost portion of the QPC due to the proximity of the pairs of SGs. Cases 1, 2, and 3 in Table B correspond to the case where the difference of potentials between gates 4 and 3, $V_{sg4} - V_{sg3}$ is equal to 0.2 (case 1), 0.1 (case 2), and 0.0V (case 3).

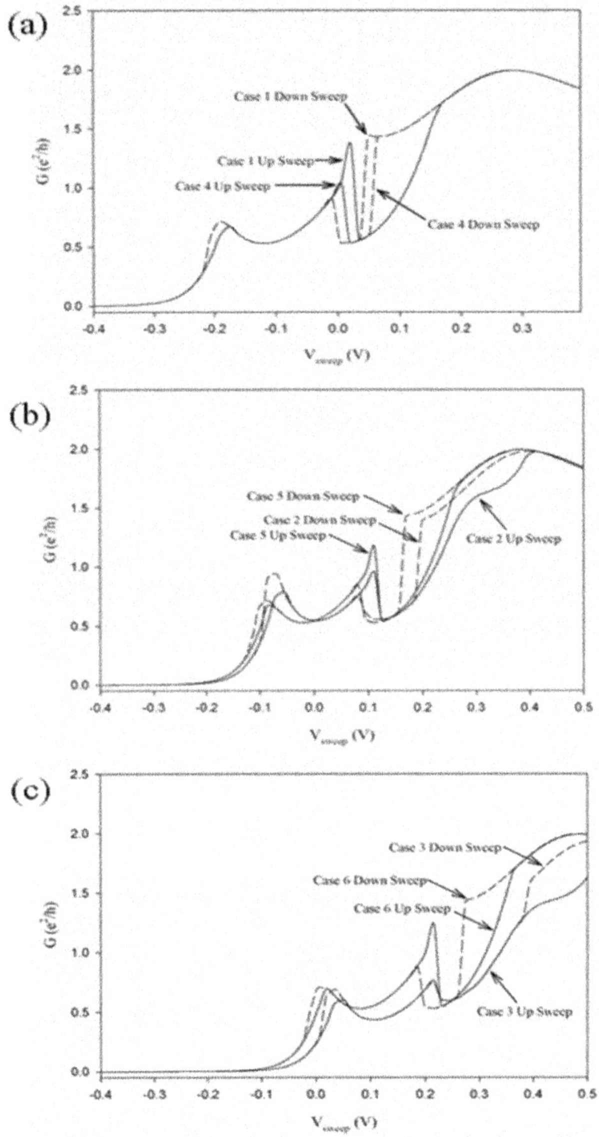

Figure 9: Conductance versus V_{sweep} of the four-gate QPC with the dimensions listed in Fig.1. In each figures, the different curves correspond to the up and down sweeps for (a) cases 1 and 4, (b) cases 2 and 5, and (c) cases 3 and 6 in Table A.

The latter corresponds to a symmetric biasing configuration on the four gates, i.e., $V_{sg1} = V_{sg2}$ and $V_{sg3} = V_{sg4}$. The conductance curves for the up and down sweep cases are shown in Figures 10(a) and (b), for $V_{sg4} - V_{sg3}$ is equal to 0.2,

0.1, and 0.0V, respectively. Figure 10(a) shows that there are two peaks (i.e., conductance anomalies) followed by a NDR region in the up sweep. The first peak in the conductance curves appears at a larger value of V_{sweep} and increase in magnitude from cases 1 to 3. On the other hand, in the up sweep, the location and value of the second peak is not affected by the difference of potential $V_{sg4} - V_{sg3}$. This feature is quite different from those observed in the conductance plots corresponding to the biasing conditions listed in Table A for which both conductance anomalies were shown to be quite sensitive to the biasing conditions on the four gates. Interestingly, the two conductance anomalies are still present in the case when $V_{sg4} - V_{sg3} = 0.$, i.e., when the biasing conditions on the four SGs are symmetric. In other words, the symmetric biasing conditions corresponding to case 3 of Table B are still triggering some spin polarization in the channel. This result will be investigated further in the next section where we analyzed the physical origin of the onset of spin polarization of a four-gate QPC device. Figure 10(b) shows that for the down sweep, it is the second conductance anomaly which is quite sensitive to the biasing conditions associated to cases 1-3 of Table B whereas the leftmost conductance anomaly stays the same. A comparison of Figures 10(a) and (b) shows the presence of hysteresis loops near both conductance anomalies but the size of the loop is much more sensitive (and increases) from cases 1 to 3. A comparison of Figures 9 and 10 show that the overall shape of the conductance curves, locations of the conductance anomalies, and size of the hysteresis loops are quite different for the biasing conditions listed in Tables A and B. *These fingerprints constitute additional evidence of the extremely high sensitivity of the amount of spin polarization in four-gate QPC devices.*

4. Discussion

All the NEGF simulations discussed above show the following trends: (a) anomalies appear simultaneously with hysteresis in the conductance curves for all the biasing conditions listed in Tables A and B; (b) There is a sharp change in the spin texture profile past the values of the common sweep voltage V_{sweep} applied to some of the SGs (note the sharp changes going from frames B to C, and D to E, in Figures 4 and 5)

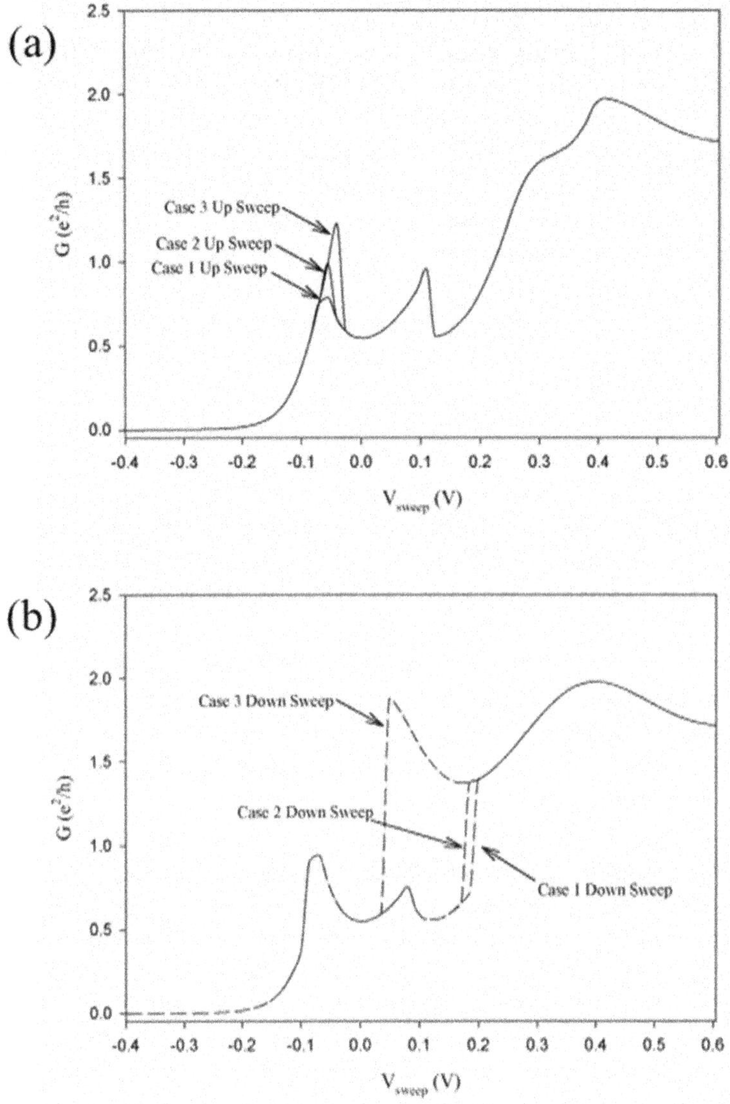

Figure 10: Conductance versus V_{sweep} of the four-gate QPC with the dimensions listed in Fig.1: (a) up sweep and (b) down sweep results corresponding for the biasing conditions listed as cases 1, 2, and 3 in Table B.

This is due to a sharp change in the potential energy profile at both ends of the narrow portion of the QPC which favors spin accumulation in the channel. It would be interesting to investigate how these potential barriers are affected by the smoothness of the transition between the wide contact regions and the narrow portion of the QPC and its influence on the spin texture on the conductance anomalies. The potential energy profile in the small region between the two sets of SGs also affects the overall spin texture in the left- and rightmost portions of the QPC; (c) Below threshold and when the conductance curves approach the first conductance plateau G_0, the spin texture is of the spin Hall type. In this case the potential barriers at the end of the narrow channel and between the two sets of SGs have little effect on the spin texture; (d) In the regions of V_{sweep} where the hysteresis appears, the spin textures associated with the up and down sweeps are drastically different, confirming some of the results obtained in refs.[24,25] for a QPC with only two SGs, where it was shown that the hysteresis is due to the existence of multistable spin textures near the conductance anomalies.

The results of Table A show that a second set of SGs of a four-gate QPC can be used to affect the amount on spin polarization in the narrow portion of the device. This tunable spin polarizer was already analyzed in ref.[25]. Here we have shown that the amount of hysteresis in the conductance of the tunable spin polarizer is also sensitive to the biasing conditions in the rightmost set of SGs. The results associated to the biasing conditions in Table B show the onset of spin polarization in a narrow QPC to be non-local, with an asymmetric bias between the rightmost SGs of a QPC can trigger a spin texture in the leftmost portion of a QPC where symmetric biasing conditions exist. The non-local aspect of the overall spin texture in the narrow portion of the QPC is due to the intricate competition between the Hartree and exchange potential energy profiles in the left- and right-most portions of the QPC.

One of the most interesting features observed in the results corresponding to case 3 of Table B is the observation of conductance anomalies and hysteresis even for the case of symmetric biasing conditions applied to the two sets of SGs. To understand this feature, NEGF simulations corresponding to case 3 were run by tuning the strength of the parameter γ characterizing the interaction self-energy associated to the strength of the electron-electron interaction and the strength of the parameter β associated to LSOC. Figure 11 shows the conductance curves vs V_{sweep} (up sweep only) for a set of simulations corresponding to the biasing conditions of case 3 in Table B, in which the value of β was set equal to zero (no LSOC) and the parameter γ was increased from 0 to $3.7\hbar^2/2m^*$. Also shown for comparison is the conductance curve for the upsweep for the case where γ = $3.7\hbar^2/2m^*$ and β = 200 Å^2. This figure clearly shows the importance of the e-e interaction in triggering the observation of conductance anomalies, even in the absence of LSOC. The conductance curves for $\gamma = 0$ has several maxima near the conductance plateau G_0. The first conductance maximum is not exactly at G_0 because of the finite value of the drain to source bias. Above the threshold value of 3.2 $\hbar^2/2m^*$, the conductance maxima obtained for the case for γ = 0 have developed into conductance anomalies. Comparing the conductance curve for $\gamma = 3.6\ \hbar^2/2m^*$ and $\beta = 0$ to the results for the case where $3.7\hbar^2/2m^*$ and 200 Å^2 suggest that the LSOC has also some influence on the size of the conductance anomalies. This is illustrated in Fig.12 where the conductance versus V_{sweep} (up sweep only) case 3 in Table B, in which γ was set equal to $3.7\hbar^2/2m^*$ and β was increased from zero to 200 Å^2 in steps of 50 Å^2. This figure indicates that the first peak is much more sensitive to the exact value of the LSOC strength than the second peak. Presumably, this is due to the difference in the spin textures near the first and second peaks. For the first peak, the spin texture has only one main lobe (see frame C in Figure 4) whereas it has two lobes with opposite polarities past the second peak (see frame E in Figure 5). The latter being composed of opposite spin polarities should be less sensitive to the LSOC.

To further understand the onset of spin polarization in the narrow portion of four-gate QPC despite the symmetric biasing conditions on the four SGs for case 3 of Table B, Figure 13 shows a plot of the spin texture for V_{sweep} values just above the first (50 mV) and second peaks (155mV) for values of $(\gamma, \beta) = (0,0)$, and of $(\gamma, \beta) = (3.7\hbar^2/2m^*, 200\ \text{Å}^2)$. The two top frames in Figure 13 shows that a spin texture of very small magnitude exists even with no LSOC and e-e interaction included in the model. This is due to the fact the scheme used to solve the NEGF equations leads to small numerical discrepancies between the contributions of spin-up and spin-down electrons. This eventually leads to the formation of small pockets of spin imbalance, especially in the narrow portion of the QPC (without the need of an external magnetic field to trigger that spin imbalance). It is worth pointing out that the role of a finite source-drain is also instrumental in this context. Because of the current induced by the bias there will be an imbalance between the population of the +k states incident from the source and –k states incident from the drain. Hence, according to Eq. (A1), a finite spin polarization can be turned on in this way. Hence, there is no need for an external magnetic field as a polarization trigger.

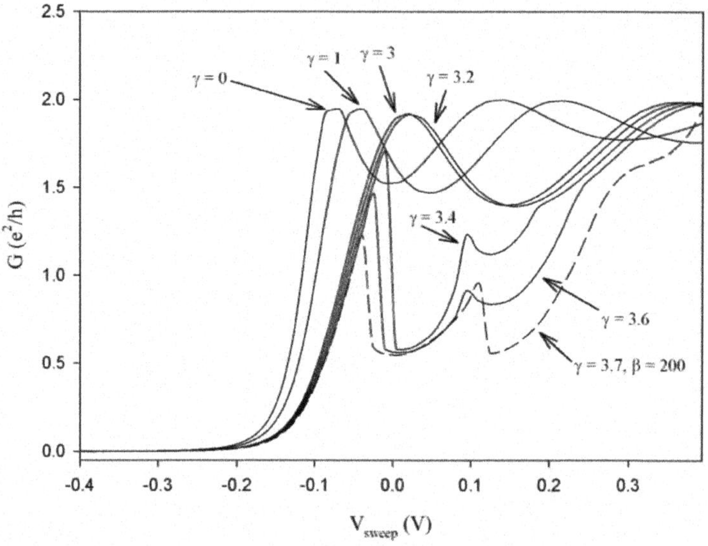

Figure 11: Conductance versus V_{sweep} for a set of simulations using the settings from Table B, Case 3. It uses the dimensions listed in Fig.1. For this specific set of simulations, the value of γ was varied from a range of 0 to $3.7\hbar^2/2m^*$, and plotted at various values. For each simulation in this set except when $\gamma = 3.7$, the value of β was 0 (for $\gamma = 3.7\ \hbar^2/2m^*$, $\beta = 200\ \AA^2$).

As the bottom frames of Fig.13 clearly shows, that small spin imbalance is enhanced drastically once the effects of e-e interaction are taken into account (notice the many orders of magnitude difference between the spin texture amplitudes between the lower and upper frames in Figure 13). This spin imbalance (triggered numerically in our simulations) would be more likely the norm in an experimental situation in which there would be always small amount of imperfections and asymmetry at the nanoscale level in a four-gate QPC device. These small imperfections can always trigger a substantial spin imbalance if the e-e interaction has the right magnitude. This only appears over a specific range of the common gate potential V_{sweep}. Bhandari et al.[23,24] recently reported the observation of the hysteresis in the conductance curves of conductance

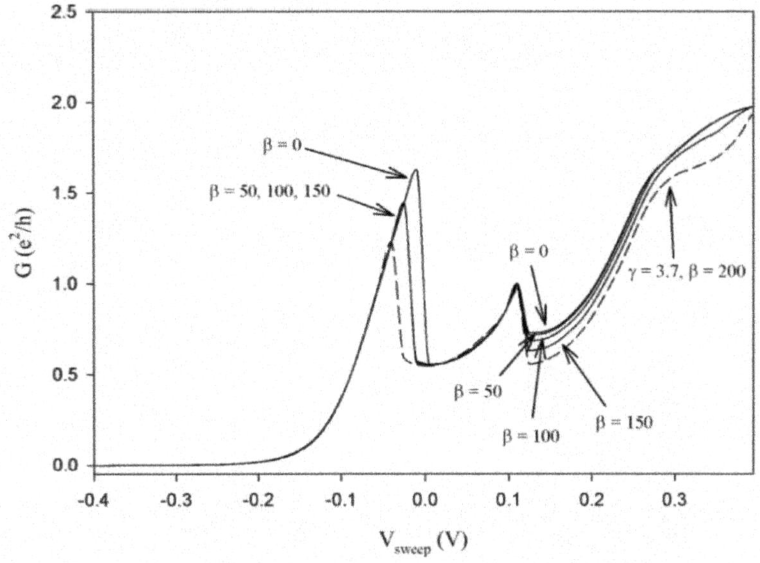

Figure 12: Conductance versus V_{sweep} for a set of simulations using the settings from Table B Case 3. It uses the dimensions listed in Fig.1. For this specific set of simulations, the value of β was varied from a range of 0 to 200 $Å^2$, and plotted at various values. For each simulation in this set, $\gamma = 3.7\ \hbar^2/2m^*$.

anomalies of four-gate QPC devices[23]. They also observed a regime of operation of the four-gate QPC devices which appear to be due to Coulomb blockaded transport, a regime of operation which cannot be reproduced by the NEGF approach.

5. Conclusions

We have studied the onset of spin polarization in a QPC device composed of two sets of SGs in close proximity in the presence of LSOC (See Fig.1). A NEGF approach was used to suggest some experimental biasing conditions of the four-gate QPC devices which lead to specific trends in their conductance anomalies. For the biasing conditions listed in Tables A and B, NEGF simulations indicate that conductance

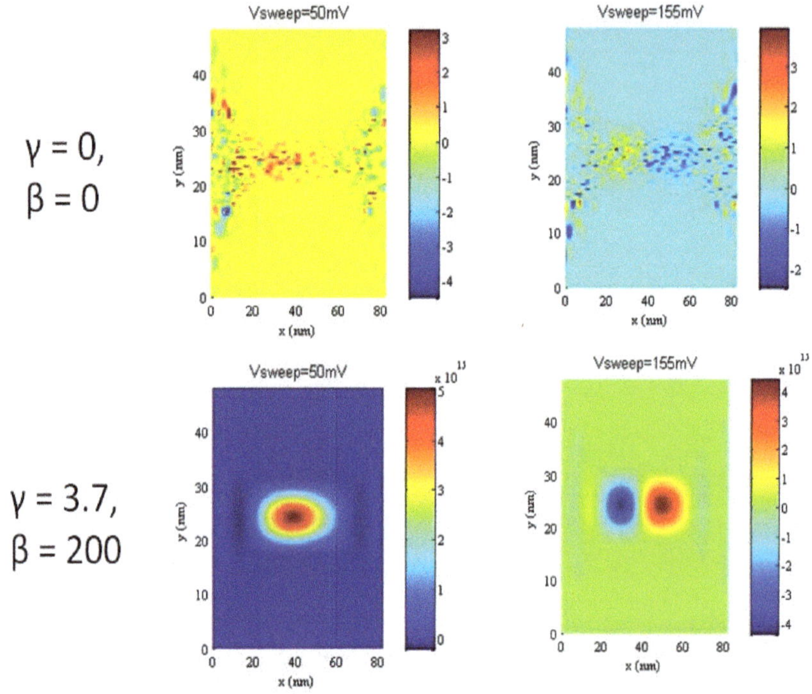

Figure 13: Two-dimensional contour plots of spin texture profile $n_\uparrow(x,y) - n_\downarrow(x,y)$ associated to some of the conductance plots shown in Fig.10 for two different values of V_{sweep}. The two top frames correspond to NEGF simulations where both β and γ are set equal to 0. For the two bottom plots, $\gamma = 3.7\hbar^2/2m^*$ and $\beta = 200$ Å^2.

anomalies (i.e., below $G_0 = 2e^2/h$) appear simultaneously with hysteresis in the conductance curves, i.e., there is a difference between the conductance curves for up sweep and down sweeps of the common potential applied to a pair of SGs. The hysteresis loops were shown to be related to the co-existence of multistable spin textures in the narrow channel of the four-gate QPC and to be very sensitive to the biasing conditions on the four SGs[24,25]. NEGF simulations show that: (a) the presence of an asymmetrically biased pair of SGs can generate a spin polarization between two adjacent SGs even when the latter are symmetrically biased; (b) There is a rectifying behavior in the conductance curves, i.e., the shape of the conductance anomalies and size of hysteresis loops are different when the biasing conditions on the leftmost and rightmost gates are swapped; (c) The amount of spin polarization in the channel is extremely sensitive to the biasing conditions on the four SGs and can be drastically different for the up and down sweeps of the common gate voltage. This is to be expected because of the quadrupole nature of the electric field in the narrow portion of the QPC and its drastic influence on the intricate relationship between the Hartree and exchange potentials. This intricate interplay is interesting from a pure physics standpoint to study the onset of spin polarization in more complicated QPC based structures but may be much harder to fully control in practical implementations of spin polarizers and detectors in a large scale QPC based circuits; (d) Conductance anomalies and accompanying hysteresis loops also appear even when the gates of opposite sides of the channel are symmetrically biased. In this case, the conductance anomalies only appear if the strength of the e-e interaction is large enough. If the latter is satisfied, the size and location of the conductance anomalies are strongly dependent on the strength of the LSOC potential. It has been shown that in this case, a small numerical imbalance between the density of spin-up and spin-down electron concentrations are amplified in specific range of the common sweep voltage on the SGs in the presence of strong e-e interaction. The overall spin texture in the four-gate QPC devices changes abruptly past the peaks (anomalies) in the conductance plots as the device enters a NDR regime of operation. This abrupt change is believed to be linked to abrupt change in the barrier profiles at both ends of the narrow portion of the QPC channel leading to enhanced space-charge built-up in specific range of V_{sweep}. The competition of the Hartree and exchange potentials due to the trapping of charge in the narrow portion of the QPC leads to very intricate metastable spin textures and the observation of the hysteresis in the conductance curves.

Hysteresis is a fundamental phenomenon associated with a bistable behavior of some physical properties of a system, typically appearing when a system undergoes phase transitions or reaches a regime of operation with multistable ground states for a certain range of parameters[29]. Intrinsic bistability has also been observed in the NDR region of double barrier resonant tunneling devices. It has been shown to be intimately related to the highly nonlinear nature of the space-charge effects in the quantum well sandwiched between the two barriers[30-34]. Additionally, several groups have reported the observation of hysteresis in the transport properties of structures based on 2DEGs, such as single layer and double layer quantum well modulation doped heterostructures used in the study of the integer and fractional quantum Hall effects[35-40]. The origin of the hysteresis observed in the double-layer structure was attributed to a phase transition between oppositely polarized ground states localized in the different layers[35]. The observations of hysteresis behavior in the modulation-doped heterostructures in the fractional quantum Hall effect has been linked to the onset of two-dimensional ferromagnetism resulting from the competition of spin-polarized and a spin-unpolarized ground states[36,37,41]. Ihnatsenka and Zozoulenko predicted hysteresis between the up and down sweeps of an external magnetic field in the transport properties of a quantum wire operating in the integer quantum Hall regime[42]. In this case, the hysteresis occurs for certain range of the external magnetic field and is due to the coexistence of two different ground states in quantum wires associated with spatially spin-polarized and spatially spin-unpolarized edge channels in the vicinity of the wire boundaries. Furthermore, it was shown that the hysteresis behavior is absent for steep and smooth confining potentials in the quantum wire and is present only for a limited range of intermediate confinement slopes. In addition, Ihnatsenka and Zozoulenko have shown that the hysteresis behavior in the integer quantum Hall effect cannot be explained using a one electron picture: it has a many-body origin resulting from the non-linear intricate interplay between the confining potential, the Coulomb interaction and the exchange interaction in the quantum wire. From our NEGF simulations, we conclude that a similar non-linear intricate interplay also exists in the transport properties of side gated four-gate QPC devices, ultimately responsible for the highly sensitive nature of the onset of spin polarization in these structures.

The various trends in the conductance curves of four-gate QPC devices are indirect evidence (as oppose to direct evidence which could be inferred from Hanle effect experiments performed on the four-gate QPCs) of the fine and non-local electrostatic tuning of the onset of spin polarization in two closely spaced QPCs with two sets of in-plane SGs in the presence of LSOC. These results show that the onset and fine tuning of conductance anomalies in QPC structures are highly sensitive to the non-local action of closely spaced gates. This effect is of practical importance and would even be more of an issue in the design of all electrical spin valves making use of middle gates to fine tune the spin precession between QPC based spin injector and detector contacts. This high-sensitivity of the conductance curves to the close proximity of gates in QPC based devices was recently demonstrated by Tse-Ming Chen and colleagues when operating an all-electric spin valve by fabricating spin injectors and detectors with an efficiency near 100% using asymmetrically biased QPCs[43].

From a more fundamental point of view, the scenario of spin-polarized currents in short (QPC based) quantum channels obviously relates to the so-called *0.7 conduction anomaly* observed in GaAlAs/GaAs QPCs with top gates in which the importance of LSOC is minute[11]. The conduction feature at 0.7 ($2e^2/h$) is named anomalous because of its anomalous temperature dependence. Many recent reports have shown that electron-electron interaction effects are obviously involved but its true origin is still highly debated[12-16]. Current models involve spontaneous spin-polarization, localized moments and Kondo dynamics, and more recently van Hove singularities[12-16,44-49]. One common objection to the model of spontaneous spin polarization in a single QPC in zero magnetic field is that there appears to be no obvious mechanism that can trigger a spin symmetry breaking. It may be introduced artificially in terms of a tiny Zeeman splitting but the direction of the polarization would be arbitrary, i.e., parallel with the direction of the applied magnetic field. One may also consider a random magnetic field or let the background numerical noise trigger the spin polarization. This ties up with systems discussed here, i.e., systems with finite LSOC. In this case, it is an imbalance in LSOC, even if it is very small, which selects the direction of magnetization which at the end is inflated by e-e interactions. Spin polarization may also be induced by spin-orbit interaction and strongly enhanced by e-e interaction. Homogeneity and order of an assembly of nanostructures are crucial for their use in large-scale electronic or optical systems[50-60]. An explicit and intriguing illustration of this technological challenge has been demonstrated recently via a statistical study of an array of 256 nominally identical QPCs[56]. This study has shown the crucial role of fluctuations in the potential profile for QPC. It was found that only 41% of total set of QPCs maintain the usual step-like conductance structure with the 0.7 anomaly. In most of the QPCs potential variations were caused by effects of disorder in spite of the fact that all QPCs were fabricated under the same conditions on the same wafer[52,53,56,58]. This seemingly insurmountable challenge could have a deleterious effect on the potential implementation of all electric spin valves which could be used as spin-based sensors, spin filters, interferometers, or other building blocks for future spin-based devices, circuits, and architectures. However, spin based devices with current-voltage characteristics that show hysteresis and/or multiple negative differential resistance could have other potential applications in multilevel logic circuits, data storage, and neuromorphic architectures. This will require a lot more work in the assessment of the reproducibility and reliability of an all-electric SpinFET and/or spin valve before their successful large scale integration.

Acknowledgment

This work was supported by NSF Awards ECCS-0725404 and ECCS-1028483. Part of this work was initiated during the sabbatical stay of M.C at the University of Linköping, Sweden, during the spring of 2014. This sabbatical leave was supported by NSF Award IIA-1341789.

Appendix

The single-particle Hamiltonian describing electron motion in the 2 DEG (selected to be the x-y plane) for the spin-up (along +z axis) and spin-down (along –z axis) electrons was taken as:

$$H_0^\sigma = H_0 I + \frac{\beta}{\hbar} \vec{\sigma} . (\vec{p} \, X \, \vec{\nabla} U), \qquad \text{(A1)}$$

where $H_0 = \frac{1}{2m^*}\left(p_x^2 + p_y^2\right) + U(x,y)$, I is the 2x2 identity matrix, $\vec{p}$ the momentum operators in the x-y plane, and U(x,y) the potential energy representing the sum of conduction band energy profile within and around the QPC and effects of space-charge through a solution of the Poisson equation. The second term in the expression of H_0^σ represents the effect of the lateral spin-orbit coupling with strength β [61-64]. In the expression of U(x,y), a term was added to simulate the interaction of electrons within the QPC at the Hartree-Fock level following Lassl et al. [26]. In this approach, the interaction between two electrons located at $\vec{r}$ and $\vec{r}\,'$ is modeled using the contact potential

$$V_{\text{int}}(\vec{r},\vec{r}\,') = \gamma\delta(\vec{r} - \vec{r}\,') \qquad \text{(A2)}$$

with an interaction strength γ is a few times expressed in units of $\hbar^2/2m^*$. In the NEGF formalism, the effective Hamiltonian for electrons in the QPC with spin σ given by

$$H^\sigma = H_0^\sigma + \Sigma_{\text{int}}^\sigma(\vec{r}), \qquad \text{(A3)}$$

i.e., a Hamiltonian operator which is the sum of a non-interacting part H_0^σ and an interaction self- energy

$$\Sigma_{\text{int}}^\sigma(\vec{r}) = \gamma n_{-\sigma}(\vec{r}), \qquad \text{(A4)}$$

where $n_{-\sigma}(\vec{r})$ is the density of electrons with spin -σ. The local potential $\Sigma_{\text{int}}^\sigma(\vec{r})$ is different for the two spin populations injected from the contacts. A spin-up electron encounters a potential which is proportional to the density of spin-down electrons, and vice versa. This leads to a repulsive interaction between electrons with opposite spin directions. Any external source leading to an imbalance between the density of spin-up and spin-down electrons is increased by the addition of the self-energy term $\Sigma_{\text{int}}^\sigma(\vec{r})$.

The Green function associated to the QPC is then a ($2N_xN_y \times 2N_xN_y$) matrix where N_x and N_y are the number of grid-points along and perpendicular to the direction of current flow:

$$G(E) = (E - H - \Sigma_S - \Sigma_D - \Sigma_{\text{int}})^{-1}, \qquad \text{(A5)}$$

where Σ_S and Σ_D are the self-energy terms representing the coupling to the source and drain contacts. For the electron-electron interaction modeled with the contact potential, the interaction self-energy in the Green function is given by

$$\Sigma_{\text{int}}(p,p) = \begin{pmatrix} \Sigma_{\text{int}}^{\uparrow}(p,p) & 0 \\ 0 & \Sigma_{\text{int}}^{\downarrow}(p,p) \end{pmatrix}, \qquad \text{(A6)}$$

where p is the node index in the two-dimensional simulation domain of the QPC, with $\Sigma_{\text{int}}^{\uparrow}(p,p) = \gamma n_{\downarrow}(p)$ and $\Sigma_{\text{int}}^{\downarrow}(p,p) = \gamma n_{\uparrow}(p)$.

To calculate the electron densities associated with the spin-up (along z axis) and spin-down (along –z axis) electrons, we use

$$n_{\uparrow}(p) = \rho_{p,p}(1,1),\ n_{\downarrow}(p) = \rho_{p,p}(2,2), \qquad \text{(A7)}$$

where the spin density is obtained from

$$\rho = \int_{-\infty}^{\infty} \frac{G_n(E)}{2\pi} dE, \qquad \text{(A8)}$$

with $G_n(E)$ is the correlation function

$$G_n(E) = A_S f_S + A_D f_D, \qquad \text{(A9)}$$

where

$$A_S(E) = G\Gamma_S G^{\dagger},\ A_D(E) = G\Gamma_D G^{\dagger}, \qquad \text{(A10)}$$

are the spectral functions of the source and drain contacts; f_S and f_D are the Fermi-Dirac distributions in the source and drain contacts, respectively.

Finally, the current through the QPC is calculated as follows

$$I = \frac{q}{h}\int_{-\infty}^{\infty} T(E)[f_S(E) - f_D(E)]dE, \qquad \text{(A11)}$$

where the transmission coefficient is equal to

$$T(E) = Trace\ [\Gamma_S G \Gamma_D G^{\dagger}]. \qquad \text{(A12)}$$

In the simulations discussed in the text, we used $N_x = 42$ and $N_y = 25$. A typical calculation of an up sweep conductance plot took about 2 days to complete on a Dell Precision T7500 64 bit dual processor workstation solely devoted to this project. The down sweep conductance calculation ran a little faster.

References

1. Awschalom, D.D. & Flatté, M.E. Challenges for semiconductor spintronics. *Nature Physics* **3**, 153-159 (2007).

2. Flatté, M.E. Solid-state physics Silicon spintronics warms up. *Nature* **462**, 419-420 (2009).

3. Williams, C.P. & Clearwater, S. H. Explorations in Quantum Computing. *Springer-Verlag*, NY (1998).

4. Awschalom, D.D. & Smarth, N. Spintronics without magnetism. *Physics* **2**, 50-54 (2009).

5. Zutic, I., Fabian, J., & Sarma, S.D. Spintronics: fundamentals and applications. *Review of Modern Physics* **76**, 323-410 (2004).

6. Kim S., Hasimoto, Y., IyeY. & Katsumoto, S. Evidence of spin-filtering in quantum constrictions with spin-orbit interaction. *Journal of the Physical Society of Japan* **81**, 054706 (2012).

7. Debray, P., Rahman, S.M.S., Wan, J., Newrock, R.S., Cahay, M., Ngo, A.T., Ulloa, S.E., Herbert, S.T., Muhammad, M. & Johnson, M. All electrical quantum point contact spin-polarizer. *Nature Nanotech* **4**, 756-764 (2009).

8. Das, P.P., Chetry, K.B., Bhandari, N., Wan, J., Cahay, M., Newrock, R.S. & Herbert, S.T. Understanding the anomalous conductance plateau in asymmetrically biased $InAs/In_{0.52}Al_{0.48}As$ quantum point contact-a step towards a tunable all electric spin valve. *Applied Physics Letters* **99**, 122105 (2011).

9. Bhandari, N., Das, P.P., Chetry, K. B., Bhandari, N., Wan, J., Cahay, M., Newrock, R.S. & Herbert, S.T. Observation of a 0.5 conductance plateau in asymmetrically biased GaAs quantum point contact. *Applied Physics Letters* **99**, 122105 (2011).

10. Das, P.P.,Bhandari, N., Wan, J., Charles, J., Cahay, M., Chetry, K. B., Newrock, R.S. & Herbert, S.T. Influence of surface scattering on the anomalous plateaus in an asymmetrically biased InAs/InAlAs quantum point contact. *Nanotechnology* **23**, 215201 (2012).

11. Thomas, K.J., Nicholls, J.T., Simmons, M.Y. *et al.* Possible spin polarization in a one-dimensional electron gas. *Physical Review Letters* **77**, 135–138 (1996).

12. Pepper, M. & Bird, J. The 0.7 feature and interactions in one-dimensional systems. *Journal of Physics: Condensed Matter* **20**, 16301–16301 (2008).

13. Micolich, A.P. What lurks below the last plateau: experimental studies of the 0.7 × 2e2/h conductance anomaly in one-dimensional systems. *Journal of Physics: Condensed Matter* **23**, 443201 (2011).

14. K.-F. Berggren and M. Pepper, M. Electrons in one dimension. *Philosophical Transactions of the Royal Society of London, Section A* **368**, 1141–1162 (2010).

15. Chen, T.-M., Pepper, M., Farrer, I., Jones, G.A.C. & Ritchie, D.A. All-electric injection and detection of a spin-polarized current using 1D conductors. *Phys. Rev. Letters* **109**, 17202 (2012).

16. Lind, H., Yakimenko, I.I. & Berggren, K.-F. Electric-field control of magnetization in biased semiconductor quantum wires and point contacts. *Phys.Rev.B* **83**, 0750308 (2011).

17. Shailos, A., Shok, A., Bird, J.P. *et al.* Linear conductance of quantum point contacts with deliberately broken symmetry. *Journal of Physics: Condensed Matter* **18**, 1715–1724 (2006).

18. Chen, J.C., Lin, Y., Lin, K.T. *et al*. Effects of impurity scattering on the quantized conductance of a quasi-one-dimensional quantum wire. *Applied Physics Letters* **94**, 01205 (2009).

19. Seo, M. & Chung, Y. Anomalous conductance quantization in a quantum point contact with an asymmetric confinement potential. *Journal of the Korean Physical Society* **60**, 1907–1910 (2012).

20. Liu, K.M., Juang, C.H., Umansky, V. & Hsu, S.Y. Effect of impurity scattering on the linear and nonlinear conductances of quasi-one-dimensional disordered quantum wires by asymmetrically lateral confinement. *Journal of Physics: Condensed Matter* **22**, 395303 (2010).

21. Wan, J., Cahay, M., Debray, P. & Newrock, R.S. On the physical original of the 0.5 plateau in the conductance of quantum point contacts. *Physical Review B* **80**, 155440 (2009).

22. Wan, J., Cahay, M., Debray, P. & Newrock, R.S. Spin texture of conductance anomalies in quantum point contacts. *J. Nanoelectron. Optoelectron.* **6**, 95 (2011).

23. Bhandari, N., Dutta, M., Charles, J. *et al*. Hysteresis in the conductance of asymmetrically biased quantum point contacts with in-plane side gates. *Journal of Applied Physics* **114**, 033702, 5 (2013).

24. Bhandari, N., Dutta, M., Charles, J., Wan, J., Cahay M. & Herbert, S.T. Potential Applications of All Electric Spin Valves Made of Asymmetrically Biased Quantum Point Contacts. Invited Book Chapter to appear in *Spintronic and nanomagnetic devices and systems*, Eds. J. Ataulasimha and S. Bandyopadhyay, Wiley (2015).

25. Charles, J., Bhandari, N., Wan, J., Cahay, M. & Newrock, R.S. Tunable All-Electric Spin Polarizer. *Applied Physics Letters* **102**, 062419 (2013).

26. Lassl, A., Schlagheck, P. & Richter, K. Effects of short-range interactions on transport through quantum point contacts. *Phyical Reiew B* **75**, 045346 (2007).

27. The Hartree-Fock approximation has the tendency to overestimate the spin-polarization, as shown by Starikov, A.A., Yakimenko, I. I. & Berggren, K.-F. Scenario for the 0.7-conductance anomaly in quantum point contacts. *Phys. Rev. B* **67**, 235319 (2003). This can be compensated by an appropriate choice of the parameter γ.

28. Sablikov, V.A. & Shchamklalova, B.S. Metastable and spin-polarized states in electron systems with localized electron-electron interaction. ArXiv:1312.6986v2 (2014).

29. Mattis, D.C. The theory of magnetism. Harper and Row, New York, 1965.

30. Goldman, V.J., Tsui, D.C. & Cunningham, J.E. Observation of intrinsic bistability in resonant tunneling structures. *Physical Review Letters* **58**, 1256 (1987).

31. Jensen, K.L. & Buot, F.A. Numerical simulation of intrinsic bistability and high-frequency current oscillations in resonant tunneling structures. *Phyical Review Letters* **66**, 1078 (1991).

32. Zou, N., Willander, M., Lennerud, I., Hanke, U., Chao, K.A. & Galperin, Y.M. Suppression of intrinsic bistability by the exchange-correlation effect in resonant-tunneling structures. *Physical Review B* **49**, 2193 (1994).

33. Dai, Z & Ni, J. Tuning bistability and current hysteresis in resonant-tunneling structures. *Physical Review B* **73**, 113309 (2006).

34. Yoder, P.D., Grupen, M. & and Smith, R.K. Demonstration of intrinsic tristability in double-barrier resonant tunneling diodes with the Wigner transport equation. *IEEE Transactions on Electron Devices* **57**, 3265 (2010).

35. Piazza, V., Pellegrini, V., Beltram, F., Wegscheider, W., Jungwirth, T. & and MacDonald, A.H. First-order phase transitions in a quantum Hall ferromagnet. *Nature* **402**, 638 (1999).

36. Cho, H., Young, J.B., Kang, W., Campman, K.L., Gossard, A.C., Bichler, M. and Wegscheider, W. Hysteresis and spin transitions in the fractional quantum Hall effect. *Physical Review Letters* **81**, 2522 (1998).

37. Eom, J., Cho, H., Kang, W., Campman, K.L., Gossard, A.C., Bichler, M. & Wegscheider, W. Quantum Hall ferromagnetism in a two-dimensional electron system. *Science* **289**, 2320 (2000).

38. De Poortere, E.P., Tutuc, E., Papadakis, S.J. & Shayegan, M., Resistance spikes at transitions between quantum Hall ferromagnets. *Science* **290**, 1546 (2000).

39. Pan, W., Reno, J.L. & and Simmons, J.A. Hysteresis in the quantum Hall regimes in electron double quantum well structures. *Physical Review B* **71**, 153307 (2005).

40. Pioro-Ladrière, M., Usher, A., Sachrajda, A.S., Lapointe, J., Gupta, J., Wasilewski, Z., Studenikin, S. & Elliott, M. Influence of the long-lived quantum Hall potential on the characteristics of quantum devices. *Physical Review B* **73**, 075309 (2006).

41. Smet, J.H., Deutschmann, R.A., Wegscheider, W., Abstreiter, G., & von Klitzing, K. Ising ferromagnetism and domain morphology in the fractional quantum Hall regime. *Physical Review Letters* **86**, 2412–2415 (2001).

42. Ihnatsenka, S. & Zozoulenko, I.V. Charge transport in two-dimensional el-ectron gas/insulator/superconductor junctions with Rashba spin-orbit coupling. *Physical Review B* **74**, 035318 (2007).

43. Chuang, P., Ho, S.-J., Smith, L. W., Sfigakis, F., Pepper, M., Chen, C.-H., Fan, J.-C., Griffiths, J.P., Farrer, I., Beere, H.E., Jones, G.A.C., Ritchie, D.A. & Chen, T.-M., *Nature Nanotechnology* **10**, 35-39 (2014).

44. Starikov, A.A., Yakimenko, I.I. & Berggren, K.-F. Scenario for the 0.7-conductance anomaly in quantum point contacts. *Physical Review B* **67**, 235319 (2003).

45. Jaksch, P., Yakimenko, I.I., & Berggren, K.-F. From quantum point contacts to quantum wires: Density-functional calculations with exchange and correlation effects. *Physical Review B* **74**, 235320 (2006).

46. Havu, P., Puska, M.J., Nieminem, R.M. & Havu, V. Electron transport through quantum wires and point contacts. *Physical Review B* **70**, 233308 (2004).

47. Shelykh, I.A., Galkin, N.G. & Bagraev, N.T. Quantum point contact with a large localized spin: Fractional quantization of the ballistic conductance. *Physical Review B* **74**, 085322 (2006).

48. Hirose, K., Meir, Y. & Wingreen, N.S., Local moment formation in quantum point contacts. *Physical Review Letters* **90**, 026804 (2003).

49. Heyder, J., Bauer, F., Schubert, E., Borowsky, D., Schuh, D., Wegscheider, W., Von Delft, J. & Ludwig, S., On the relation between the 0.7-anomaly and the Kondo effect: geometric crossover between a quantum point contact and a Kondo quantum dot. *Physical Review B* **92**, 19504 (2015).

50. Yang, Q.-Z., Kelly, M.J., Farrer, I., Beere, H.E. & Jones, G.A.C. The potential of split-gate transistors as one-dimensional electron waveguides revealed through the testing and analysis of yield and reproducibility. *Applied Physics Letters* **94,** 033502 (2009).

51. Burke, A.M., Klochan, O., Farrer, I., Ritchie, D.A., Hamilton, A.R., and Micolich, A.P. Extreme sensitivity of the spin-splitting in one-dimensional nanoelectronic devices. *Nano Letters* **12**, 4495-4502 (2012).

52. Smith, L. W., Al-Taie, H., Sfigakis, F., See, P., Lesage, A.A.J., Xu, B., Griffiths, J.P., Beere, H.E., Jones, G.A.C., Ritchie, D.A., Kelly, M.J. & Smith, C.G. Statistical study of conductance properties in one-dimensional quantum wires focusing on the 0.7 anomaly. *Physical Review B* **90**, 045426 (2014).

53. Smith, L.W., Al-Taie, H., Lesage, A.A.J., Sfigakis, F., See, P., Griffiths, J.P., Farrer, I., Jones, G.A.C., Ritchie, D.A., Kelly, M.J., & Smith, C.G. Dependence of the 0.7 anomaly on the curvature of the potential barrier in quantum wires. *Physical Review B* **91**, 235402, (2015).

54. Al-Taie, H., Smith, L.W., Xu, B., See, P., Griffiths, J.P., Beere, H.E., Jones, G.A.C., Ritchie, D.A., Kelly, M.J. & Smith, G.C. Cryogenic on-chip multiplexer for the study of quantum transport in 256 split-gate devices. *Applied Physics Letters* **102**, 243102 (2013).

55. Hornibrook, J.M., Colless, J.I., Conway Lamb, I.D., Paula, S.J., H. Lu, Gossard, A.C., Watson, J.D.,mGardner, G.C., Fallahi, S., Manfra, M.J. & Reilly, D.J. Cryogenic control architecture for large-scale quantum computing. *Physical Review Applied* **3**, 024010 (2015).

56. Smith, L.W., Al-Taie, H., Sfigakis, F., See, P., Lesage, A.A.J., Xu, B., Griffiths, J.P., Beere, H.E., Jones, G.A.C. D., Ritchie, A., Kelly, M.J., and Smith, G.C. Statistical study of conductance properties in one-dimensional quantum wires focusing on the 0.7 anomaly. *Physical Review B* **102**, 045426 (2014).

57. Conway Lamb, I.D., Colless, J.I., Hornibrook, J.M., Paula, S.J., Waddy, S.J., Frechtling, M.K. & Reilly, D.J. An FPGA-based instrumentation platform for use at deep cryogenic temperatures. *The Review of Scientific Instruments* **87**, 014701 (2016).

58. Al-Taie, H., Smith, L.W., Lesage, A.A.J., See, P., Griffiths, J.P., Beere, H.E., Jones, G.A.C., Ritchie, D.A., Kelly, M.J. & Smith, C.G. Spatial mapping and statistical reproducibility of an array of 256 one-dimensional quantum wires. *Journal of Applied Physics* **118**, 075703 (2015).

59. Puddy, R.K., Smith, L.W., Al-Taie, H., Chong, C.H., Farrer, I., Griffiths, J.P., Ritchie, D.A., Kelly, M.J., Pepper, M. & Smith, C.G. Multiplexed charge-locking device for large arrays of quantum devices. *Applied Physics Letters* **107**, 143501 (2015).

60. Smith, L.W., Al-Taie, H., Lesage, A.A.J., Thomas, K.J., Sfigakis, F., P. See, Griffiths, J.P., Farrer, I., Jones, G.A.C., Ritchie, D.A., Kelly, M.J & Smith, C.G. The effect of split gate size on the electrostatic potential and 0.7 anomaly within one-dimensional quantum wires on a modulation doped GaAs/AlGaAs heterostructure. arXiv:1508.03085 (2015).

61. Winkler, R. Spin-orbit coupling effects in two dimensional electron and hole systems. Springer, Berlin (2003).

62. Engel, H.-A., Halperin, B.I. & Rashba, E.I. Theory of spin Hall conductivity in n-doped GaAs. *Physical Review Letters* **95**, 166605 (2005).

63. Bokes P. & Horváth, F. Edge-induced spin polarization in two-dimensional electron gas. *Physical Review B* **81**, 125302 (2010).

64. In vacuum, the strength of the spin-orbit interaction β is negative and is much smaller than in semiconductors by several orders of magnitude, as shown explicitly in ref.[62].

Chapter 6

Introduction of All-electric Spin Field-Effect Transistor

Sheng-Chin Ho

Department of Physics, National Cheng Kung University, Tainan 701, Taiwan
b45612398@hotmail.com

The spin field-effect transistor envisioned by Datta and Das opens a gateway to spin information processing. A functional spin field-effect transistor for information processing has yet to be realized owing to several fundamental challenges. Such are the low spin-injection efficiency due to resistance mismatch, the spin relaxation, and the spread of spin precession angles. A realization of an all-electric all-semiconductor spin-field effect transistor in which these obstacles are overcome by using two quantum points contact as spin injector and detector. The spin polarization in QPCs are created by lateral spin-orbit coupling which an asymmetric one-dimensional confinement is formed and the 0.25 structure which a dc source-drain bias is applied. The polarized spins that travel from injector to detector experience a Rashba spin-orbit coupling (which can be controlled by the middle gate voltage) and rotate. This causes the detector voltage to oscillate when the spin orientation continuously vary between parallel to and antiparallel to the polarization direction of the QPCs, giving rise to an on/off switching with respect to middle gate voltage. This device exploits distinct engineering architectures of spin-orbit coupling to achieve complete control of the electron spins in a purely electrical manner and provides an all-semiconductor design for exploring the possibility of future spintronics computing.

1. Introduction

With the advent of second-generation computing devices it becomes increasingly important to understand and control spin transport in semiconductors [1, 40]. Spin transistors are new devices based on ordinary transistor concepts but using spin-related functionalities to realize logic operations and storage. The remarkable features of spin transistors include higher speed in logic computation and nonvolatile information storage. Efforts on realizing spintronics involve two approaches. The first approach is already in used in the industry: giant-magnetoresistive (GMR) devices which apply to information storage [23]. The second approach is still in development as a spin field-effect transistor(FET) used in logic computation. Spin transistors analogous to conventional FETs are more interesting than two terminals devices because they are suitable for integrated circuits.

The spin FET, proposed in 1990 by Datta and Das [11], has since become a quintessential example of spintronic devices. Such devices have a similar design to the ordinary FETs, which have a source, a drain, and a top gate. Two ferromagnetic electrodes are applied into two contacts, which are used to inject and detect the spin-polarized current in a semiconductor channel. The top gate is used to rotate the spin orientation instead of to create a potential barrier to block the driven current in a channel. The logic switch between 1 and 0 still replies on current modulation, but is instead achieved by manipulating the spin orientation: electrons can sink into the drain if their spin is parallel to the magnetization of the drain contact, and cannot if their spin is antiparallel. The voltage applying to flip spins to achieve such a logic switch is much lower than the threshold voltage of conventional FETs, giving many benefits. Although the control of spin orientation in a Datta-Das type spin FET has successfully been demonstrated by a South Korean scientist Hyun Cheol Koo and his colleagues in 2009 [20], the on/off current modulation in such a device is extremely small and far from practical application. So far a functional spin FET is still absent, thus and how to realize a functional spin transistor becomes one of the most important challenges in spintronics. One must first understand the problems before trying to solve them. In the next section, I will briefly describe three major obstacles to achieving a functional spin transistor. The reasons for realization of spin-FET ascribe to three obstacles: spin life time, spin injection efficiency and phase spread. We widely introduce these three obstacles.

2. Obstacles to realizing spin FET

2.1. *Spin relaxation*

Since spin relaxation mainly results in a loss of spin information, the length of spin information carried by the transport electrons is generally limited by the electron's mean free path. It is because every each collisions between particles may result in the orientation no longer predictable, especially in a spin-orbit coupled system. To maintain completely the spin information during transport, sufficient mean free path is demanded. The mean free path l_m can be written as $l_m = v_F \tau = v_F e\mu/m^*$ where v_F is Fermi velocity, τ is momentum relaxation time which can be expressed by elementary charge e times material mobility μ and divided by the electron's effective mass m^*. Following this relation, one can find that l_m is basically determined by electron mobility.

2.2. *Impedance mismatch*

The semiconductor heterostructure, even for strong spin-orbit interaction(SOI) materials (e.g., InAs, p-type GaAs, etc.), has a substantial degree of equilibrium carrier spin polarization. To create the spin transportation, it is necessary to generate non-equilibrium spin, which can be done by a couple of different methods [28]. Generally,

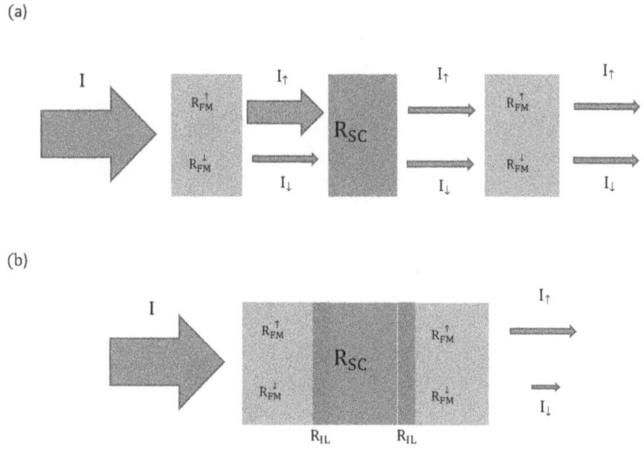

Fig. 1. resistance network was generated in different spin circuit. (a) Ferromagnetic material/semiconductor/ Ferromagnetic material configuration. (b) Ferromagnetic material/insulating layer/semiconductor/insulating layer/ Ferromagnetic material configuration. A electric current (which is illustrated by large green arrow) inject into these configuration and then can be polarized (which is illustrated by two small green arrow).

spin-injection through the ferromagnetic contacts into the semiconductor, but the efficiency of spin polarization is normally very low owing to an impedance mismatch problem [28]. The spin transport in ferromagnetic material exhibits a very high degree of spin polarization but significantly reduces when ferromagnetic material is combined with a semiconductor. This problem can be appreciated by adjusting the current of the spin-up and spin-down circuit in the parallel magnetization configuration. As shown in Figure 1(a), the upper and lower circuits obtain different resistances, in which $R^{\uparrow}_{FM}$ and $R^{\downarrow}_{FM}$ represent the resistances for spin-up and spin-down circuit in ferromagnetic material, respectively. Although the resistance of upper circuit is smaller than lower, the resistance ratio $I^{\uparrow}_{FM}/I^{\downarrow}_{FM}$ still approach to 1 because of $R^{\uparrow}_{FM}$ and $R^{\downarrow}_{FM}$ can be neglected, comparable with semiconductor resistance. In order to avoid the impedance mismatch problem, insulating layer is applied between the ferromagnetic material and semiconductor, as shown in Figure 1(b), in which $R^{\uparrow}_{IL}$ and $R^{\downarrow}_{IL}$ represent the schottky resistances for spin-up and spin-down circuit in insulating layer, respectively. The resistance of spin-up and spin-down circuit should be sufficiently different because equivalent resistance increase obviously comparable with semiconductor resistance after spin through the insulating layer. Therefore, the spin-injection efficiency can be improved.

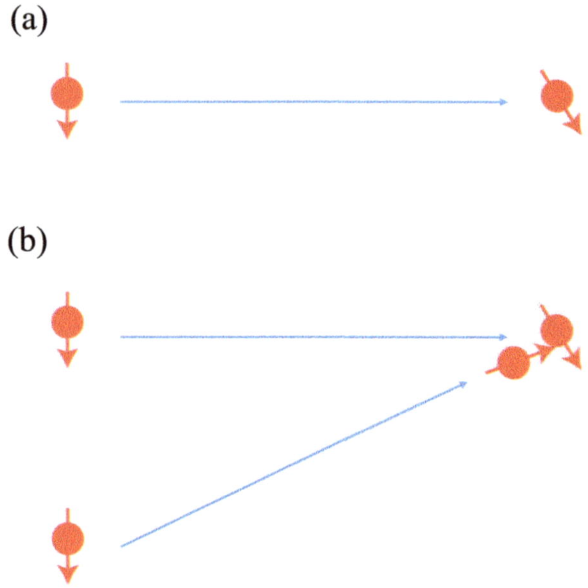

Fig. 2. The schematic diagram of spin phase spread. (a) 1D description. (b) 2D description.

2.3. *Spin spread*

It has been suggested that the spin information in the quasi-one-dimensional type spin FET is easier to maintain even if there are a few scattering events [32]. In the quasi one-dimensional channel, as shown in Figure 2(a), the spin precession angle from source to drain is $\phi = (2\alpha m^* L)\hbar$, where α is Rashba spin orbit parameter, m^* is effective mass for material and L is the propagation length for spin. The phase can be easily predicted because there is only one transport path. However, when the case is extended to two dimensions, transport momentum is no longer limited to one direction and thus the arrival electrons coming from different source points will travel through different paths and acquire different spin precession angles, resulting in complication and unpredictable (Figure 2(b)). Spin phase is averaged out due to different traveling length L from source to drain. For this reason, spin-FET base on 1D-wire as a propagation channel is suggested, also noted in [11]. A common 1D-wire, for example, nano-tube, nano-wire and quantum point contact are easily to create 1D transport channel which are potentially used to create spin-FET.

These three issues are generally believed to be the bottlenecks in the development of spin FETs over the past 25 years. In short, to operate spin-FET even to create a functional spin-FET, it is crucial to find a new route to overcome all these obstacles.

3. Spin-Orbit interaction

A main feature of the spin-FET presents a periodic modulation of conductance as controlled by voltage-induced precession of the electron spins owing to spin-orbit interaction. In atomic physics, orbital electrons exhibit the spin properties in orbital motion. Because the electromagnetic interaction between the electron's spin and the effective magnetic field generated by the electron's orbit around the nucleus result in shifts in an electron's energy levels. In condensed matter physics, electrons moving in an electric field experience an effective magnetic field with direction perpendicular to the momentum and electric field. The Hamiltonian of spin-orbit interaction (SOI) is described as:

$$H_{so} = -\frac{\hbar}{4m_e^2c^2}\sigma \cdot \mathbf{p} \times \mathbf{E} \tag{1}$$

and effective magnetic field

$$B_{so} = -\frac{\hbar}{m_e c^2}\mathbf{k} \times \mathbf{E} \tag{2}$$

where $\hbar$ is the Planck constant divided by 2π, c is the speed of light, m_e is an electron mass, σ is pauli matrix $\mathbf{p}$ is momentum, k is wavevector and E is electric field. In the $eq2$, the patterns of the electric field determine the types of SOI. Two kinds of spin-orbit interactions, Dresslhaus SOI and Rashba SOI, are present in two-dimensional materials.

The Dresslhaus SOI results from the lack of inversion center that exists in the bulk crystal of zinc-blende structure-based materials; so-called bulk inversion asymmetry. It is influenced by the internal structure configuration. Generally, its strength is determined by a materials constant parameter while the value is difficult to modulate. On the other hand, Rashba SOI is induced by additional symmetry reduction which results from the structure asymmetry of the potential. Thus, it is possible to vary the spin-orbit interaction by manipulating the external gate voltage. In experiment, spin-orbit parameter α can be easily estimated by Shubnikovde Haas oscillation [33].

4. Quantum point contact

As discussed in the previous section, a one-dimensional(1D) channel, which avoids the spin phase spread, is important for the realization of spin FET. There are a lot of ways to form a one-dimensional systems, for example, a carbon nanotube, the edge state of topological insulator or a quantum point contact created by a pair of metallic gates. The first two systems have rich physics but may be difficult to be utilized for spin FET, while a quantum point contact (QPC), a mature technology

Fig. 3. Schematic of a quantum wire formd at interface of a GaAs/AlGaAs heterostructure. Split metallic gate locating at the top of whole material provide a electric potential confining electron moving in the 2DEG. Adapted from [18].

can easily be used in integrated circuits, is suitable to be applied in spin FET manufacturing process.

The most common way to form a one-dimensional system is to first confine elections within a surface of an object or an interface between two substances. A two-dimensional electron gas (2DEG) systems is formed within such a interface, i.e., a semiconductor heterostructure. A further reduction of dimensions was achieved by applying different fabrication techniques such as etching or electron lithography. The 2DEG may be scraped narrowly to confine electrons moving in one direction. However, scraped may induce lattice imperfections and consquently carrier mobility reduction. Alternative one-dimensional lateral confinement scheme uses metallic split gates. Applying a negative voltage V_g to split gates, a depleted zone is formed under the gated regions, which gives rise to a narrow channel between them, as shown schematically in Figure 3. Although the split gate geometry create a extremely simple one-dimensional system, it contains rich physics. In the following sections, I will outline some 1D features that are relevant to spin polarization, connecting with necessary issues of all-electric spin-FET.

4.1. *Quantized plateau*

The split gate, as usual as called quantum point contact (QPC), can vary the width of the 1D channel when change the gate voltages, leading to a feature of stepped plateaus in the linear differential conductance $G = dI/dV$ at low temperatures.

Such stepped conductance plateaus occur in steps of $2e^2/h$ due to the formation of one-dimensional energy subbands. The first quantized transport evidence was obtained in 1988, as shown in Figure 4 [36, 38]. This quantized resistance value $h/2e^2$ had been explained by Landauer-Buttiker formula within a simple model for non-interacting electrons. Consider an electron confined in a QPC formed in a 2DEG channel, i.e., allowed to move freely in the x-direction only, the Hamiltonian can be written as:

$$(-\frac{p^2}{2m^*} + V(y))\Psi = E\Psi \tag{3}$$

where the $V(y)$ term is the lateral confinement potential, which gives the eigenenergy:

$$E_{k_x,n} = E_n + \frac{\hbar^2 k_x^2}{2m^*}, n = 0, 1... \tag{4}$$

As illustrated in Figure 5, a 1D channel is connected to two 2D reservoirs. An incoming electron incident from the left channel j tunnels though the potential barrier into the right i channel with a tunneling probability T_{ij}^L, and reflects back into the left channel i with a reflectivity R_{ij}^L. Also, an incoming electron incident from the right channel j tunnels though the barrier into the left i channel and reflect back into the right i channel with a tunneling probability T_{ij}^R and reflectivity probability T_{ij}^R. Note that the unitary condition requires:

$$\sum R_{ij}^R + \sum T_{ij}^L = 1, \sum R_{ij}^L + \sum T_{ij}^R = 1 \tag{5}$$

The current of the ith channel on the right-hand side is given by:

$$I_i = \int dEev_i\rho_i^{1D}(E)\sum(f^L T_{ij}^L + f^R R_{ij}^R - f^R) \tag{6}$$

where f^L and f^R are the Fermi-Dirac distributions on the left-hand and right-hand side, respectively, and v_i is the group velocity defined by:

$$v_i = \frac{1}{\hbar}\frac{dE_{k_x}, n}{dk_x} \tag{7}$$

Combine Eq. 5, Eq. 6 and Eq. 7, we obtain the ith of electron current:

$$I_i = \frac{2e}{h}\int dE(f^L - f^R)T_i^L = \frac{2e(\mu_L - \mu_R)}{h}\int dE(-\frac{\partial f}{\partial E})T_i^L \tag{8}$$

The conductance measured between the outside reservoirs is given by:

$$G = \frac{I}{V_{sd}} = \frac{eI}{\mu_L - \mu_R} = \frac{2e^2}{h}\int dE(-\frac{\partial f}{\partial E})\sum T_{ij} \approx \frac{2e^2}{h}\sum T_{ij} \text{ if } k_B T \to 0 \tag{9}$$

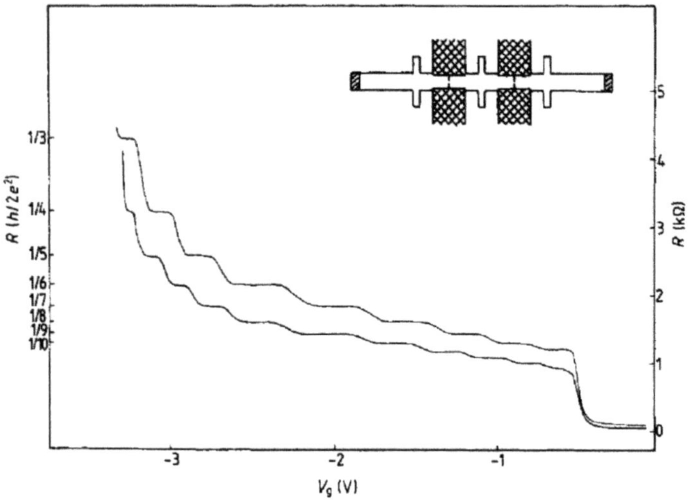

Fig. 4. The resistance as a function with split gate voltage is present. The value is quantized at each $h/2Ne^2$, where N is integer.Adapted from [38].

Each occupied subband provide the conductance $2e^2/h$ provide the transmission equal to 1, and the prefactor 2 accounts for spin degeneracy. The total conductance is therefore $G = N(2e^2/h)$ if there are N subbands being occupied.

Figure 6 [7] illustrates the relative positions between the 1D subbands and Fermi energy and their corresponding conductances. When the electrons from the reservoirs do not have enough energy to occupy 1D subbands (i.e., when the chemical potential μ is below the lowest subbandas shown in Fig. 6(a)) there is no transport electron and thus the conductance maintains zero. Electrons start to transmit through the quantum wire when μ is passing through the bottom of the 1D subband since some electrons will now enough energy to occupy lowest subband (Figure 6(b) and (c)). Once μ moves across the subband and lies between two subsequent subbands as shown in Figure 6(d), the conductance reaches a quantized plateau, in which the width of plateaus depend on subband spacing. The quantumed plateaus appear in steps of $N \times 2e^2/h$, as shown in Figure 6(e) and 6(f).

The application of magnetic field B in 1D transport measurement, spin polarization is exhibited because of obviously spacing of 1D subband leading beneficial observation of Zeeman effect [18]. By solving in quantum mechanism, one obtains the eigenenergies of the 1D subband with Zeeman effect as:

$$E_{n,\uparrow} = E_{K_x,n} + g^* \mu_B B/2 \tag{10}$$

$$E_{n,\downarrow} = E_{K_x,n} - g^* \mu_B B/2 \tag{11}$$

where g^* is the effective g-factor and μ_B is Bohr magneton. We note that the Zeeman effect splits the spin-degenerate subband into two, raising and lowering

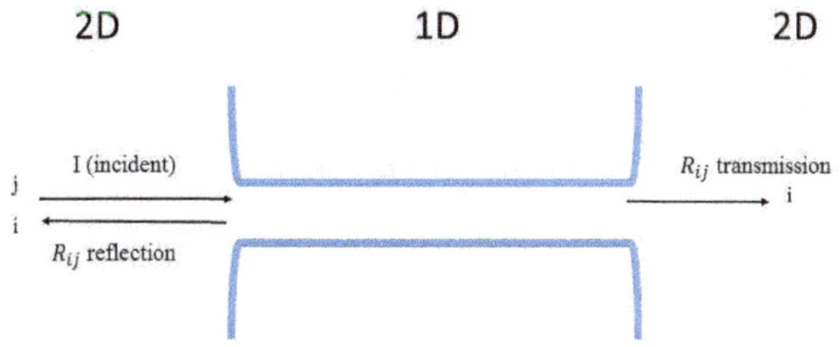

Fig. 5. Schematic diagram of a multichannel tunnelling situation. The incident electron tunnels to 1D wire from the left j channel and then yield reflective and transmissive situation, in which the probability is expressed as R^L_{ij} and T^L_{ij}, respectively.

the spin-up and spin-down subband by $g\mu_B B/2$, respectively. The 1D plateaus therefore split into two as well. The conductance plateaus now locate at half integer multiples of e^2/h, as shown in Figure 7. This result shows that electron spins can be polarization even up to fully polarized in the magnetic field. However, such a magnetic field is not necessary for spin polarization in 1D channel. Strong electron-electron(e-e) interactions and/or spin-orbit interactions can also be used to generate spin polarization, which will be discussed in the following subsections.

4.1.1. *0.7 structure*

The first possible observation of spontaneous spin polarization at zero magnetic field can be linked to the so-called 0.7 anomaly. It was a surprising feature studied by Thomas *et al.* [34, 35] in 1996 which shows that a sub-plateau appears at $0.7(2e^2/h)$. This anomalous feature has a distinct physical origin and is not related to the scattering events. The initial investigation presented that in the presence of a magnetic field the 0.7 plateau decreases to $0.5(2e^2/h)$, i.e. spin polarized plateau. This evolution suggested that it is the spin-imbalanced transport that results in the appeared of 0.7 feature. Exchange interactions can significantly enhance spin splitting of energy levels once a small spin imbalance is created. This mechanism was suggested to give rise to a partially spin-polarized 0.7 anomaly and also the observed evolution from a partially spin-polarized 0.7 anomaly into a fully spin-polarized 0.5 plateau with increasing magnetic field [17, 25]. Temperature investigations(as shown in Figure 8), however, exhibited a weak connection between the 0.7 anomaly

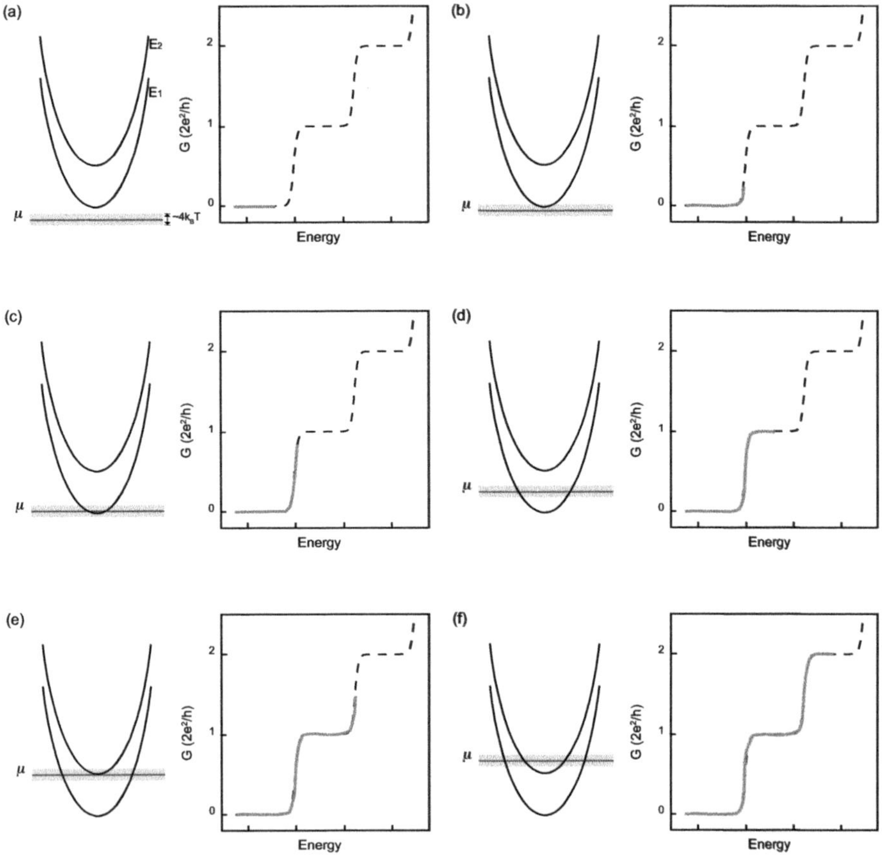

Fig. 6. Schematic diagrams of the dispersion relation of 1D subband. We consider the thermal broadening $4K_BT$ for the chemical potential μ and assume the transmission probability T in the Landauer-buttiker formula which is a step function. When the subband decline with the negative gate voltage decreases, the chemical potential situation corresponding to energy space is: (a) far below the bottom of the lowest subband, (b) and (c) moving across the lowest subband edge but not crossover it, (d) completely across the lowest subband and under second lowest, (e) moving across second lowest subband and (f) completely across the second lowest and under third lowest subband. Resulting the conductance variation are corresponded red solid traces which are shown on the right-hand side next to the schematics of energy levels. Adapted from [7].

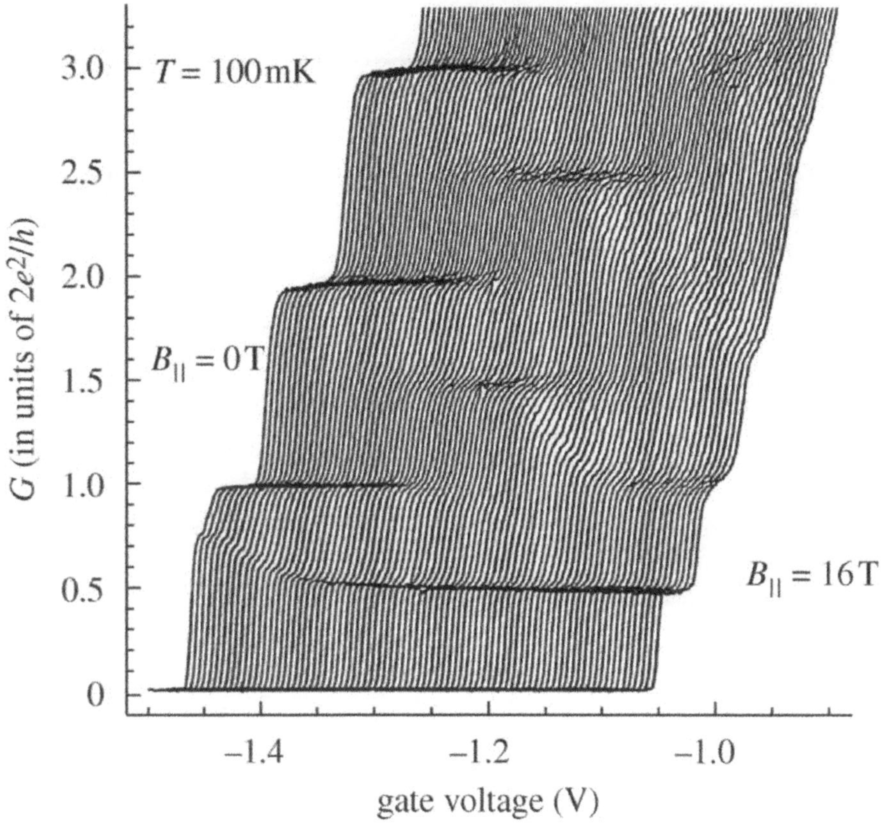

Fig. 7. Conductance trace evolution as a function of magnetic field. The clear spin degenerate plateau is lifted , resulting that the half integer plateaus$N \times 2e^2/h$ appear from 0 to 16T. Adapted from [18].

and the spin polarzation model, an enhancement of 0.7 plateaus with increasing temperature from 0.07K to 1.5K, distinct from most quantum phenomena (e.g., quantized plateau) that are suppressed with increasing temperature, was observed. Some explanations were suggested trying to link this temperature tendency to the spin polarization model, but are still under debate [20]. Zero bias anomaly, which was found by Cronenwett *et al.* in 2002 [9], provided a different route to investigate the 0.7 anomaly. Zero-bias anomaly, a conductance peak which appears at zero source-drain bias V_{sd}, appear a maximum value at zero V_{sd}, was found to be suppressed with increasing temperature and split into two peaks with increasing magnetic field. All of these are typical signature of Kondo physics normally observed in quantum dots [10, 14, 15]. The observation of Kondo-like phenomenon in QPC is very peculiar because in principle Kondo effect should not occur in an open system like this. Although there is still no consensus on the origin of the 0.7 anomaly, the 0.7 anomaly is generally believed to be related to spins and opens a

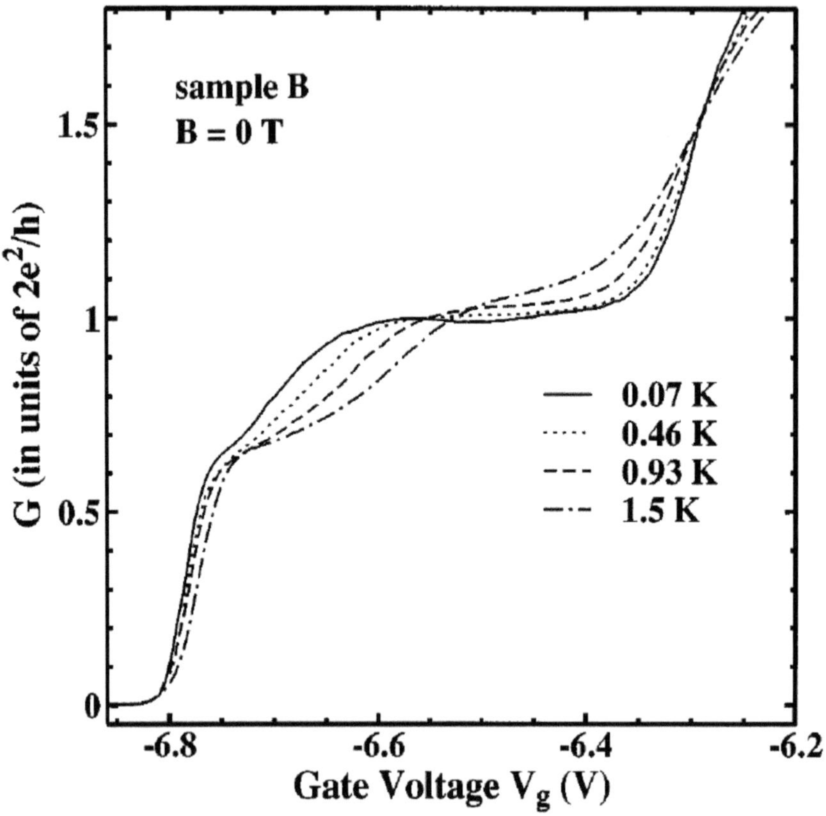

Fig. 8. The temperature dependence of 0.7 anomaly. The temperature change from 70mK to 1.5K. Adapted from [34].

gateway to investigate many-body spin dynamics and in QPC system.

4.1.2. *0.25 structure*

The 0.7 structure is not the only one in QPC which can be related to spin physics. In general, the 1D conductance measurement is performed using the standard lock-in technique, which means that a small ac excitation bias, typically 10-100 μV at 10-400 Hz, is applied to the source. Based on this measurement, a dc source-drain bias can be added in, resulting in a signal consisting of a small ac bias riding on a dc bias from the source to the drain. More details of dc measurement can be found in Refs [2–5]. When a finite dc bias is applied, the conductance plateau should appear in an integer multiple of $0.5(2e^2/h)$ instead of $2e^2/h$ because of separation of chemical potential crossing to energy subbands successively. The chemical potential separate to source potential μ_s and drain potential μ^d, while the theoretical feature of integer quantized plateau drop from $N(2e^2/h)$ to $N \times 0.5(2e^2/h)$ due to μ_s cross the spin-degenerate subband. In practice, however, the separation appears not only in the chemical potential but also in the lowest energy subband in large dc bias. This phenomenon suggests an additional spin splitting in 1D subbands apart from the splitting of chemical potentials. The underlying mechanism was suggested to be the strong e-e interactions [29]. However, this interpretation is still under debate, similar to the case of the 0.7 anomaly. Although up to date there is no consensus on whether the spin polarization is indeed the origin of the 0.25 anomaly, the evidence emerging in experiments [2, 5, 6] is highly consistent with spin polarization. Two different experiment investigated by Chen *et al.* provide strong evidence to sustain the suggestion of spontaneous spin polarization in the 0.25 structure.

Figure 9(a) shows the differental conductance evolution of QPC for applying dc bias from 0 to 4 mV in steps of 0.1 mV. Two obvious plateaus exhibit below $2e^2/h$, the 0.85 and 0.25 structure when dc source-drain bias is applied. Here we will only discuss 0.25 structure and its mechanism of spin polarization. In Figure 9(a), the 0.25 structure exist strongly for a finite dc bias applied. More investigation as shown in Figure 9(b), this 0.25 plateau is relatively unaffected for $B_{\|}$ from $B = 0$T to $B = 12$T. As the dc source-drain bias is larger than the energy subband spacing, one of the spin-subbands is always lying between μ_s and μ_d, thus the quantized 0.25 plateau is present in G_{ac}.

In order to verify whether the 0.25 plateau is indeed spin-polarized, a direct measurement of the spin content of the electrons passing through the QPC is necessary. A technique known as magnetic focusing is therefore used to probe the degree of spin polarization of the QPC. Figure 10 shows the schematic diagram of magnetic focusing geometry; the charged particles are injected from a QPC emitter passing, through the base regime, and then can be detected by a QPC collector once these electrons are focused into it by a transverse magnetic field [8]. The focusing spectrum is obtained by measuring the collector-base voltage as a function of the transverse magnetic field. In the semi-classical picture, the Lorentz force experienced by the electron undergoing the cyclotron motion is given by:

$$F_c = m^* v_F^2 / r = e v_F B_\perp \tag{12}$$

where v_F is the Fermi velocity of the electrons, r is the radius of the cyclotron motion and $B_\perp$ is the applied magnetic field. The focusing condition is satisfied when $L = 2r$, where L is the distance between the emitter and the collector. The focusing spectrum is expected to show voltage peak at:

$$B_{peak} = 2hk_F / 2\pi e L \tag{13}$$

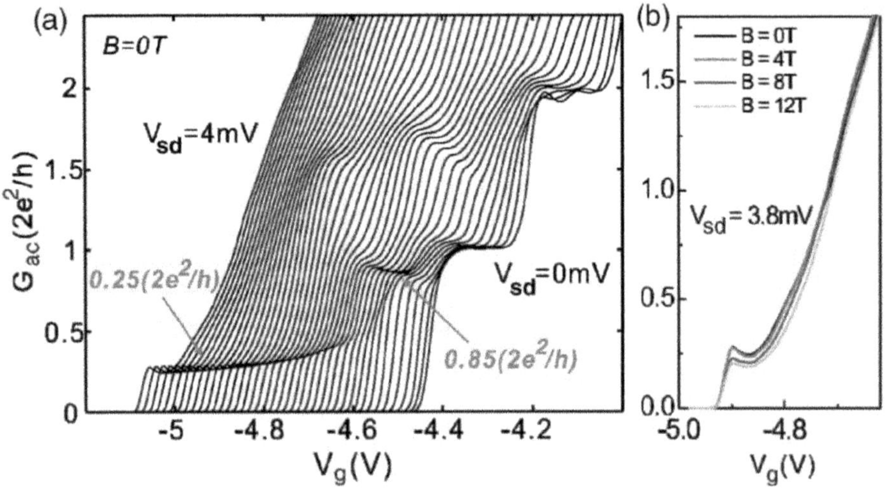

Fig. 9. (a) The evolution of conductance trace as a function of dc source-drain bias from 0 mV to 4 mV. The clearly 0.25 structure is appeared for dc bias increasing. (b) The conductance trace as a function of magnetic field changing from 0 T to 12 T, showing that the 0.25 $(2e^2/h)$ strongly against magnetic field. Adapted from [2].

i.e., when electrons are focused into the collector and build up a base-collector voltage. Magnetic focusing is a technique that was used extensively in the studies of mesoscopic physics [6, 16, 22]. This technique can be further extended to probe the degree of spin polarization. Based on this technique, Chen *et al* further demonstrate the spin polarization of 0.25 feature by magnetic focusing measurements [6]. Potok *et al.* has shown that when both emitter and collector are spin polarized, the peak voltage will be doubled compared to the situation when either the emitter or the collector QPC is spin denerate. The height of the peaks in collector-base voltagefollows the relation:

$$V_c = \alpha\frac{h}{2e^2}(I_\downarrow + I_\uparrow)(1 + P_eP_c) \tag{14}$$

where $P_e = (I_\downarrow - I_\uparrow/I_\downarrow + I_\uparrow)$ is degree of spin polarization for the emitter and $P_c = (T_\downarrow + T_\uparrow)/(T_\downarrow + T_\uparrow)$ is the spin selective for the collector, $I_{\uparrow\ \downarrow}$ are different types of spin current and $T_{\uparrow\ \downarrow}$ are transmission rates of the collector for spin-up and-down. It can be cleary seen that the doubled focusing signal appears when both the emitter and the collector are fully spin-polarized. Chen *et al.* employed this method to measure Pe and Pc in different regions, aiming to demonstrate that the 0.25 region is indeed spin polarized. As shown in Figure 11(c) and 11(d), the focusing peaks appear periodically, at intervals of $B = 0.05T$, consistent with the cyclotron motion relation. The height of the focusing peaks barely changes with

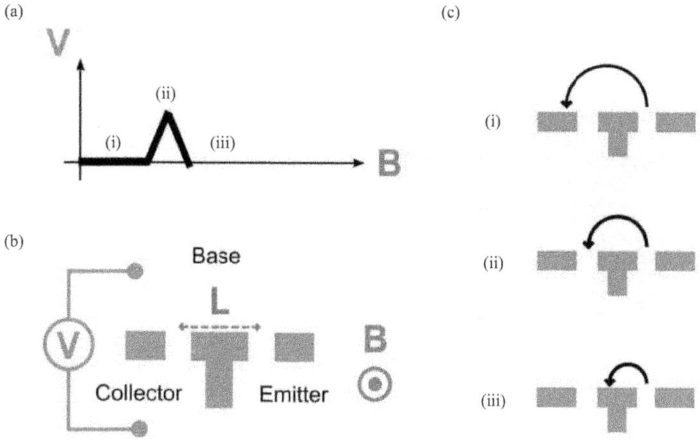

Fig. 10. Schematic diagram of the magnetic focusing. (a) The diagram illustrates the focusing spectrum which is obtained from the collector-base voltage as a function of the transverse magnetic field. (b) This diagram present the circuit of magnetic focusing measurement. The electrons travel from the emitter, through the base region into the collector. (c) The focusing peak evolution with different magnetic field are shown, which three different radius conditions (i), (ii) and (iii) correspond to (a), respectively. Adapted from [8].

the decreasing conductance of both emitter and collector from $2e^2/h$ to $0.25(2e^2/h)$, indicating that there is no spin polarization. In contrast to zero bias case, the focusing peak height exhibits very different behaviour when the source-drain bias was applied. It was observed that the peak height barely changes when the collector conductance decreases from $1.5(2e^2/h)$ to $0.5(2e^2/h)$, but rises doubly at $0.25(2e^2/h)$, corresponding to the 0.25 anomaly shown in Figure 11(b). This phenomenon, according to that both the emitter and collector QPCs are spin polarized when they are tunned to the 0.25 region. They used this method to measure the degree of spin polarization of the QPC as a function of both conductances and source-drain biases, concluding that the 0.25 anomaly is a fully spin-polarized state.

4.1.3. *0.5 structure*

In 2009, Debray *et al.* [12] present experimentally a spin polarization technique which the electron current can be polarize when a lateral confinement is formed by applying asymmetric gate voltage in QPC. The measurement perform in wet etch InAs quantum wires, a high mobility with strong SOI and e-e interaction. To introduce the effort of [12], I will start to introduce the spin-orbit interaction considering in quantum wire system. The transverse electric potential gradient resulting in generation of SOI, so-called lateral spin-orbit interaction(LSOI), was first proposed in 2006 [19, 39]. Consider that a quantum wire formed with the

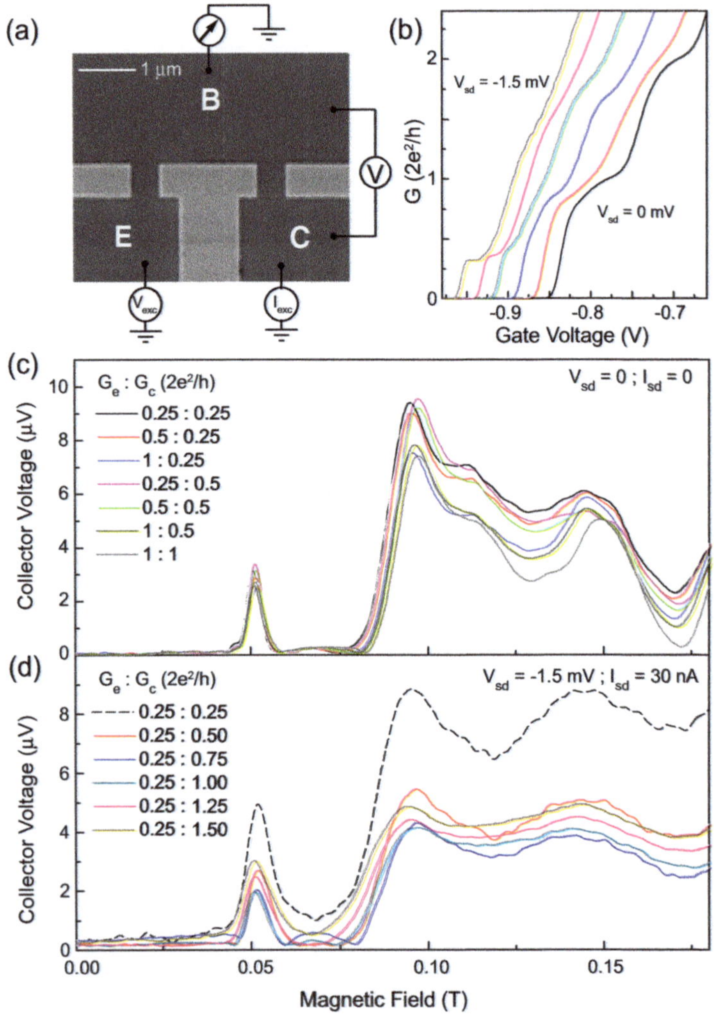

Fig. 11. (a) The diagram is shown the scanning electron micrograph of electron focusing device which the circuit include the emitter-collector configuration. Electrons are focusing in to collector via cyclotron motion by a small perpendicular magnetic field. (b) The differential conductance trace of the emitter with finite source-drain dc bias applied at $B = 0$. (c) The collector focusing voltage for $V_{sd} = I_{sd} = 0$. (d) Same as (c), but $V_{sd} = -1.5mV$ and $I_{sd} = 30nA$ are applied. The focusing voltage as a function of conductance increasing either emitter's or collector's, which is applied from 0.25 $(2e^2/h)$ to $2e^2/h$.

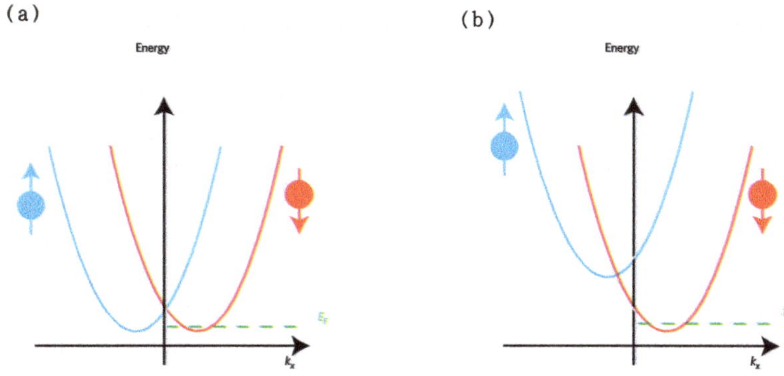

Fig. 12. The dispersion relaxation of 1D subband with SO interaction. (a) The 1D subband consider SOI, in which spin subband lift in momentum space. (b) The 1D subband considering SOI and e-e interaction, in which spin subband lift not only in momentum but also in energy space.

confinement potential $U(y)$ rises steeply at the edge, then the in-plane electric field $E = \nabla U$ lso points to the y-axis. Once the electron moves into the quantum wire, it would be affected by the lateral electric field. The LSOI Hamiltonian can be written as :

$$H_{LSO} = -\hbar/(4m_e^2c^2)\sigma \cdot (k_x \times \nabla U(y)) \tag{15}$$

and

$$B_{LSO} = -(\frac{\hbar}{m_e c^2}(k \times \nabla U(y)) \tag{16}$$

where k_x is the propagation vector along the x-direction and $\nabla U(y)$ is a lateral potential gradient. This Hamiltonian is analogous to Rashba Hamiltonian but the spin direct different direction. Assuming an electron is travelling along x-direction, the orientation of the effective magnetic field of RSOI points to y-dimension and LSOI points to z-direction. Figure 12 shows the dispersion relation of nondengerenrate spin subband according to this relation:

$$E_\pm = \frac{\hbar^2}{2m}((k \mp \alpha_z)^2 - \frac{\hbar^2}{2m}\alpha_z^2 \tag{17}$$

in which the first part provides the separation of momentum and second part is the energy shift. According to dispersion relation, left-moving electrons and right-moving electrons are in different spin states, meaning that spin has been polarized in the momentum space. For the electrons with momentum $+K_x$, spin-up have lower effective potential than the spin-down. For an electron with momentum $-K_x$, the spin orientation on both sides of the channel is reversed. However, one still observes the degeneracy in the energy space. It is difficult to lift the spin subband only existence of SOI in energy space. It has to further consider the e-e interaction in the quantum wire system. The transport spins unbalance gradually due to e-e interactions and thus lift the spin subband in the energy space. The evidence of spin polarization is present in [12]. The well-defined 0.5 plateau was observed in one-dimensional quantum transports when applying asymmetrical split gate voltage without an external magnetic field, as shown in Figure 13. On the other hand, the numerical calculation presents that a 0.5 plateau was not observed in the free electron Hamiltonian but did appear when considering e-e interaction. This evidence indicate that the e-e interaction plays a important role to lift the spin subband and thus provide spin polarization. However, the 0.5 structure has been observed requiring strong e-e interaction. The QPC structures of normal geometry have not exhibited evidence of spin polarization in the conductance measurement, which is shown in Figure 14. Nevertheless, the spin polarization mechanism still exists in the QPC structure without e-e interaction. When the Fermi energy is below the crossing point between two spin subbands, intermediate e-e interaction still lift spin subband, resulting that chemical potential can only cross one of the spin subband. Thus, spin is still polarized in QPC structure. Moreover, nearly full polarization efficiency can be provided.

To confirm the spin polarization creating by e-e interaction and SOI, magnetic focusing measurement still can be utilized. Figure 15 shows the focusing voltage spectrum as a function of a transverse magnetic field. The QPC injector applied asymmetrical voltage and varied the conductance from 2 $(2e^2/h)$ to 0.3 $(2e^2/h)$. When the QPC injector is set at high conductance values(over $(2e^2/h)$), the doubled focusing spectrum appeared, in which the high-field and low-field correspond to the spin-down and spin-up particles, respectively. This implies that two types of electron spins appeared with different momentums. As conductance decreases, however, the spin-up peak vanishes gradually, while the spin-down peak remains constant from 2 $(2e^2/h)$ to 0.3 $(2e^2/h)$. The data shows that spin is indeed polarized after passing through the emitter only controlling the conductance to 0.3 $(2e^2/h)$. Applying asymmetry voltage bias is a common method, usually use to investigate the disorder in GaAs/AlGaAs heterostructure. However, the spin polarization evidence present in the InGaAs/InAlAs heterostructure because of the strong SOI demanded. In this condition, a lateral potential is caused by asymmetry gate voltage and thus induce the lateral spin orbit interaction. This also implies that spontaneous spin polarization is formed.

5. Realization of All-electric Spin FET

In 2015, a new type of spin FET was proposed and demonstrated [24, 26]. This device overcomes the obstacles discussed above by employing two quantum point

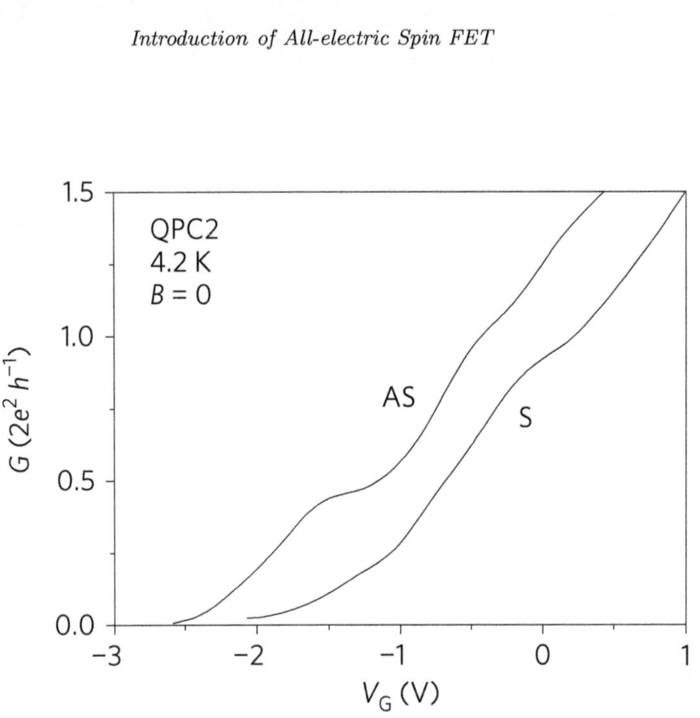

Fig. 13. The transport evidence of spin lift in QPC structure. Only the clearly first plateau appear at $(2e^2/h)$ for symmetry(s) saturation. Two plateau appear at 0.5 $(2e^2/h)$ and $(2e^2/h)$ for asymmetry saturation. Two conductance traces measure without magnetic field. The environment temperature fixed at 4.2K. Adapted from [12].

contacts. The ferromagnetic materials are replaced by using QPCs as a spin injector and detector which not only provides a 1D propagation channel but also improves the spin injection efficiency. The lateral SOI, and consequently, spin polarization in QPCs created when an asymmetric one-dimensional confinement is formed [12]. Figure 16 illustrates the all-electric spin FET and its operating principle. To utilize the QPCs as spin injectors/detectors in the all-electric spin FET device, asymmetrical split gate voltage was set to generate a lateral inversion asymmetry and thus formed an effective LSOI magnetic field resulting that the spin degenerate subbands is lifted in the momentum space. After passing through the QPC, only the electron spins that remain ballistic transport can be detected. On the other hand, these spins experience the Rashba field only and precess when travelling in the 2D region. The detector voltage depends sensitively on the difference between spin orientation of the arrival electrons and the spin polarization of QPC. When the polarization direction of the detector QPC align to the injectors polarization direction because of accumulated spins at the drain and vice versa. Therefore oscillating signals is formed by the two situations: the electron spins pass through the QPC detector when their orientation rotates to parallel to the polarization direction, and do not for the anti-parallel case.

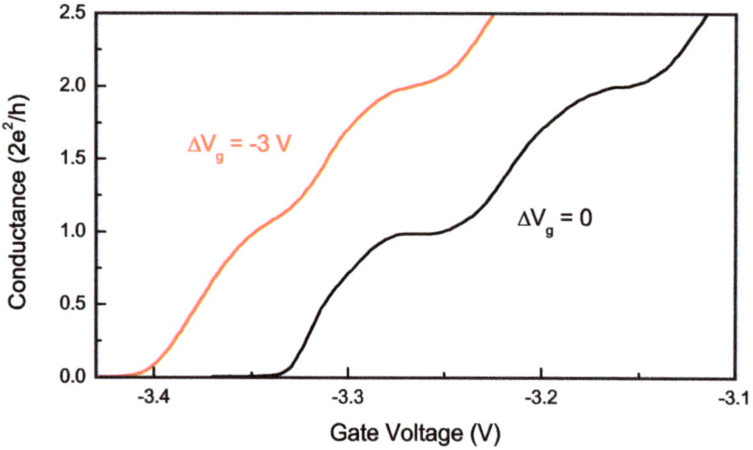

Fig. 14. QPC conductance. The QPC conductance as a function of QPC gate voltage with symmetry (black trace, $V_g = 0$) and asymmetry (red trace, $V_g = -3V$) confinement potential.

Figure 17 shows the oscillatory of the spin FET. The conductance value of both QPCs are initially fixed at $G_{QPC} = 0.3(2e^2/h)$, below the crossing point which is the point of intersection of spin-up and spin-down subbands, confirming the presence of a complete spin polarization. The oscillatory signal appears (disappears) with asymmetrical (symmetrical) lateral confinement potential. For the symmetry saturation, both spins are allowed to travel to the detector, resulting in a constant voltage signal. On the other hand, sinusoidal voltage is yielded due to the mechanism of the detector filter, which is operated when the lateral asymmetry confinement potential is applied (i.e. $\nabla V_g = -3\,\text{V}$). Three obvious wave packets exhibit the asymmetry condition, and its periods simultaneously increase with the bias of increasing the top gate voltage. The Rashba spin-orbit parameter α can be tuned by the z-direction external electric field, in which the electric field is parallel or antiparallel with α. Each maximum (minimum) implies an on (off) state and thus determines the on/off ratio, which is an important parameter affecting the power consumed.

This spin-FET geometry is complexed to estimate an accurate value of α because the potential is distributed to three part by two QPCs and the middle gate. The estimation of α is not appropriately fit by Shubnikov de Haas measurement [21]. In contrast to the relation obtained through the Shubnikov de Haas measurement, this fitting is consistent. The fitting formula of oscillatory voltage follows from a

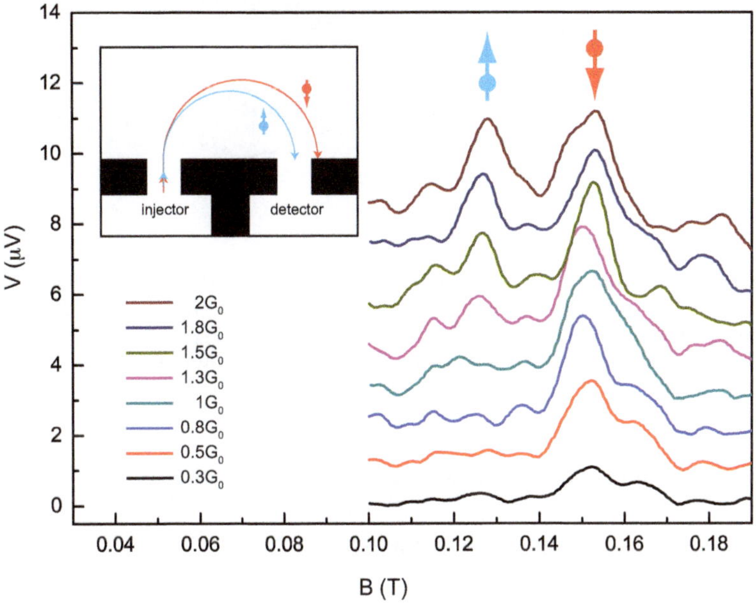

Fig. 15. The magnetic focusing spectrum shows the spin-dependent double peaks evolution with conductance of the QPC injector from 2 $(2e^2/h)$ to 0.3 $(2e^2/h)$. The appeared left peak (right peak) correspond to the spin-up (spin-down) focusing electrons. The left peak vanishes while the right peak still remains when the emitter conductance is tuned below $(2e^2/h)$ indicating that the current injected from emitter is spin-polarized. The data are vertically offset by 1 μV for clarity. The inset shows the schematic of a magnetic focusing device. The QPC collector is fixed at 1.5 $(2e^2/h)$ to receive both spins.

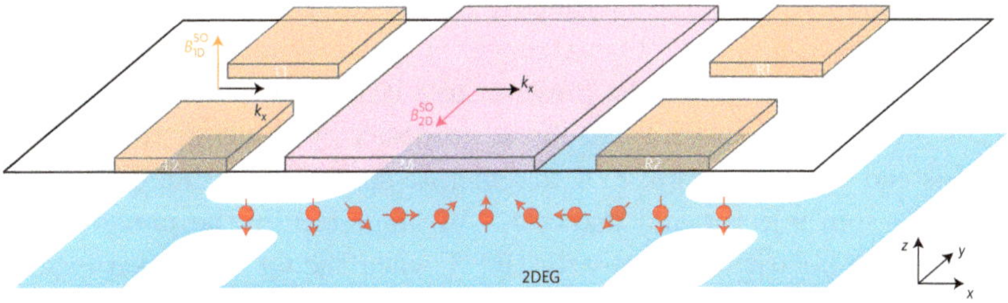

Fig. 16. The illustrates of operation principle of all-electric spin FET. A pair of origin bulk illustrate the QPCs which act as emitter and detector. The purple bulk illustrates the middle gate. The bottom layer with light green illustrate the carrier channel. The white regime with different size under the QPCs illustrate the depletion zone when QPCs apply asymmetry negative voltage. The orange arrows indicate the polarization direction of QPCs and consequent rotation red arrows indicate the evolution travelling spin with experienced RSOI.

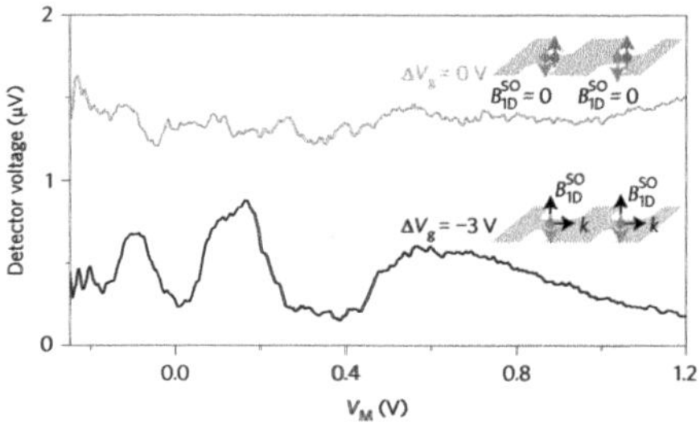

Fig. 17. The On/off oscillatory state of the spin-FET. The detector voltage as a function of middle gate voltage V_m. The measurement environment present at 30 mK and $G_{QPC} = 0.3\ (2e^2/h)$. The on switch (black line) occurs when a voltage difference $V_g = V_{l1} - V_{l2} = V_{r1} - V_{r2} = -3V$ is applied to QPCs. The off state (oscillation disappear) at $V_g = 0$. Data are vertically offset by $1\mu V$ for clarity.

Datta-Das proposal and is described as:

$$V = A\cos(2m^*\alpha L/\hbar^2) \tag{18}$$

where m^* is effective mass and L is the length between the two QPCs. The expression ignores the arbitrary phase shift. The fitting for the oscillatory voltage is shown in Figure 18(a), which results in the external electric field dependence of the strength of α [Figure 18(b)]. Inserting curve into Equation 11 results in the fitting curve [Figure 18(b)]. Fittings that coincide well indicate that the operations characteristic of the all-electric spin FET is consistent with the original spin-FET.

The working temperature is also the important problem for the realizing of spin FET. For this all-electric spin FET device, the oscillatory signal exist up to 17K, as shown in Figure 19. Considering the transport remain ballistic regime, the spin relaxation which is induced by momentum scattering can be negligible. Thereby the decrease of the oscillation amplitude mainly results from the thermal reduction of the QPC polarization efficiency. Thus stronger SOI and e-e interaction can increase the working temperature due to the presence of a larger 1D spin splitting.

Another all-electric spin-FET has been proposed by utilizing the 0.25 structure [30]. Instead of using lateral SOI to generate spin polarization 0.25 structure is now used for spin injector and detector. As suggested by the previous experiments [6], the spin polarization is nearly 100 % when the QPC conductance drops below $0.25(2e^2/h)$. Owing to different voltage biases applied to the injector and detector, the chemical potential differs in the three different regimes (source reservoir, middle regime, and drain reservoir), which is shown in Figure 19. The spin-precession-driven voltage oscillation is observed as a function of middle gate voltage, as shown in Figure 20. The period of voltage signal increase gradually as the top gate voltage larger than 0.4V. No variation for the voltage larger 0.4V the injector and detector conductances are higher than $0.25(2e^2/h)$ resulting in the absence of

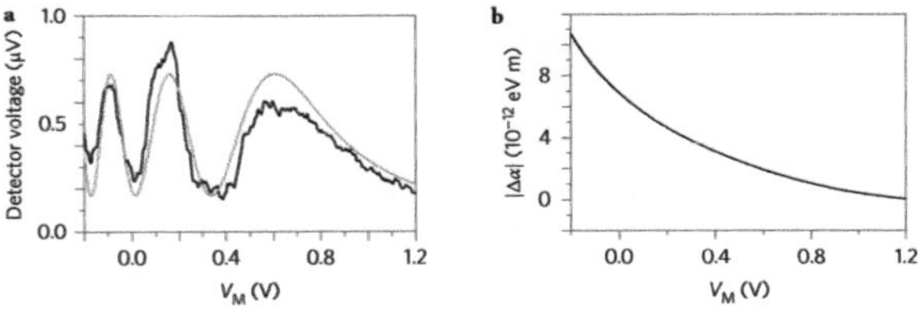

Fig. 18. Theoretical calculation of alpha variation. (a)Experimental data for the detector voltage is fitted by using eq 18. (b) The absolute value of spin-orbit parameter variation $|\triangle\alpha|$ as a function of middle gate voltage V_M which obtain from (b).

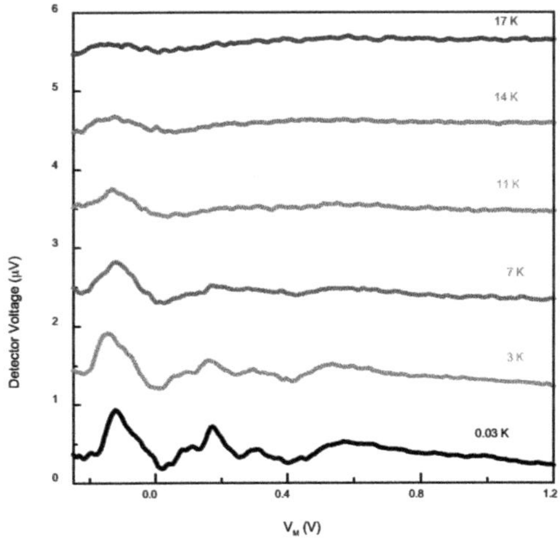

Fig. 19. The temperature dependence of oscillating signal. The temperature variation raise from 0.03 to 17 K, for $G = 0.3(2e^2/h)$.

spin polarization. Simultaneously, the spin precession is not able to precess a period due to weaker SOI. Moreover, on both sides of the QPC, the chemical potentials are

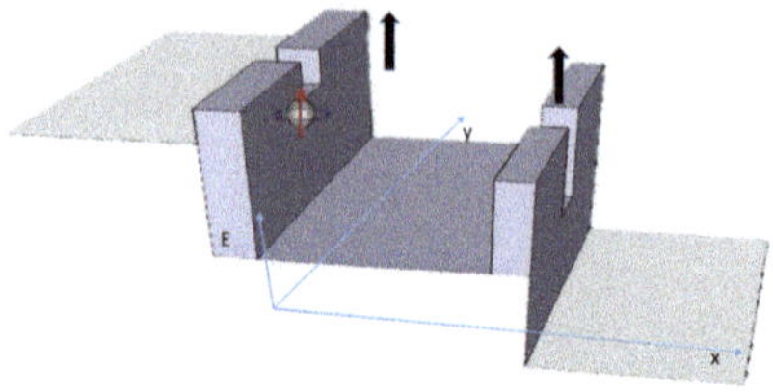

Fig. 20. The illustrate of electric potential distribution consider dc source-drain bias applying in spin FET device. The potential is formed steps from spurce to drain and is split by each QPC.

unequal (i.e., nonlinear transport) with the result that partial spins forcibly pass through the QPC injector and detector even if their direction is incorrect but also the variation of α is difficult to estimate the accurate value.

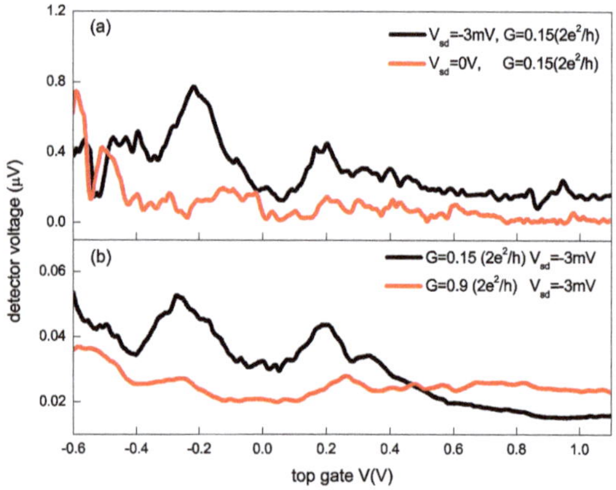

Fig. 21. The cross voltage as a function of top gate voltage. (a) shows the signal for the emitter dc bias at $V_{sd} = 3\text{mV}$ and the dc bias at $V_{sd} = 0\text{mV}$ with fixed conductance $G = 0.15(2e^2/h)$. (b) shows the signal for the emitter conductance at $G = 0.15(2e^2/h)$ and the conductance at $G = 0.9(2e^2/h)$.

6. Conclusion

The Datta-Das FET opens a gateway to the post-CMOS information processing; however, ferromagnetic elements, which are required to generate spin current for most of the spintronic devices, cannot be easily incorporated into the very large-scale integrated circuit, thereby making spin FET difficult for practical application. The QPCs have been proposed and utilized to set a 1D channel to overcome those obstacles and consequently provides fully spin polarization efficiency and remain a collision-free regime in the transport, resulting in a large on/off switch in the spin FET. Moreover, the all-electric control and all-semiconductor properties offer a viable route for spin information processing to connected industries.

References

[1] Awschalom, D. D. & M. E. Flatté, M. E. Challenges for semiconductor spintronics. *Nat. Phys.* **3**, 153-159 (2007).

[2] Chen, T. M., Graham, A. C., Pepper, M., Farrer, I. & Ritchie, D. A. Bias-controlled spin polarization in quantum wires. *Appl. Phys. Lett.* **93**, 032102 (2008).

[3] Chen, T. M., Graham, A. C., Pepper, M., Farrer, I. & Ritchie, D. A. Spontaneous spin polarisation in one dimension under finite DC-bias. *Phyical E* **40**, 1295 (2008).

[4] Chen, T. M., *et al.* Odd-even spin effects and variation of g factor in a quasi-one-dimensional subband *Phys. Rev. B* **79**, 081301 (2009).

[5] Chen, T. M., *et al.* Direct Observation of Nonequilibrium Spin Population in Quasi-One-Dimensional Nanostructures. *Nano Lett* **10**, 2330 (2010).

[6] Chen, T. M., Pepper, M., Farrer, I., Jones, G. A. C. & Ritchie, D. A. All-electrical injection and detection of a spin-polarized current using 1D conductors. *Phys. Rev. Lett.* **109**, 177202 (2012).

[7] Chen, T. M., Electron-electron interactions in GaAs quantum wires, Ph D thesis, Cambridge university (2009).

[8] Chin-Hung Chen, Study of spin-orbit interaction with magnetic focusing, Master thesis, Cheng-Kung University (2014).

[9] Cronenwett, S. M., *et al.* Low-Temperature Fate of the 0.7 Structure in a Point Contact: A Kondo-like Correlated State in an Open System *Phys. Rev. Lett* **88**, 226805 (2002).

[10] Cronenwett, S. M.,A Tunable Kondo Effect in Quantum Dots *et al.* Kondo effect in a single-electron transistor *science* **281**, 540 (1998).

[11] Datta, S. & Das, B. Electronic analog of the electro-optic modulator. *Appl. Phys. Lett.* **56**, 665-667 (1990).

[12] Debray, P. *et al.* All-electric quantum point contact spin-polarizer. *Nature Nanotechnology* **4**, 759-764 (2009).

[13] D'yakonov, M. I. & Perel', V. I. On spin orientation of electrons in interband absorption of light in semiconductors. *Zh. Eksp. Teor. Fiz.* **60**, 1954-1965 (1971) [*Sov. Phys. JETP* **33**, 1053 (1971)].

[14] Goldhaber-Gordon D, *et al.* Kondo effect in a single-electron transistor *science* **391**, 156 (1998).

[15] Hershfield S, Davies J. H. and Wilkins J. W. Probing the Kondo resonance by resonant tunneling through an Anderson impurity *science* **67**, 3720 (1991).

[16] Katherine E, *et al.* Imaging magnetic focusing of coherent electron waves *Nature Phys* **3**, 464 (2007).
[17] K.-F. Berggren and I. I. Yakimenko Effects of exchange and electron correlation on conductance and nanomagnetism in ballistic semiconductor quantum point contacts *Phys. Rev. B* **66**, 085323 (2002).
[18] K.-F. Berggren and M. Pepper, Electrons in one dimension *Phil. Trans. R. Soc. A* **368**, 1141-1162 (2010).
[19] Kiminori Hattori and Hiroaki Okamoto, Spin separation and spin Hall effect in quantum wires due to lateral-confinement-induced spin-orbit-coupling. *Phys. Rev. B* **74**, 155321 (2006).
[20] Kristensen, A *et al.* Bias and temperature dependence of the 0.7 conductance anomaly in quantum point contact *Phys. Rev. B* **62**, 10950-10957 (2000).
[21] Koo, H. C. *et al.* Control of spin precession in a spin-injected field effect transistor. *Science* **325**, 1515-1518 (2009).
[22] L. P. Rokhinson, *et al.* Spin separation in cyclotron motion. *Phys. Rev. Lett.* **93**, 146601 (2004).
[23] M. N. Baibich, *et al.* Giant Magnetoresistance of (001)Fe/(001)Cr Magnetic Superlattices.*Phys. Rev. Lett.* **61**, 2472 (1988).
[24] P. Chuang *et al.* all-electric all-semiconductor spin field-effect transistor. *Nature Nanotechnology* **10** 39 (2015).
[25] P. Jaksch, I. Yakimenko and K.-F. Berggren From quantum point contacts to quantum wires: Density-functional calculations with exchange and correlation effects *Phys. Rev. B* **74**, 235320 (2006).
[26] Po-Jen Chuang, All-electric spin field-effect transistor using lateral spin-orbit coupling, Master thesis, Cheng-Kung University (2013).
[27] R. M. Potok *et al.* Detecting spin-polarized currents in ballistic nanostrutures. *Phys. Rev. Lett* **89**, 266602 (2002).
[28] Schmidt, G., Ferrand, D., Molenkamp, L. W., Filip, A. T. & van Wees, B. J. Fundamental obstacle for electrical spin injection from a ferromagnetic metal into a diffusive semiconductor. *Phys. Rev. B* **62**, R4790 (2000).
[29] H. Lind, I. I. Yakimenko, and K.-F. Berggren, Electric-field control of magnetization in biased semiconductor quantum wires and point contacts. *Phys. Rev. B* **83**, 075308 (2011).
[30] Sheng-Chin Ho, All-electrical spin field effect transistors using source-drain bias induced spin filter, Master thesis, Cheng-Kung University (2013).
[31] Simmonds, P. J. *et al.* Quantum transport in $In_{0.75}Ga_{0.25}As$ quantum wires. *Appl. Phys. Lett.* **92**, 152108 (2008).
[32] Sugahara, S. & Nitta, J. Spin-Transistor Electronics: An Overview and Outlook. *Proceedings of the IEEE* **98**, 2124-2154 (2010).
[33] Takaaki Koga,*et al.* Rashba Spin-Orbit Coupling Probed by the Weak Antilocalization Analysis in InAlAs/InGaAs/InAlAs Quantum Wells as a Function of QuantumWell Asymmetry. *Phys. Rev. Lett.* **89** 046801 (2002).
[34] Thomas K J, *et al.* Possible spin polarization in a one-dimensional electron gas. *Phys. Rev. Lett.* **77**, 135-138 (1996)
[35] Thomas K J, *et al.* Interaction effects in a one-dimensional constriction. *Phys. Rev. B* **58**, 4846-4852 (1996)
[36] van Wees, B. J., *et al.* Quantized conductance of point contacts in a two-dimensional electron gas. *Phys. Rev. Lett* **60**, 848-850 (1998).
[37] Vera Lenchner, Bulk and Structure Inversion Asymmetry in Semiconductor Quantum Well Structures. *University of Regensburg.*

[38] Wharam, D. A., *et al.* One-dimensional transport and the quantisation of the ballistic resistance. *J. Phhys. C. Solid State Phys.* **21**, L209-L214 (1998).
[39] Yanxia Xing, Qing *et al.* Accumulation of opposite spins on the transverse edges of a two-dimensional electron gas in a longitudinal electric field. *Phys. Rev. B* **74**, 155313 (2006).
[40] Zutic, I., Fabian, J. & Das Sarma, S. Spintronics: Fundamentals and applications. *Rev. Mod. Phys.* **76**, 323-410 (2004).

Chapter 7

Semiconductor Spintronic Devices and Circuits

Bruce W. Wessels, *Northwestern University, Evanston, IL 60208*

With the discovery of large magnetoresistance effects in InMnAs heterojunction diodes and bipolar heterojunction transistors at room temperature there is interest in exploiting these devices for spintronic logic. An emitter coupled transistor (ECT) logic family has been proposed using the bipolar magnetic junction transistor (MJT). The logic circuit is predicted to have improved speed, area and power characteristics compared to conventional ECT making it potentially suitable for very high speed logic circuits.

1. Introduction

With the discovery of III-V dilute magnetic semiconductor there has been widespread interest in device development using these materials for both information processing and storage.[1] Early on Datta-Das proposed a spin FET device using a ferromagnetic source and drain.[2] To date this device has not been realized due to in part the difficulty in making efficient spin-polarized injection contacts. An alternative approach we have taken is to fabricate bipolar transistors using a narrow gap magnetic semiconductor. InAs is alloyed with Mn to form InMnAs. The material is p-type with hole concentrations of the order of $1 \times 10^{19} cm^{-3}$. Using InAs as the substrate p-n heterojunctions have been realized[3] [4]. Furthermore, epitaxial magnetic heterojunction transistors have been demonstrated.[5] The devices have been operated over a wide range of temperatures up to and including room temperature. Since InMnAs is ferromagnetic both processing and storage devices can be potentially realized. We demonstrated that this semiconductor is multifunctional in that by applying an external magnetic field the electronic properties of the transistor could be controlled. The transistor showed magnetoamplification, whereby the gain of the device decreased with applied magnetic field. The magnetoamplification results from the magnetic semiconductor having a giant magnetoresistance (MR) even in the paramagnetic state. Indeed a large positive MR of 1300% was observed in InMnAs/InAs p-n heterojunctions at a field of 9T and room temperature.[6] To explain the magnetoresistance, a Mott model for the origin of ferromagnetism was applied whereby current transport involves spin selective scattering of carriers by the ions in this case manganese. The spin transport can be described in terms of a two band model consisting of aligned and antialigned electron spin. The resulting current through the device will be spin polarized. The spin polarization p depends on the channel conductances G_1 and G_2 and $p = (G_1 - G_2)/G_1 + G_2)$. This spin polarized current leads to the observed giant magnetoresistance. The magnetoresistance depends on both the population of the two bands and the carrier mobilities. It was proposed that the conductance depends on the quantity $\exp[g\mu_B H/ kT]$ whereby g is the effective Lande g factor, μ_B is the Bohr magneton and H is the magnetic field. The total conductance G is given by

$$G = \frac{1}{2cosh\,(\frac{\mu_B gH}{k_B T})}\left(G_1 \exp(\frac{-\,\mu_B gH}{k_B T}\right) + \,G_2 \exp\left(\frac{\mu_B gH}{k_B T}\right) \quad (1)$$

The conductance depends strongly on the magnetic field depending on the value of g. The larger the value of g, the stronger the field dependence of the conductance. For magnetic semiconductors an effective g-factor (g_{eff}) needs to be considered where

$$g_{eff} = g + \frac{N_{Mn}\,\beta\langle S_z\rangle}{6\mu_B H} \quad (2)$$

is the value determined by the band structure and the second term is associated with the giant Zeeman effect that depends on the exchange integral β, the number of manganese atoms N_{Mn} and $<S_z>$ the average value of the Mn spin. For high temperature the Zeeman term g^* is approximated by

$$g^* = \frac{N_{Mn}\,\beta\; g_{Mn} S}{k_B(T+T_{AF})} \quad (3)$$

Where $T_{AF}(x) < 0$ describes the antiferromagnetic interactions between Mn^{2+} spins. The equation indicates that the giant Zeeman term is an explicit function of both N_{Mn} *and* T. Thus by increasing N_{Mn} and decreasing T, g_{eff} should become larger. Consequently a larger positive magnetoresistance will result.

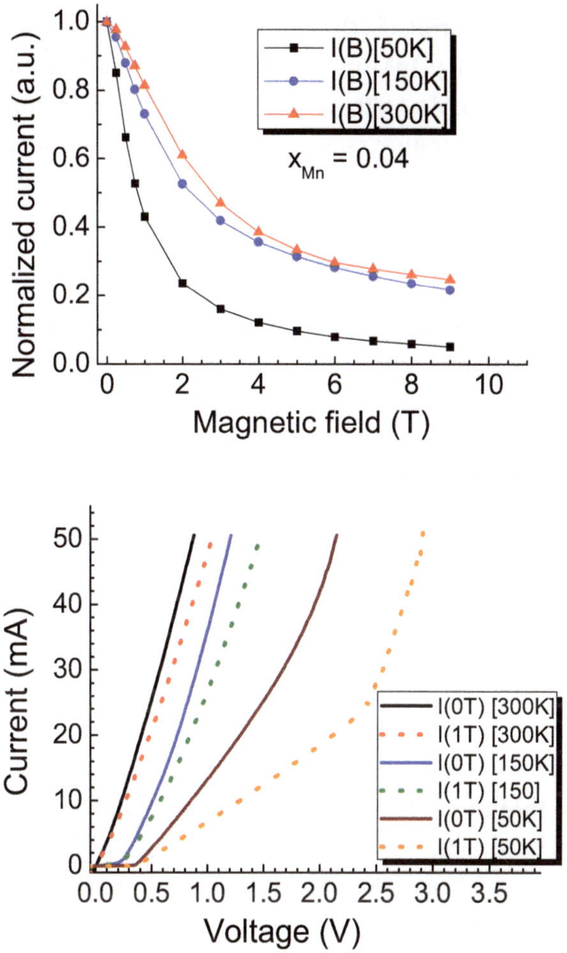

Fig.1a) Current-voltage characteristics of an InMnAs p-n heterojunction at 295 K. b) Normalized junction current vs. magnetic field after ref 7.

Experimentally the measured g_{eff} values showed a weak dependence on N_{Mn} but a strong dependence on T.[7] Fig 1a shows this temperature dependence in the I-V characteristics of InMnAs / InAs p-n heterojunction. As can be seen in Fig 1b the field required to reduce the current by a factor of two is four Tesla at 300K and 0.9 Tesla at 50K. Further reduction in the required field is expected as the temperature is decreased.

InMnSb devices

InMnSb has a large spin-orbit (SO) coupling and a g-factor of 50 in undoped materials. Heterodiodes of p-InMnSb/n-InSb were characterized in a magnetic field.[8] As in the case of InMnAs/InAs a giant positive magnetoresistance effect was observed over the temperature range of 75 to 298K. By fitting the conductance data to Eq (1) the g_{eff} factor was found to increase from 68 at 298K to 140 at 75K. The large g-factors are consistent with prior studies on both III-V and II-VI DMS systems where magnetic impurities induce large g-factors.

The initial data indicates that there may be an advantage to operating these materials at cryogenic temperatures. Recently we have studied low temperature behavior of InMnSb/InSb diodes at temperatures of 7.2 to 100K at magnetic fields of 0 to 0.88 Tesla.[9] A positive magnetoresistance was measured at all temperatures. It was observed that the forward current decreased with decreasing temperature. An appreciable magnetoresistance of 50% was observed at 7.2K,and 0.5 T. Thus for a five percent magnetoresistance a field of 500 Gauss would be required.

Bipolar Magnetic Junction Transistor

InMnAs is a p-type semiconductor with a hole concentration of the order of 10^{18}–$10^{20}cm^{-3}$. To fabricate heterojunction transistors, either n type or p-type InAs is required. Consequently both npn and pnp heterojunction transistors can be fabricated. A p–InMnAs/n–InAs/p–InAs heterojunction magnetic junction transistor was previously reported by our group.[5] A gain of twenty was initially reported and gains as high as 100 were subsequently obtained. The transistors showed large positive magnetoresistive effects. Unlike a conventional transistor, due to the positive magnetoresistance the gain of the magnetic junction transistor decreases under an applied magnetic field. We refer to this behaviour as a negative magnetoamplification. The amplification β_{dc} in the common emitter mode of a bipolar MJT is given by[10]

$$\beta_{dc} = \frac{\gamma \propto_T}{1-\gamma \propto_T} \tag{4}$$

For a gain of twenty the emitter injection efficiency γ is 0.95 for a 0 T field and a base transport factor $\propto_T$ of 0.9998. Fig. 2 shows the measured amplification for a bipolar MJT at different magnetic fields. The amplification is plotted as a function of collector current. For small currents the amplification is small and in the cut-off region. As the current increases the amplification increases and is in the active region. It reaches a maximum. As can be seen when a magnetic field is applied, the gain decreases. For a field of 8T, the amplification drops from 20 to 4 at the maximum. For a current of 15 mA the gain drops from 7 to less 1T.

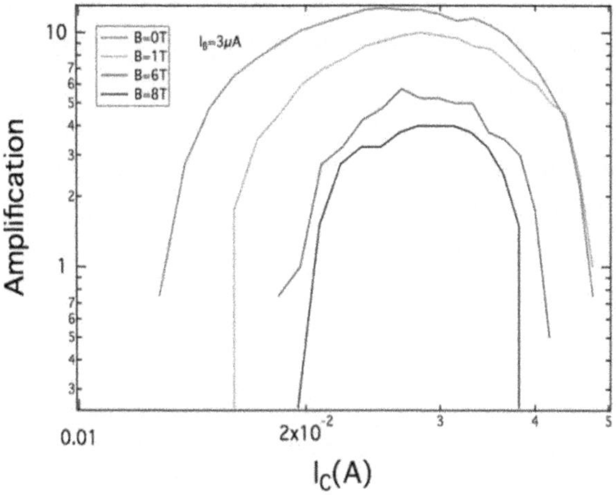

Fig. 2. Amplification versus collector current for an InMnAs transistor at 295 K for external magnetic fields up to 8 T. The base current is 3 microamps, after Ref 10

While the mechanism for the observed magnetoamplification behavior is still under debate, we attribute the observed charge transport characteristics to the positive magnetoresistance of the InMnAs collector. The emitter-base voltage V_{EB} decreases with magnetic field due to the magnetoresistance of the InMnAs collector. In that case the hole diffusion current decreases where

$$I_{pE} = I_{p_0}^{*} \exp(\frac{qV_{EB}}{kT}) \tag{5}$$

where I_{p0}^{*} is a constant which is determined by the doping and doesn't depend on magnetic field. A decrease in V_{EB} results in a decrease in the emitter injection current.

To increase the magnetoresistance of the collector the magnetic semiconductor requires a higher g factor. According to Eq (3) the g factor can either be increased by increasing the Mn concentration or operating at a lower temperature. Alternatively a semiconductor with a larger g-factor can be used.

Applications to logic circuits

Several spin logic families have been proposed based on the magnetic heterojunction diode or the bipolar magnetic junction transistor.[11 12 13] For the heterodiode at low magnetic fields the diode is conductive and at a high magnetic field the diode is highly resistive due to the positive magnetoresistance. Thus at low field the conductive state is a digital "1" and at high field a digital "0" is obtained due to the resistive state. Using high and low resistance states a binary logic family with a full range of logic functions is obtained. By routing magnetic control wires next to the diodes, the diode can be switched from the low to high resistivity state. The control wires are configured so as to control the magnetoresistance of the other diodes. For a low resistance the diode is propagating a "1" thus creating a magnetic field in the next diode. When the diode is in the high resistivity state "0", a low magnetic field will be present.

Key to utilization of the magnetic heterodiode logic family is that it exhibits a large magnetoresistance. In order to differentiate between the two digital states 0 and 1 a high I_{on}/I_{off} ratio is needed. Thus the larger the magnetic field the lower the current. Fig 1b shows a plot of normalized current versus magnetic field for a InMnAs/InAs diode. As can be seen for normalized current of 0.5, a field of three Tesla at 300K is required, whereas it is one Tesla at 50K.

Emitter-coupled transistors logic (ECL) uses bipolar junction transistors for very high speed electronics. Instead of the conventional transistor a bipolar magnetic junction transistor (MJT) is used for ECSTL. Fig. 3 shows a schematic of the bipolar MJT including control wires.

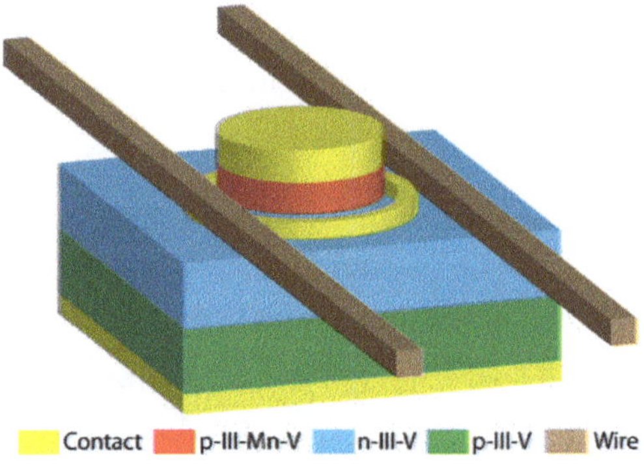

Fig. 3 A schematic of a bipolar MJT. Two parallel control wires are shown after Ref 12.

Conventional ECL dissipates minimal dynamic power, resulting in the power dissipation to be nearly frequency independent. This is in contrast to conventional CMOS circuits where the dissipated power increases proportionally with frequency. It is proposed that ECL using the bipolar magnetic junction transistor the power dissipation is much less. Thus for high speed applications there is a potential advantage using ECSTL at frequencies of 10 GHz and beyond.

To perform logic operations control wires are used. By passing current through the wire, a magnetic field results. This magnetic field, in turn, controls the amplification of the bipolar MJT. A control wire tangential to the transistor can be used to perform logic. When metal wires are placed parallel to the plane of the junction and isolated by a dielectric the resulting magnetic field is perpendicular to the plane of the junction. For a low net field the junction is in the conductive state, whereas for high fields the junction is in the resistive state. Depending on the direction of the currents through the control wires, the magnetic fields can either combine constructively or destructively. If the fields add then the emitter base junction will be in the resistive state. If the fields cancel the junction will be in the conductive state. For operation the bipolar MJT is forward biased. At low bias the transistor is in the cut off operation region. Application of a magnetic field will further decrease the current due to the negative amplification and lower current will flow between the emitter and the collector. For a higher voltage between the emitter and collector the transistor is in the active region. If the current through the control wires is zero or equal and opposite, the transistor is in the active state. For a high emitter-collector voltage or a large net magnetic field the transistor is in the secondary cut off region.

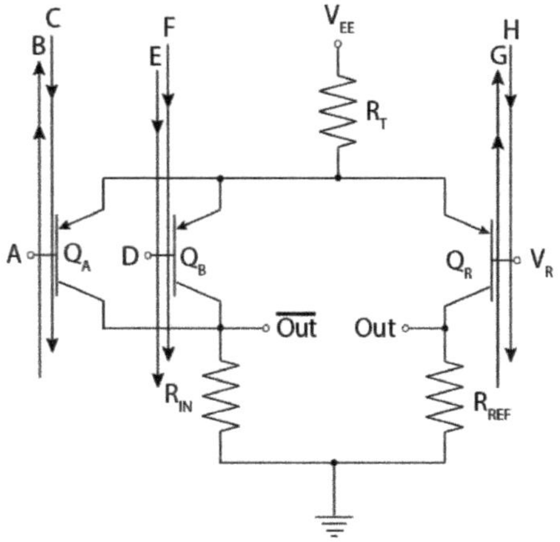

Fig.4. ECSTL basis logic gate. The lines labeled B,C,E,F,G and H are control wires after Ref 13

The general ECSTL basis gate is shown in Fig 4; it shows two control wires for each transistor. There are three transistors and thus six current input and two voltage inputs. When there is a net current through the control wires, it results in the increase in the negative magneto amplification and operation in the cut-off region. The truth table for the ECSTL basis gate is given in ref 13.

The inverter/buffer circuit demonstrates the operating principle as shown in Fig (5).When there is no magnetic field the circuit serves as a differential amplifier. When input A is a low voltage and B (control wire) produces a small magnetic field, Q_{IN} is in the active region and draws more current then Q_{REF}. The current through R_T thus flows through R_{IN} resulting in a "1" and "0" current through R_{REF}. These currents result in a large voltage across the base of Q_{INV} and a small voltage on the base of Q_{REF}. A "1" voltage is propagated to the output labeled INV and a "0" voltage is propagated to the output labeled BUF. When there is a large current through the B input, Q_{IN} is in the cut off mode. The current then flows and R_{REF} produces opposite value for the output voltages and currents.

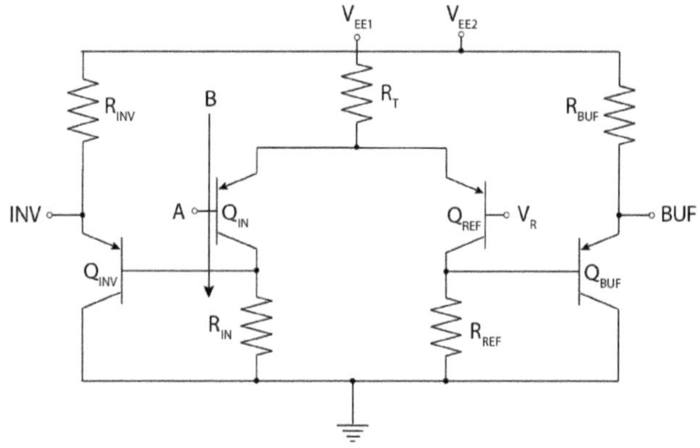

Fig. 5. ECSTL inverter/buffer circuit with voltage followers. There are two inputs: voltage A and field B. The currents through R_{IN} and R_{REF} flow in control wires for other magnetic junction transistors and are the magnetic field outputs. Q_{INV}, Q_{BUF}, R_{INV}, and R_{REF} are used for voltage followers to create outputs that can be connected to the gates of other magnetic junction transistors after Ref. 13.

As an example when the input voltage at A is held constant at 0.7V and the input field B is varied between 0 and 5 T, at low field the bipolar MJT is in the cut off region. Consequently the current flows through R_{REF}.

By using the bipolar MJT in multiplexer circuits they are much more compact than conventional ECL multiplexes circuits. Conventional circuits require multiple stages of logic and 20 or more transistors. The ECSTL circuits are also more compact than standard CMOS circuits which require 12 transistors and two stages. A 2:1 multiplexer is shown in Fig. 6. When the control wire labeled *Sel* carries current, Q_1 is in the cut off condition. When –*Sel* wire carries current, Q_2 is cut off. Thus Sel chooses which transistor responds to the input at A and B. The chosen signal is propagated to the outputs labeled – *Out* and *Out* in the inverted and non-inverted forms. HSPICE simulations have been performed on the 2:1 multiplexer (Ref.13)

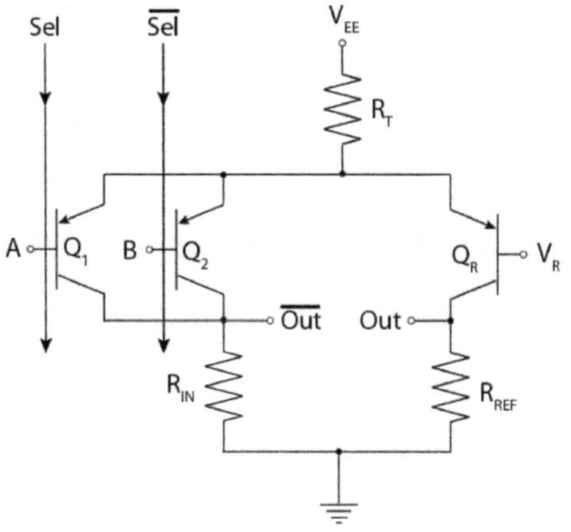

Fig 6..ECSTL 2:1 multiplexer after ref. 13.

A 4:1 multiplexer has been designed as shown in Fig 7. There are four bipolar MJT, and each having two control wires. The control wires implement a NOR function to select which transistor is in operation. If either of the two control

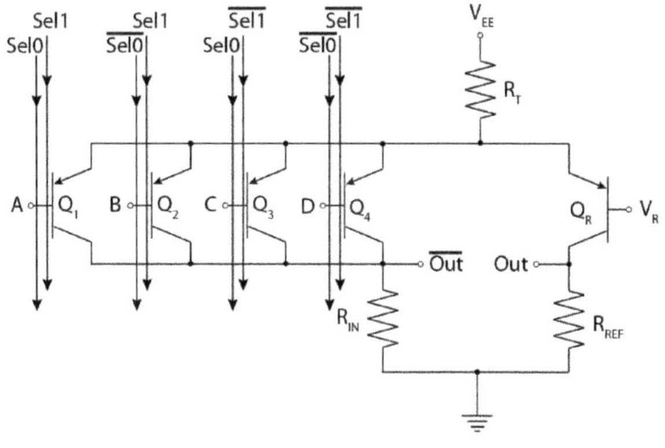

Fig. 7. ECSTL 4:1 multiplexer after Ref 13 .

wires carry current the transistor is in cut-off. There is one transistor in operation at all times and the selected input is propagated to the output along with its complement.

For logic circuits the important metrics are performance, power, area, scalability and cascading characteristics. Other factors include cost and reliability. The mechanisms for power dissipation in ECSTL are current flow from the high voltage rail to the ground and the current in the control wires. For a conventional ECL, gate dissipation is about 50 microwatts and the power dissipation is almost frequency independent. The power dissipation is proportional to the number of logic gates. In the case of the multiplexor and adder circuits the power dissipation is proportional to the number of logic stages. ECSTL can execute logic functions using only 20-30 % of ECL thus there is a 70 – 80 % decrease in power dissipation.

As to the bandwidth, the total switching time is the cumulative switching time of the transistors comprising the gates and the magnetic components associated with changing the current in the control wires. Now heterojunction transistors used in ECL have bandwidths approaching 200 GHz. The magnetic switching time will be determined by the spin relaxation and the magnetic induction time for changing the current in the control wires. The spin relaxation time is of the order of 10 picoseconds. The magnetic induction time is expected to be faster than the base-voltage switching time. Thus the gate delay is presumably determined by the cumulative switching time of the transistors. Due to the decrease in the number of devices in the ECSTL by 3x to 5x, there is a proportional decrease in the delay time, potentially enabling a logic family with capabilities of very high speed computing.

As to the device area, like conventional ECL, ECSTL uses a large number of resistors which take up area. Furthermore the control wires used in ECSTL will need additional area. However there are fewer transistors so ECSTL can be smaller than ECL circuits. The circuits using the magnetic semiconductors can operate at room temperature since they are in the paramagnetic state. However there are advantages to operating at low temperature including lower current and power requirements.

The ECSTL can be scaled to smaller dimensions. This scaling can lead to both speed-up and decrease in power dissipation. The control wire current depends on the current necessary to switch the transistors from the 0 to 1 state. The magnetic field created by the current wire is proportional to the device diameter D, thus the device can be scaled. A constant I/D produces a constant magnetic field. This makes it possible decrease the current I and power. Finally the match between the input and the voltage enables cascading.[13]

While the basic semiconductor devices for ECSTL have been demonstrated and their characteristics measured, the next step is to design and fabricate logic elements using these magnetic heterodiodes and bipolar MJTs. The logic circuits are predicted to have both improved speed and power requirements compared to conventional ECL as well as CMOS when operated at high frequencies,

The author would like to thank Dr. J. S. Friedman for extensive collaboration on logic circuits using magnetic semiconductor devices. The author acknowledges the collaboration of the electronic, magnetic and photonics group at Northwestern University. This work was supported by the NSF under grants DMR 0804479 and DMR 1305666.

References

1 H. Munekata, H. Ohno, S. von Molnar, A. Segmuller, L.L. Chang and L. Esaki, Phys. Rev. Lett. 63 1849 (1989).
2 S.Datta and B. Das, Appl.Phys. Lett. 56 665 (1990).
3 S. J. May and B. W. Wessels, Appl. Phys. Lett. 88 072105 (2006).
4 S. May, A. Blattner,and B. W. Wessels Phys. Rev. B 70 073303 (2004).
5 N. Rangaraju, J. Peters and B. W. Wessels, Phys. Rev. Lett. , 105 117202 (2012)
6 S. May, and B. W. Wessels, JVSTB 23 1769 (2005).
7 J. A. Peters, C. Garcia, and B. W. Wessels, Appl. Phys. Lett. 103 053503 (2012).
8 J. A. Peters , N. Rangaraju, C. Feeser, and B. W. Wessels, Appl. Phys. Lett 98 193506 (2011).
9 J. A. Peters, W. Huang and B. W. Wessels, unpublished
10 J. Liu, J. A. Peters, N. Rangaraju, and B. W.Wessels, IEEE Trans. on Electron. Devices, 62 2470 (2015).
11 J. S. Friedman, N. Ranagaraju, Y. I Ismail and B. W. Wessels, IEEE Trans.Nanotech. 11 1026 (2012).
12 J. S. Friedman, J. A. Peters, G. Memik, B. W. Wessels, and A. V. Sahakian, J. Parrallel Distributed Computing 74 2461 (2014)
13 J. S. Friedman, B. W. Wessels, G. Memik, and A. V. Sahakian, IEEE J. on Emerging and Selected Topics in Circuits and Systems, 5 17 (2015).

Chapter 8

Organic Spintronics: The First Decade and Beyond

Sayani Majumdar

Department of Applied Physics, Aalto University School of Science, FI – 00076, Aalto, Finland

sayani.majumdar@aalto.fi

The field of organic spintronics gained momentum since the last decade where the motivation was mainly to explore the advantage of low spin orbit and hyperfine coupling strengths of organic semiconductors for achieving long spin diffusion length and time in the spintronic devices. Initial years saw both promising results and substantial challenges of fabricating high quality devices showing giant and tunneling magnetoresistance and reproducibility of the device performance. However, recent years witnessed a more promising route for organic semiconductor based spintronic components through molecular engineering at the inorganic-organic interface. A high degree of spin polarization was reported from such interfaces that lead to term them as "spinterface". Also modified interfacial magnetism is being reported by different groups that can eventually lead to tailoring of device properties at the first molecular level. The chapter discusses the progress of the field in the last decade in terms of spin valves and magnetic tunnel junction performances, introduction of new materials and device architectures for improved spin injection and transport, significant challenges in terms of low performance of the devices at room temperature, reproducibility, speed and the more recent route towards controlling spins at the single molecule level. Finally the chapter discusses the future prospects for improved molecule based spintronic components for future memory, sensor and logic operations.

1. Introduction

The first decade of the field of organic spintronics has been phenomenal where many breakthrough experiments have advanced the frontiers of our knowledge and conceptual understanding on spin-dependent phenomenon in organic semiconductor small molecules and polymers. On one hand we observed several exciting and promising results regarding the spin injection and transport (or tunneling) in organic spin valves and magnetic tunnel junctions, however, at the same time several technical problems of device fabrication, reproducibility, etc. have also emerged. Also several experimental results challenged our present understanding of the subject and there are strong debates and discussions on those issues in the organic spintronics community. Several comprehensive review articles have covered the initial years of this field quite thoroughly. [1-5] The current chapter will discuss some of the more recent findings focusing on breakthrough experiments, their theoretical understanding and the open questions in this field. The chapter is divided in two main sections. The first section will discuss experiments on organic spin valves, magnetic tunnel junctions and single molecular devices and the second section of the chapter will mainly focus on the spin polarization and magnetic properties of the interface between a ferromagnetic material and an organic semiconductor. In general the aim is to provide the readers with an overview of the major findings in the field of organic spintronics and the future research directions.

2. Spin valves and magnetic tunnel junctions

Before going into the details on the organic spin valves and magnetic tunnel junctions, let us first discuss a little about the fundamental aspects of spin devices. Spin valve architectures where a non-magnetic material is sandwiched between two ferromagnetic electrodes with different coercive fields H_{C1} and H_{C2}, form the very basic architecture of spin-electronic devices (schematics shown in figure 1). When this device is placed in an external sweeping magnetic field, the magnetic moments of the two ferromagnetic electrodes attain either parallel or antiparallel alignment depending on the coercive fields of the ferromagnets and also the applied field value (H). When the applied field value is in between the

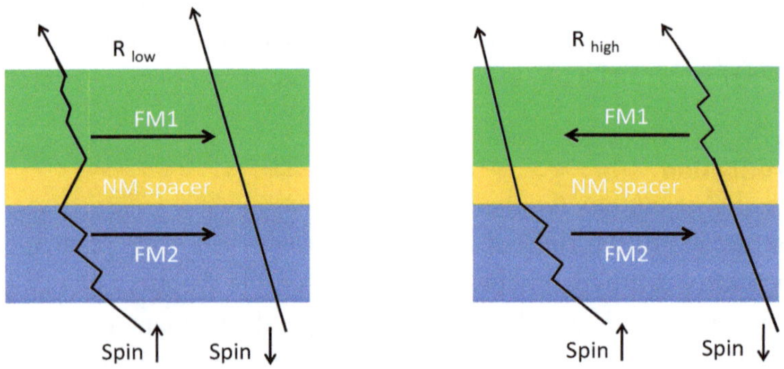

Fig.1. Schematic diagram of a spin-valve architecture for parallel and antiparallel alignments of the two ferromagnetic electrodes showing spin dependent scattering at the two electrodes.

coercive fields of the electrodes ($H_{C1} < H < H_{C2}$), the two electrodes attain an antiparallel orientation and when the applied field is higher than both the coercive fields, $H > H_{C1}$ and H_{C2}, the two electrodes are in parallel configuration. The spin polarized carriers entering the device through the ferromagnetic (FM) electrode undergo spin dependent scattering in the two electrodes. For parallel alignment of the spin moments, one of the spin directions is passed without much scattering whereas for antiparallel alignment, both spin directions are scattered. Scattering of both spin channels for antiparallel alignment gives rise to a high resistance (OFF) state and a low resistance (ON) state is achieved for the parallel orientation of the electrodes. Thus the magnetoresistance (MR) arising from the resistance bi-stability due to sweeping magnetic field is referred to as giant magnetoresistance (GMR). In spin-valves, the roles of the two FM electrodes are to inject and detect spin polarized carriers whereas the role of the non-magnetic spacer is to decouple the two ferromagnets so that they can switch their magnetic moments independently. At the same time, the spacer should allow the transport of spin polarized carriers without too much scattering for maximizing the device magnetoresistance. The up and down spin carriers suffering a spin dependent scattering in the two FM electrodes based on their parallel and antiparallel orientation leading to a high and low resistance state, is shown schematically in Figure 1. Hence based on the applied magnetic field, electrical resistance of a spin valve device switches between high and low states leading to GMR effect. GMR is defined as,

$$GMR = \frac{R_{high} - R_{low}}{R_{low}} * 100\% \tag{1}$$

where the R_{high} and R_{low} is the high and low device resistance respectively. In magnetic tunnel junction (MTJ), a similar architecture is used, however the

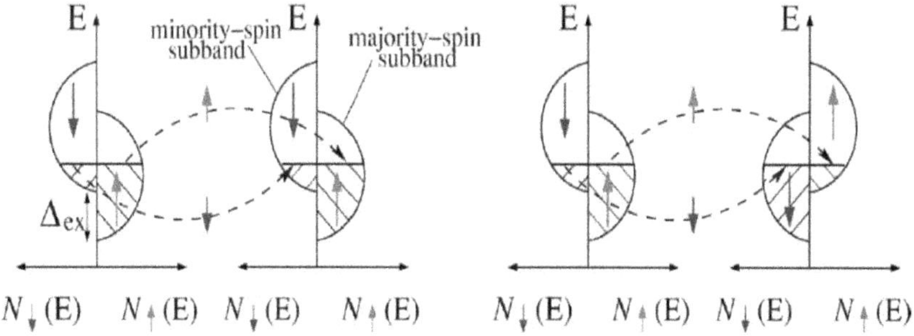

Fig. 2. Schematic diagram showing spin polarized tunneling of carriers between the two ferromagnetic electrodes with different spin resolved density of states. Reprinted with permission from Ref. 12, copyright (2004) by the American Physical Society. Reprinted with permission from Ref. 12, copyright (2004) by the American Physical Society.

non-magnetic spacer in this case is a very thin layer of an insulating material (a few nm) called a barrier through which spin polarized carriers can quantum mechanically tunnel from one electrode to the other. The resulting magnetoresistance is termed as tunneling magnetoresistance (TMR). Although the physics of GMR and TMR was known since Sir Neville Mott's theory of conduction of spin polarized electrons and Jullier's experiment in the early 70's, the field could not gain enough momentum due to the lack of technology of fabricating a few nanometer thick, continuous films with good material properties. GMR was first reported in metallic multilayers of Fe and Cr in 1988 separately by Albert Fert's group [6] and Peter Grundberg's group [7] which subsequently lead to the Nobel prize in Spintronics in 2007. Tunneling devices were reported by Meservey and Tedrow already in 1971 [8] and TMR was first reported by Julliere in his seminal paper in 1975 showing TMR response in Fe/Ge (10-15 nm)/Co device based on the parallel or antiparallel alignment.[9] However, instead of this promising finding, the field did not progress at that time. Since the reporting of magnetic tunnel junctions by Moodera et al. [10] and Miyazaki et al. [11] separately in 1995 showing that the tunneling current between two FM films separated by a thin oxide layer strongly depends on an external magnetic field that aligns the spin orientation of the two FM electrodes (as shown in Fig. 2), the research on MTJs became the most important branch of spintronics. The TMR is calculated similar as GMR as described in equation (1). In Jullier's formalism the TMR can be written as

$$TMR = \frac{2P_1P_2}{1-P_1P_2} \tag{2}$$

In this formalism it was assumed that spin travels from one spin polarized electrode to the other without any spin flip and P_1 and P_2 represent spin polarization of electrode 1 and 2. P is defined as

$$P = \frac{N(\uparrow)-N(\downarrow)}{N(\uparrow)+N(\downarrow)} \tag{3}$$

where $N(\uparrow)$ and $N(\downarrow)$ represents density of states at the Fermi level of the up ($\uparrow$) and down ($\downarrow$) spin carriers. Also TMR can be calculated using the tunneling currents in the parallel and antiparallel configuration of the devices or the spin polarizations of the two FM electrodes using the following formula, where $G_{\uparrow\uparrow}$ and $G_{\uparrow\downarrow}$ are the junction conductance for the parallel and anti-parallel configuration of the electrodes respectively.

$$TMR = \frac{G_{\uparrow\uparrow}-G_{\uparrow\downarrow}}{G_{\uparrow\downarrow}} \tag{4}$$

However, it is important to note that in all these formulas, the effect of the electrodes have been taken into account only. In practice, the interface between the electrodes and the barrier is of utmost importance. Spin scattering, trapping, filtering effects at the interfaces are significant in most practical devices and often determine the device properties.

The last one and half decade in the field of spintronics witnessed many fascinating materials, device architectures and exciting science that lead to miniaturization of computer memories to an unprecedented level. Introduction of semiconductors in the field of spintronics in the late 90s [12] and eventually introduction of organic semiconductors (OSC) in the last decade has also substantially contributed to more fascinating results improving our fundamental understanding of spin injection and transport leading to versatile multifunctional spin devices for future memory and logic operations.

2.1. *Organic semiconductor based spin valves*

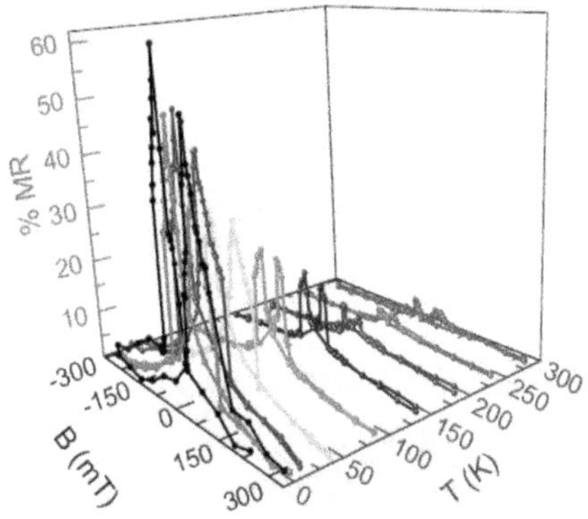

Fig. 3. GMR response of a typical LSMO/RRP3HT/Co as a function of temperature measured with 100 nA current. Reprinted from Ref. 21 with permission from Elsevier.

Hybrid inorganic-organic spin valves: The field of organic spintronics was pioneered by the first report of spin polarized injection and transport in OSC molecule Sexythiophene (T_6) in 2002, where the authors reported a spin diffusion length of 200 nm at room temperature in a planar spin valve structure with $La_{2/3}Sr_{1/3}MnO_3$ (LSMO) as both spin injecting and detecting electrodes. [13] In 2004, Xiong et al. reported the first vertical spin valve structure using organic semiconductor tris-(8,hydroxyquinoline) aluminum (Alq_3) as the spacer layer between FM electrodes LSMO and Co. [14] At 11K, a GMR response of 40% was reported from these devices. However, in these spin valve devices, the GMR response decayed quite quickly with increasing temperature leaving no MR response above 180 K. This loss of GMR response with increasing temperature is found to be universal later on with the reports of similar loss of GMR for different FM electrodes and different organic semiconductors. The effect was initially dedicated to the loss of spin polarization at the FM electrodes, especially LSMO with Curie temperature (T_C) of 350 K.

However, in next few years several other groups reported MR response from organic spin valves even at room temperature although the effect merely existed.[15, 16] By spin ½ photoluminescence detected magnetic resonance (PLDMR) studies, Wang et al. showed [17] that the spin diffusion length in organic semiconductors are more or less temperature independent and hence the observed decay of GMR response of the organic spin valves with increasing temperature solely appears due to loss of spin polarization of the FM electrode. However it is important to mention that reported spin-valve GMR effect showed similar decreasing tendency with increasing temperature in devices with low T_C half-metallic LSMO or high- T_C transition metal electrodes. It was shown that LSMO

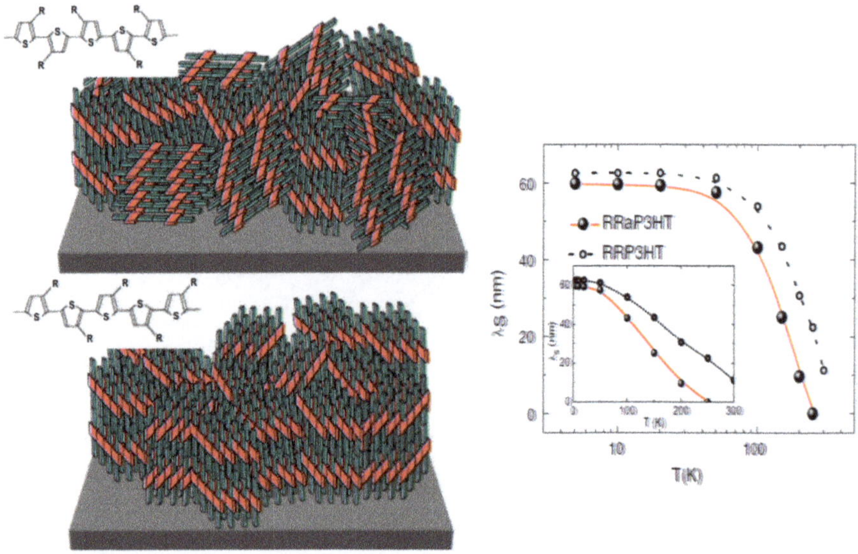

Fig. 4. (Left) Schematic diagram (reproduced from R. J. Kline et al., Nature Mater. 5 (2006) 222–228) of the structure and organization of region-random (top) and regio regular poly (3-hexyl thiophene) (bottom) forming either a poor orientation or a preferentially oriented crystal structure on the substrate. Conducting π-stacking planes are shown in red while the insulating hexyl chains are in green. (Right) Spin diffusion length (λ_S) of RRaP3HT and RRP3HT, calculated using modified Júlliere formula, as a function of temperature. Reprinted from Ref. 22 with permission from Elsevier.

on different substrates with varying strain states and hence different T_C show similar decrease in GMR response with increasing temperature. [18] Even $La_{2/3}Ca_{1/3}MnO_3$ (LCMO) with significantly reduced T_C, also showed similar loss of GMR response with increasing temperature. [19] In 2009, Drew et al. [20] studied the spin diffusion length (λ_S) of Alq_3 in a spin-valve structure by muon spin rotation technique and reported that spin diffusion length in Alq_3 decays sharply with temperature. In 2012, similar temperature dependent spin diffusion length is reported in regio-regular poly (3-hexyl thiophene) (RRP3HT) although the decay of λ_S with increasing temperature was not as sharp as in Alq_3. [21] Also comparison with regio-random P3HT (RRaP3HT) showed that depending on the polymer structure and charge carrier mobility, λ_S could be different. The more ordered RRP3HT showed larger spin diffusion length with lesser temperature dependence. [22] In 2009, Bobbert et al. [23] presented a theory for spin diffusion in disordered OSCs, based on incoherent hopping of a charge carrier and coherent precession of its spin in presence of a random hyperfine field and an applied magnetic field. From Monte Carlo simulations and an analysis of the waiting-time distribution of the carriers they predicted a weak temperature dependence, but a considerable magnetic-field dependence of the λ_S. It was shown that spin

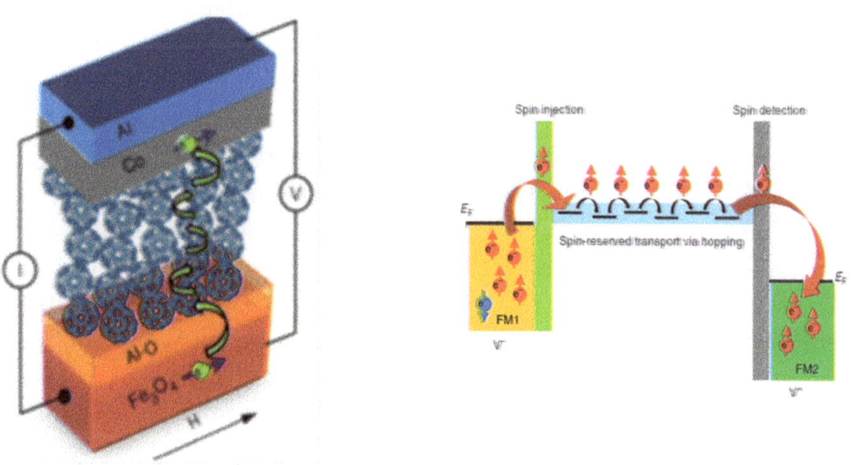

Fig.5. (Left) Schematic diagram of the device structure with an MgO substrate/Fe_3O_4/Al-O/C_{60}/Co/Al stacking structure. Excellent matching of energy levels within the C_{60} layer for spin-reserved transport via a hopping mechanism, and the interfaces between the FM electrodes and C_{60} layer result in high efficiencies for spin injection/detection under optimized fabrication conditions. Reprinted from Ref. 28 by permission from Macmillan Publishers Ltd.

polarized transport in organic semiconductor can get hampered in presence of hyperfine interaction. One of the major source of the hyperfine interaction being the Hydrogen atoms of the hydrocarbons. The hyperfine fields could act as local magnetic fields and the spin polarized carriers could start precessing around them losing their initial spin orientation. Therefore it was suggested that molecules without any hydrogean atoms i.e. fullerene molecules and Graphene could be ideal for spin polarized transport. Experimentally this was also verified that λ_S of hundreds of nanometers in C_{60} atoms [24] and in micrometers in Graphene can be obtained [25] where no hydrogen atoms are present. In subsequent years substantial experimental investigation was carried out to check the possibility of achieving long spin diffusion length from C_{60} based spin valve devices. In 2011, Lin et al. [27] experimentally demonstrated organic spin-valves based on C_{60}, however, a significantly increased spin-diffusion length could not be observed in those devices. The authors explained their observations by three possible reasons (a) a mechanism other than hyperfine coupling causes the loss of spin-polarization; (b) the observed magnetoresistance is really due to tunneling magnetoresistance which would disappear for C_{60} thicknesses beyond the tunneling range; (c) in thick devices an increasing conductivity mismatch limits spin-injection. Based on an analysis of the devices' current–voltage curves, and expectations resulting from the close match of the electrodes' work function with the lowest unoccupied molecular orbital of C_{60}, the authors proposed that the conductivity mismatch is a likely cause of the loss of spin-valve signal with increasing C_{60} layer thickness. Gobbi et al. [27] have obtained spin coherent transport in fullerenes at room temperature, as observed from the sizable values (> 5%) of RT MR in relatively thick (>25 nm) fullerene based spin valves. To interpret their experimental results the authors presented a multi-step tunneling model for explaining both electronic and spin coherent transport. They concluded that both the large MR values and the small decrease of this MR with applied bias are related to the robust intrinsic properties of fullerenes for spin transport. Zhang et al. [28] also reported organic spin valve devices with C_{60} as the spacer layer where a magnetoresistance ratio of over 5% was observed at room temperature. Also, a large spin-dependent transport length of approximately 110□nm was experimentally observed for the C_{60} layer at room temperature which is much higher compared to that observed for small molecule Alq_3 [20] or polymers.[22]

Recently, Li et al. have reported more than 3% magnetoresistance at room temperature from the spin valves based on n-type semiconducting polymer P(NDI2OD-T2) with better charge carrier mobility.[29] The improvement of interface properties via introducing a thin AlO_x barrier at the Co/polymer interface together with optimal annealing of bottom LSMO electrode, the authors reported MR ratio of 54 % at low temperature and more than 3% at room temperature. Furthermore, it is found that P(NDI2OD-T2) has λ_S of 64.5 nm, similar to that as RRP3HT however with a weaker temperature dependence. Better spin transport properties of P(NDI2OD-T2) could prompt it as a promising material for spin electronic devices and in future this result might lead to exploration for more novel organic materials for the development of high-performance spintronic devices. Another interesting way was tried by Sun et al. [30] This report showed substantial room-temperature spin transport from spin valves with an interlayer of barrier layer between the bottom FM electrode and BCP. A leaky and and a non-leaky barrier of AlO_x was used. The non-leaky AlO_x seed layer formed a true inorganic tunnel barrier, with extremely high device resistance even at room temperature and showed weak, tunneling-like temperature dependence. In this case, the MR was only measured in SVs with thin BCP layers, as expected for transport in the tunneling regime. The reference junction showed a 2% MR. These devices became extremely resistive when the BCP thickness is increased above 10 nm, preventing the measurement of any clear MR signal. Long-distance spin transport could not be achieved in the devices with non-leaky AlO_x layer. Instead, when leaky seed layers are employed, a strong temperature dependence of the junction resistance is found. Under these conditions, MR is measured even for thick BCP layers (up to 60 nm) which is far higher than the recorded for the non-leaky-AlO_x spin valves with 5 nm BCP. In this case, MR for different thicknesses of the BCP layer for 5, 10, 30 and 60 nm of BCP, was measured. Even for the case of the 60-nm-thick BCP film the MR signal was clearly distinguishable, although the curves became noisy because of the high junction resistance (in the GigaOhm range). A maximum 7% MR value at ambient conditions were reported and it was conjectured that in these devices spin transport takes place in the molecular levels of BCP.

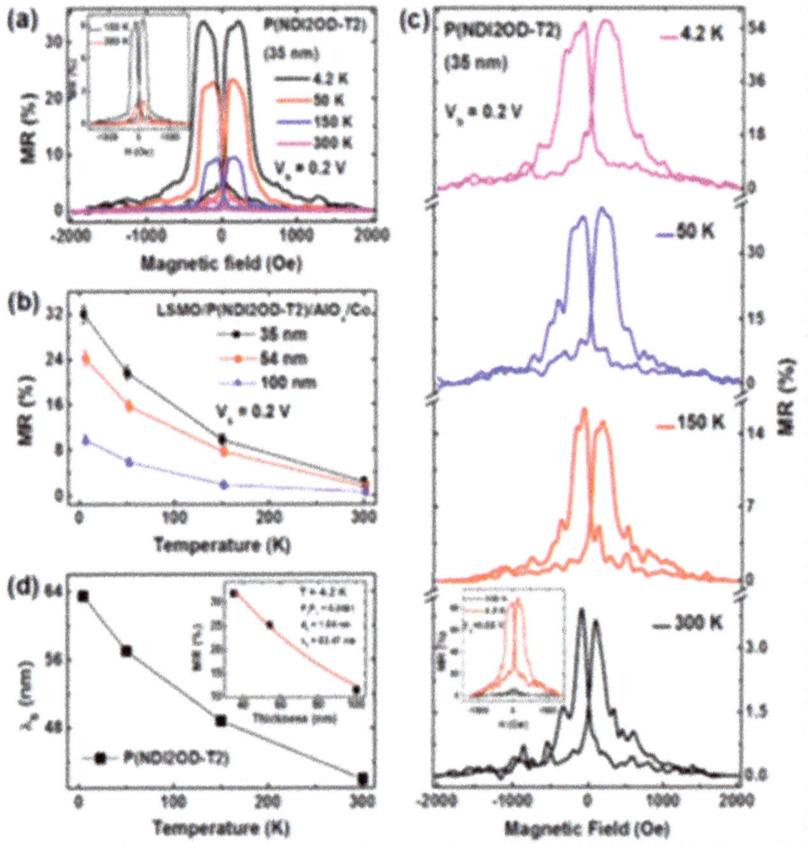

Fig. 6. (a) MR curves for the LSMO/P(NDI2OD-T2)/AlOx/Co device with a 35-nm P(NDI2OD-T2) at different temperatures. Inset shows the MR curves at higher temperatures. (b) The temperature dependent MR with different thicknesses of P(NDI2OD-T2). (c) MR curves for the SV device of a 35-nm P(NDI2OD-T2). The inset shows the MR of the same device measured at a bias of 0.02 V. (d) The temperature dependence of spin diffusion length λ_S of P(NDI2OD-T2). Inset shows the fitting of the MR values measured at 4.2 K. [Reproduced from Ref. 29]

All organic spin valves: The attempt of replacing the traditional inorganic ferromagnets from the organic spin valves was made by Li et al. [31] who showed spin injection and detection in an all-organic-based spin valve using the organic magnetic semiconductor V[TCNE]$_x$ as both the spin injector and analyzer. V[TCNE]$_x$ is the first reported room-temperature molecule-based magnet with magnetic ordering temperature $T_C \approx 400$ K. [32] The magnetic order originates from the antiferromagnetic (AFM) coupling between the three unpaired electrons of the V^{2+} ion in $3d(t_{2g})$ orbitals with a total spin $S = 3/2$ and the unpaired electrons of the [TCNE] $-$ anion in π^* orbital with $s = 1/2$. [33] V[TCNE]$_x$ has a unique half-semiconductor electronic structure with fully spin-polarized valence and conduction bands. An inverted spin valve effect was observed with maximum of 0.4% MR value at 100 K and the authors proposed that the negative magnetoresistance originates from the spin-dependent tunneling between highly spin-polarized bands split by Coulomb interaction.

Bipolar spin valve devices – On a more versatile application side, a spin polarized organic light emitting diode (spin-OLED) was presented as a prototype bipolar organic spintronics device by the Utah scientists. [34] Similar to ordinary OLEDs, electrically injected positive and negative charge carriers fuse together to form triplet (spin $S = 1$) and singlet ($S = 0$) excitons, where the latter subsequently recombine and produce electroluminescence emission. However, in a spin-OLED device spin polarized charge carriers alter the generated singlet to triplet exciton ratio, and this, in turn, changes the singlet exciton electroluminescence intensity. This effect can, in principle, be used to detect the degree of spin polarization of the injected carriers. Furthermore, in OLEDs where both electro-luminescence and longer wavelength electro-phosphorescence are substantial, a spin-OLED may be used to control the device emission color. Ehrenfreund et al. [34] discussed the mechanisms that dominate the spin relaxation of injected spin polarized carriers and limit their spin diffusion length in unipolar and bipolar devices. A space charge limited current model with

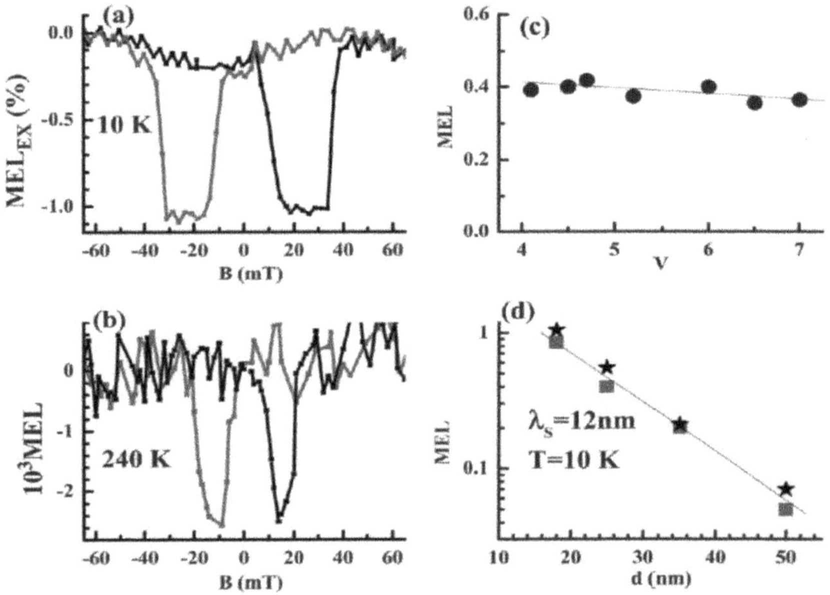

Fig. 7. Magneto-electroluminescence of a D-DOO-PPV spin-OLED device. (a) As obtained $MEL_{EX}(B)$ response for upward (black) and downward (blue) B-sweeps, measured at $V_b = 4.5$ V and T = 10 K, for device thickness of 18 nm and LiF thickness 0.8 nm. (b) The hysteretic MEL response at 240 K. (c) The bias voltage dependence of the maximum MEL_{SV} value measured at 10 K. (d) The maximum MEL_{SV} response of spin-OLED devices at various polymer thicknesses and the LiF buffer layer thickness, 0.8 nm (blue symbols) and 1.5 nm (black symbols), respectively, measured at 10 K and V_b= 4.5 V. Reproduced from Ref. 34 with permission of The Royal Society of Chemistry.

$$MC_{SV} = \frac{2p_1p_2e^{-d/\lambda}}{1+p_1p_2e^{-d/\lambda}} \quad (5)$$

FM electrodes is used to describe the bipolar organic spin valve operation. For unipolar spin valves, finite spin diffusion length was taken into account to obtain a general expression that is valid for devices with thick organic interlayers. For bipolar spin valves, where the injected spin polarized electrons and holes selectively form polaron pairs having singlet or triplet spin configurations, modified expressions for the magneto-resistance (or magneto-conductance, *MC*) and magneto-electroluminescence (*MEL*) was obtained (Equation 5 and 6 respectively) that show that they are proportional to the difference in singlet–triplet

$$MEL_{SV} = \frac{2p_1p_2\Delta b/2b}{1+ p_1p_2\Delta b/2b} \qquad (6)$$

generation rates and the polarization degree of spin injection of the FM electrodes. The spin-polarized organic light-emitting diode (spin-OLED) based on a deuterated derivative of poly(phenylene-vinylene) with small hyperfine interaction was reported by Nguyen et al. [35] In the double-injection limit, the device showed ~1% spin valve MEL response, which follows the FM electrode coercive fields and originated from the bipolar spin-polarized space charge–limited current. In stark contrast to the response properties of homopolar spin valve devices, the MEL response in the double-injection device was found to be practically independent of bias voltage, and its temperature dependence follows that of the FM electrode magnetization. These findings provided an interesting pathway for organic displays controlled by external magnetic fields.

The top interface: One of the major practical challenges of the organic spintronic devices being the ill-defined interface between the organic spacer layer and the top electrode where the heavy magnetic metals like Co, Fe, Ni etc. can penetrate through the soft molecules making either a shorted device or a mixture of a clean spacer layer and a layer with metal pin-holes. [36] This layer with metal inclusion can act as additional source of scattering for spin polarized carriers. This leads to modification of the Julliers's formula where the effect of the ill-defined layer also needs to be taken into account (as shown in equation 7, where $d\text{-}d_0$ is the thickness of the clean organic layer).

$$\frac{\Delta R}{R} = \frac{R_{AP} - R_P}{R_{AP}} = \frac{2p_1p_2e^{-(d-d0)/\lambda s}}{1+ p_1p_2e^{-(d-d0)/\lambda s}} \qquad (7)$$

Experimentally the effect of Co inclusion on the spin transport has been demonstrated in many works.[36, 37] In 2008, Vinzelberg et al. [37] presented a detailed study of irreproducibility of device resistance and magnetoresistance in organic spin valves. This paper studied the GMR effects in LSMO-based junctions with Alq_3-spacer and FM top layers where device resistance showed a broad resistance interval from 50 Ω to MΩ range without systematic dependence on spacer layer thickness and device area. However, spin-valve effects at 4.2 K have been observed in all of them. In some samples, the MR changed sign as a function of the bias voltage. The observed similarity in the bias voltages dependences of the MR in comparison with conventional magnetic tunnel junctions with oxide barriers suggests a description of the found effects within the classical tunneling concept. The proposed model implied the realization of the transport via local Co chains embedded in the Alq_3 layer and spin dependent tunneling over barriers at the interface Co grains/Alq_3/LSMO.

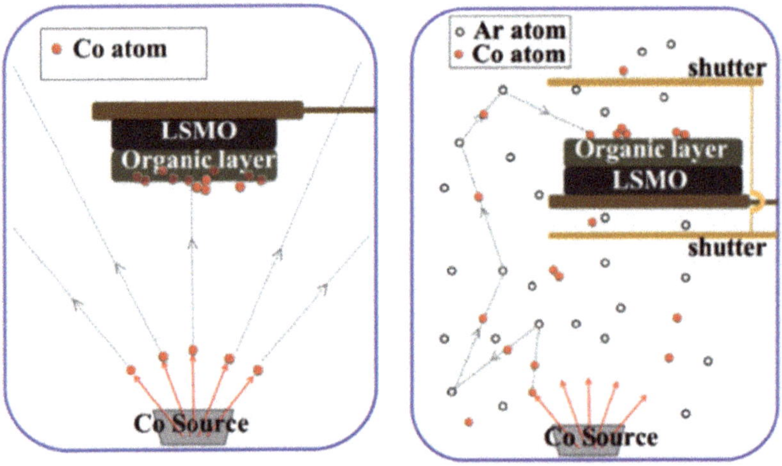

Fig 8. Schematic diagram of (left) direct and (right) indirect deposition. Reprinted from Ref. 39 with permission from Institute of Physics.

The existence of conducting Co chains within the organics is supported by transmission electron microscopic/electron energy loss spectroscopic studies on cross-sectional samples from analogous layer stacks. Also Pang et al. [38] reported a positive spin valve effect in spin valves with the similar half-metal perovskite manganites LSMO as the bottom electrode, Co as the top electrode, and Alq_3 as the spacer. Several groups tried to improve the top interface by inserting an additional Aluminium oxide (AlO_x) layer that helped in achieving the improved spin valve performance. Dediu et al. [16] showed room-temperature operation for organic spin valve devices with AlO_x or Lithium Fluoride (LiF) tunnel barrier layer between the Alq_3 spacer and the top electrode. Through control and engineering of the interfaces between organics and the spin polarized electrodes by insertion of AlO_x and LiF layers, a 0.15% MR was obtained at room temperature from the LSMO/Alq_3/Co spin valves and the authors concluded that the organic semiconductor does not pose any limitation for spin transport at least up to room temperature and the improvement in device performance achieved by the introduction of tunnel barriers in organic spin valves could pave the way for future development of such devices.

However, later results confirmed that the interface spin polarization at the FM/organic interface is very important for any spin injection devices and any additional layer between the organic and the ferromagnet can modifying the interfacial magnetism and spin polarization properties. This issue is discussed in the following section of the article. Also another new technique of indirect evaporation method (as schematically shown in Fig. 8) for depositing the top FM electrode without much penetration was used by Chen et al. [39] with encouraging results. Unlike the regular direct deposition method, in which the deposition process is accomplished in a high vacuum chamber, the vacuum chamber is intentionally filled with inert Ar gas to lower the vacuum level to about 3×10^{-3} Torr. The vaporized atoms collided with the inert gas to release their kinetic energy. To ensure the vaporized atoms be scattered by the Ar atoms several times, the organic layer was covered by a shadow mask that faced away from the electron beam evaporation source. A shutter above the sample was put to let the Co atoms scattered more times. Eventually, the scattered atoms softly landed on the organic surface layer and the metal penetration was reduced. The deposition rate was calibrated to be about 0.004 nm/s, which is ~ 25 times lower than that without Ar gas with the sample facing the evaporation source and only a fraction of the evaporated metal atoms are deposited on the organic surface. Devices prepared in this way showed sharper switching and more reproducible performance which is a promising step forward for organic spin injection devices with vertical device geometry.

Sign of magnetoresistance: Since the birth of organic spintronics, the debate over the sign of the observed MR consistently remained. This puzzling issue has its origin in the fact that the device structure with same materials combination, namely LSMO/Alq_3/Co, can produce either negative [14,16, 40] or positive [41] MR that is incompatible with the Julliere's formula, unless it is assumed that the sign of the spin-polarization of the injected electrons at one of the two interfaces may change from device to device. It has been reported recently in several articles (discussed in detail in the following sections) that chemical bonding between the magnetic electrode and the organics can modify the interface in the following ways. The spin-polarization (p_1 and p_2) in the Julliere's formula for effective tunneling corresponds to the electrodes' density of states (DOS) of spin polarized carriers. However the chemical interaction between the molecular orbitals of the organics and the magnetic surfaces of the electrodes forms hybrid surface states and this hybridization at the Fermi level, can alter the spin- polarization of the tunnel electrons. In 2010, Sanvito [42] has theoretically shown that the DOS of the electrodes and that of the organic molecules in isolation (where it is assumed that only the HOMO is responsible for the conduction) shows no resonance at the Fermi Energy (E_F) of the electrodes and any molecule states. In this situation, the transport is tunnel-like and the spin-polarization of the tunnel carriers, are same as that of the electrodes. However, when the two materials are brought in contact, chemical interaction takes place leading to new features like broadening of molecular DOS that is spin selective. This broadening depends on the interaction strength of a particular molecular orbital with the extended electronic states of the metal. Since the Fermi surfaces of majority and minority spins of FM metals are different and also the interaction with the molecular orbitals may be different, the minority consequently the broadening is larger for that spin direction. Such a spin-dependent level broadening brings only the minority component of the HOMO at the electrodes' E_F, so that polarization will change sign with respect to the situation. Another effect could be a shift of the molecule spin-band interacts more strongly and DOS with respect to the electrodes' E_F that can also be spin dependent leading to a new spin-polarized molecular orbital at E_F. Now the largest DOS at the Fermi level for the hybrid state has majority character and the sign of polarization changes again. If polarization is determined by a hybrid state, the strength of the interaction between the molecule and the electrodes also sets the energy scale of the MR. The prospect of changing the sign of MR of an organic device by engineering the molecular bonding is exciting and is capable of paving new ways for devices with tailored properties. This opportunity has given birth to the new field of "Spinterface" science, described in detail later.

Recently Ciudad et al. [43] showed that the sign of the MR in organic spin valves with intercalated LiF layers can be tuned by simply changing the order of deposition of the different layers. It was shown that depositing LiF on a FM electrode changes its chemical and magnetic nature. An AFM layer is formed at the interface which allows inversion of the sign of the magneto-resistance of organic and inorganic spintronic devices. OSVs were produced with bathophenantroline (BPhen –4,7-diphenyl-1,10-phenanthroline – CHN) as the organic layer and cobalt and $Ni_{80}Fe_{20}$ electrodes. Choice of BPhen was motivated due to close matching of its LUMO energy with the Fermi energy of typical transition metals doped with LiF. [43] The authors showed that for two different spin valve structures (from bottom to top), namely sample A Co(8)/Al_2O_3(1)/BPhen(10)/LiF(0.9)/ $Ni_{80}Fe_{20}$(12) and sample B Co(8)/LiF(0.9)/BPhen(10)/Al_2O_3 (1)/$Ni_{80}Fe_{20}$(12), differing only in the position of the LiF, placed either above or below the organic layer, different MR signs can be observed. Sample A exhibited a positive MR (2.8%), while sample B showed negative MR (−0.7%). The different signs of the MR in the two structures suggest that the LiF layer affects the magnetic polarization of the FM electrodes, since otherwise samples A and B would show a similar sign in the magnetoresistance response. It was also shown that reactive interlayers can not only control the spin transport properties in devices but could also be used to determine the Néel temperature of ultra-thin AFM films with thicknesses within

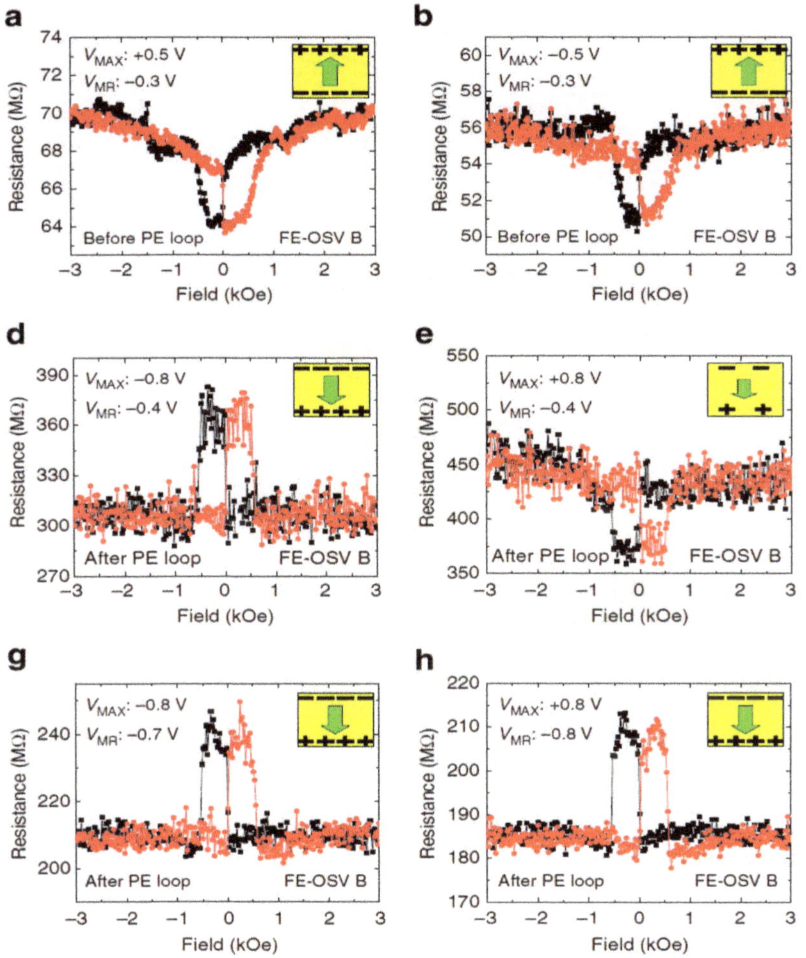

Fig.9. Reversal of MR sign at T = 11 K. (a,b) MR scans for the 'as-grown' state of the FE-OSV device B acquired at V_{MR} = - 0.3 V/V_{MAX} = + 0.5 V and V_{MR} = - 0.3 V/V_{MAX} = - 0.5 V, respectively. (d,e,g,h) MR scans taken when the polarization of the PZT is reversed. The red (blue) circles indicate the V_{MR} values at which the MR loops (a,b,d,e,g,h) are taken with positive (negative) V_{MAX} Reprinted from Ref. 44 by permission from Macmillan Publishers Ltd.

1–2 nm. This measurement could be achieved by inserting a thin AFM film on the bottom magnetic electrode, just before the tunnel barrier and the Néel temperature should mark the transition from no MR to negative MR when decreasing the temperature. This result reconfirmed the importance of an accurate and complete characterization of the chemical nature and the properties

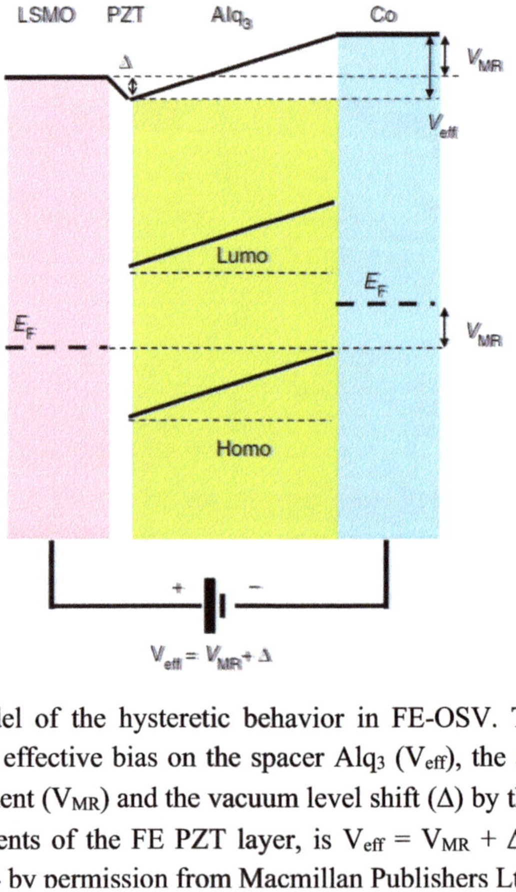

Fig.10. Model of the hysteretic behavior in FE-OSV. The relation between the effective bias on the spacer Alq_3 (V_{eff}), the applied bias at measurement (V_{MR}) and the vacuum level shift (Δ) by the remanent dipole moments of the FE PZT layer, is $V_{eff} = V_{MR} + \Delta$. Reprinted from Ref. 44 by permission from Macmillan Publishers Ltd.

of the electrodes used in spintronics, and particularly in organic spintronics and spinterface-based devices. The authors concluded that in these devices, it should always be shown whether the magneto-transport properties are arising from a chemical change, or from hybridization of molecules or energy-level depletions and/or vacuum level changes, particularly when inversions of the MR are measured. Chemical reactions produced by the molecules themselves could also lead to inversion of the MR.

Another new approach for controlling the sign of magnctoresistance was investigated by D. Sun et al.[45] The authors studied the tuning of the resistance of organic spin valves by adding a thin ferroelectric interfacial layer between the FM electrode and the organic spacer. It was found that the magnetoresistance of the spin valve depends strongly on the history of the bias voltage, which is correlated with the polarization of the ferroelectric layer; the magnetoresistance even changes sign when the electric polarization of the ferroelectric layer is reversed (Fig. 9). This finding is significant and could enable active control of resistance using both electric and magnetic fields, opening up possibility for multi-state organic spin valves. A schematic model is illustrated in Fig. 10 to explain this hysteretic behavior of MR in the FE-OSV. When a V_{MAX} is applied to the junction, the polarization of the PZT gets modified. Owing to the dipole moment created by the electric polarization of PZT, the effective voltage V_{eff} applied on the Alq_3 layer equals $V_{MR}+\Delta$, where Δ is the vacuum level shift caused by the remnant dipole moment of PZT. When a positive (negative) V_{MAX} is applied, Δ becomes larger (smaller), which explains why the MR (V_{MR}) profile can be shifted by V_{MAX}.

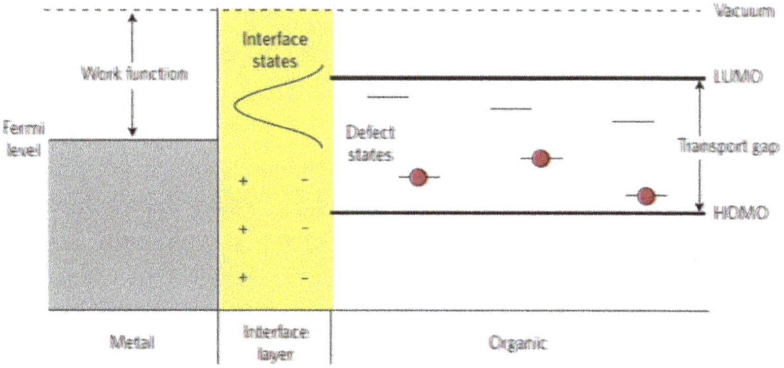

Fig. 11. Schematic representation of the interface between a metal and an organic layer. Dipoles are induced at the interface, which modify the work function of the metal. There are also interface states created by chemical reaction, and defect/impurity states distributed across the HOMO–LUMO gap. Reprinted from Ref. 45 by permission from Macmillan Publishers Ltd.

To summarize the above discussion it can be said that despite different efforts with engineered interfaces and materials, the observed GMR effect at room temperature is still quite modest from organic spintronic devices. Model calculations predicted that the incoherent hopping transport of spin polarized carriers through thick organic spacer layers can cause significant spin loss despite their low spin-orbit and hyperfine coupling strengths. Szulczewski et al. [45] have predicted the major reason for loss of spin information inside organic semiconductors in the following way. The traps due to impurities and defects can accommodate up to two electrons with opposite spin, according to the Pauli principle, but their energy levels are scattered across the HOMO–LUMO gap (Fig. 11). An electron located in an unoccupied trap can hop over to another empty trap retaining its spin direction, however, if a spin up electron hops into a trap already occupied by a spin-down one, the electron that hops out can have either spin direction, and the polarization is destroyed provided there is an effective spin-relaxation mechanism. This can be seen as the localized version of 'spin–spin scattering' in band semiconductors. As the defect and trap states inside the organic layer is plenty, effective spin relaxation mechanism is substantial in organic spin devices, however, it can be controlled by optimizing the vacuum conditions, growth rates, film morphology, surface roughness and density of impurities influencing the performance of organic-based magnetic devices.

By combining experiments with model calculations, Schoonuss et al. [46] have shown that intrinsic loss of magnetoresistance occurs at the onset of multiple-step tunneling from direct tunneling regime. In the regime of multiple-step tunneling, under the condition of low hopping rates, spin precession in the presence of hyperfine coupling is conjectured to be the relevant source of spin relaxation. The authors shown that from the direct tunneling to two-step tunneling, the MR reduces at least by a factor of 4, which matches well with the maximum experimentally found MR. The factor 4 was explained by considering that tunneling via an intermediate site results in an intrinsic loss of MR by a factor 2. Additionally, there are factors like inequal transmission through two barriers Al_2O_3 (1.5 nm) and Alq_3 in their device structure, a strong hybridization at the Alq_3/Co interface that results in more interface sites, can force tunneling via sites with unequal transmission of both barriers, permitting more than two tunnel steps, which may lead to a further decrease of MR. Also lowering of the spin polarization of Co adjacent to Alq_3, and spin precession due to hyperfine coupling could contribute in the loss of MR response in thick devices. Therefore, obtaining significant GMR response at room temperature from organic spintronic devices with thick spacers remained a distant dream.

2.2. *Organic magnetic tunnel junctions*

As in many thicker devices the magnetoresistance response decays fast with increasing temperature, one best possible way to achieve significant room temperature response could be the organic magnetic tunnel junctions although achieving an MTJ without short is quite challenging in organic devices due to the top electrode penetration problem. First magnetic tunnel junction using OSC Alq_3 was reported by Santos et al. [47] where a 6% positive TMR response was observed at room temperature. Subsequently other OSC small molecules, like rubrene [48, 49] was used as the tunnel barriers in organic magnetic tunnel junctions. Using the Meservey-Tedrow technique [50] Santos et al. directly determined the spin polarization of Co, Fe, and Permalloy (Py) electrodes through the Alq_3 barrier junctions with an Al counter electrode. For the Co electrode and Al_2O_3/Alq_3 barrier, a P value of 27% was determined. Similarly, P values of 30% for Fe and 38% for Py were determined. This measurement demonstrated polarization of the tunnel current from a FM through an organic semiconductor. However, the value of P was down to 6% for the junction without Al_2O_3 at the Al/Alq_3 interface. For both junction structures, with and without Al_2O_3 at the

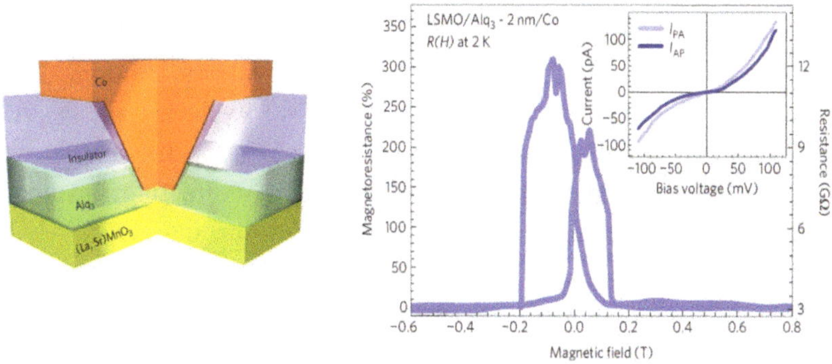

Fig 12. (Left) The device schematics of a $La_{0.7}Sr_{0.3}MnO_3/Alq_3$ /Co nanometric-size MTJ. The nanoindent in the Alq_3 layer is realized by a CT-AFM, allowing the control of the organic tunnel barrier thickness. This nanohole is then filled with cobalt, leading to the nanometric-size MTJ. (Right) Magnetoresistance curve of the organic MTJ obtained at 2 K and −5 mV. The lower coercive field corresponds to the LSMO magnetic reversal and the higher coercive field to the Co magnetic switching. Inset: *I(V)* curves recorded at 2 K in the parallel (I_{PA}) and antiparallel (I_{AP}) magnetic configurations. Reprinted from Ref. 41 by permission from Macmillan Publishers Ltd.

Al/Alq_3 interface, a positive P for Co, Fe, and Py electrodes were measured, which justified the positive TMR values. However, this is in contrast to the inverse GMR observed by Xiong et al. [14] and others [16, 40] in $LSMO/Alq_3/Co$ spin valves. The authors mentioned that tunneling behavior observed in this study is similar to that of both amorphous Al_2O_3 and amorphous $SrTiO_3$ tunnel barriers, [51] in which positively polarized, itinerant *sp* electrons dominate spin transport.[52]
In 2010, Barraud et al. [41] reported a nanosized magnetic tunnel junction using the AFM tip that exhibited a positive TMR response of up to 300% at 2K (Fig. 12). The nanosized junctions were realized using contact atomic force microscopy tips avoiding defect prone areas of the bottom LSMO and the Alq_3 layer. The MTJs obtained in this way showed 65% of working devices, among which 20% showed measurable magnetoresistance from 10 to 300%. The authors also proposed a model for explaining the sign inversion of magnetoresistance response called the spin-hybridization-induced polarized states (SHIPS) model. SHIPS model took into account the spin injection into molecules and all of the apparent discrepancies on the observed magnetoresistance signs in $LSMO/Alq_3/Co$ spin valves, discussed previously. Earlier negative magnetoresistance sign in large-area junctions were explained using the Jullière model with LSMO as a spin analyzer with $P_{LSMO}/Alq_3 * > 0$ and $P_{Co}/Alq_3 * < 0$ at the Co/Alq_3 interface. However, SHIPS model proposed that, oppositely, a SHIPS inversion occurs in large-area samples at the bottom $LSMO/Alq_3$ interface, leading to an effective $P_{LSMO}/Alq_3 * <0$ while having $P_{Co}/Alq_3 * >0$. The $P_{Co}/Alq_3 * > 0$ sign, suggesting a weak coupling in the model, which was supported by the experimental results of $P_{Co}/Alq_3 > 0$ directly measured [48] by the Meservey–Tedrow technique. Also insertion of an inorganic Al_2O_3 spacer between Alq_3 and Co in a $LSMO/Alq_3/Co$ spin valve did not change the sign of the MR [16] whereas an inversion would have been expected because of the commonly reported $P_{Co}/Al_2O_3 > 0$. The $P_{LSMO}/Alq_3 * < 0$ sign suggested a large enough intermediate coupling in the model in addition to disorder contribution. With the disorder contribution being distributed over space and energy, it was predicted that the inversion should not happen for all of the interfacial molecular states but only locally for those brought close enough to the electrode's Fermi level that are also dominating the injection step. The local character of this SHIPS-induced bottom-interface inversion explained why it is not observed in the spin-polarized scanning-tunneling-microscopy-like experiment where a single outcome of the energetic disorder distribution is probed that mainly remain off resonance. This model supported the results of Vinzelberg et al.,[37] where negative MR on millimetre-wide $LSMO/Alq_3/Co$ spin valves turned to positive with bias voltages for a few samples. This MR inversion as a function of bias is

ascribed to a signature of spin-dependent resonant tunneling through localized states. For a single merged donor– acceptor state in resonance with the Fermi levels, the spin-dependent Breit–Wigner formula is used. The proposed SHIPS mechanism suggested that through modification of metal/molecule coupling, it is possible to tailor the properties of spintronics devices ultimately controlling the spin injection sign and amplitude by an applied voltage. Use of self-assembled monolayers that could be grafted on any surface and where one could easily play with the anchoring groups and the backbone of the molecules, is suggested as one particularly interesting way that could bring new properties beyond the now recognized low-cost, flexibility and long spin lifetime interest, chemistry hardly available in conventional inorganic spintronics.

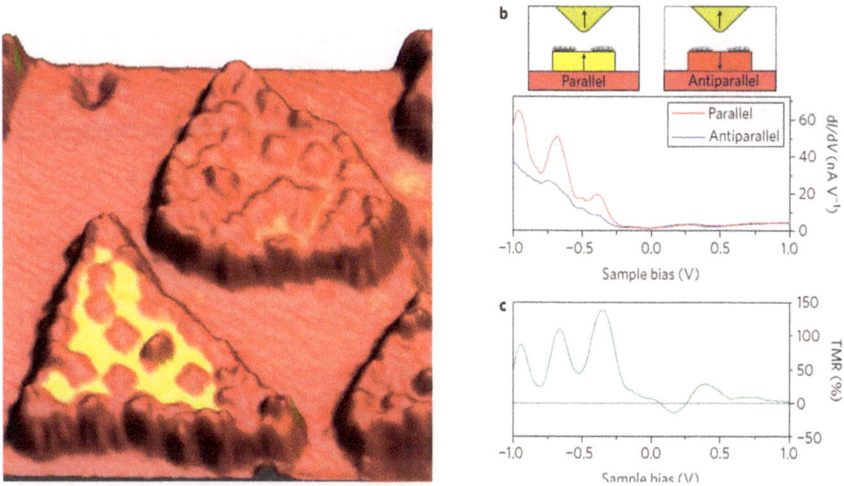

Fig 13. (Left) H_2Pc molecules adsorbed on cobalt islands with different out-of-plane magnetic orientations. a, Topographic image of H_2Pc molecules adsorbed onto two cobalt islands on the Cu(111) surface. Colour code: measured dI/dV at -310 mV. The two island species can be distinguished by the magnetization parallel (in yellow) and antiparallel (in red) to the tip magnetization. b, Typical dI/dV spectra taken on parallel and antiparallel oriented cobalt islands (marked by red and blue crosses in a) clearly reveal spin-polarized density of states below the Fermi edge. c, Energy dependence of the optimistic TMR ratio calculated from the dI/dV spectra. The highest value is measured at approximately -350 meV, and is used to distinguish between the magnetic orientation of the islands. Reprinted from Ref. 53 by permission from Macmillan Publishers Ltd.

Almost at the same time Yoo et al.[49] reported first organic magnetic semiconductor $V(TCNE)_x$-based MTJs with the structure LSMO/$LaAlO_3$/rubrene/$V(TCNE)_x$, In this report $V(TCNE)_x$ was used as spin polarized top electrode and LSMO as spin analyzer in a standard MTJ device geometry. The results demonstrated the spin-polarizing nature of the organic-based magnetic semiconductor, $V(TCNE)_x$ (x ~ 2; Tc ~ 400 K), and its function as a spin injector/detector in hybrid magnetic multilayer devices. The TMR response showed close to 2% value at 90 K with strong temperature and bias dependence and the device performance being limited by the high device resistance at low temperature, especially due to increased resistance of the semiconductor electron spin polarizer.

Towards Single Molecular Devices: In recent times spin-polarized transport through a single molecule is being seen as the future of spintronics because it combines the goals of downward size scaling of electronic components (to lower

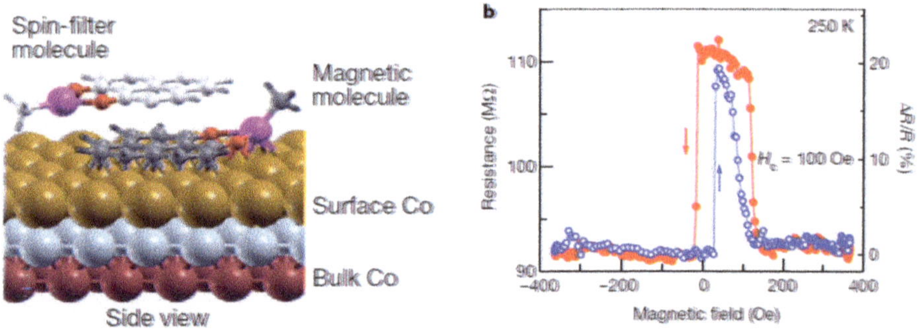

Fig 14. (Left) Ab initio model: side view of the relaxed molecules' configuration on a Co(111) surface; grey, carbon; red, oxygen; purple, zinc. The first-layer ZMP molecule ('magnetic molecule') absorbs flat, with six α-carbons and the central carbon of the phenalenyl sitting directly on top of Co atoms, while the β-carbons of the phenalenyl sit in a bridge position between Co atoms. The molecular plane is ~2.1–2.2 Å above the Co(111) surface (the C atoms are ~2.1 Å above, the O and Zn atoms are ~2.2 Å above). The second molecule ('spin filter molecule') sits in a staggered configuration over the first adsorbed molecule with a spacing of ~3.26 Å, forming a molecular π-dimer. (Right) Interface switching response from a Co (8 nm)/ZMP (35 nm)/Py (12 nm) device. at 250 K, showing large IMR of 22% with reduction in switching field (~100 Oe). Reprinted from Ref. 54 by permission from Macmillan Publishers Ltd.

power consumption while increasing speed and density) and increasing computational bandwidth by manipulating spin in addition to charge. Since the recent promising results from highly spin polarized interfaces between a ferromagnet and the organic molecules, the idea of extreme downsizing of spin devices to single-molecular ones have gathered momentum. One such promising result is reported by Schmaus et al. [53] across a single, nonmagnetic hydrogen phthalocyanine molecule (H_2Pc) with chemical formula $C_{32}H_{18}N_8$ contacted by the FM tip, giant magnetoresistance was measured by using spin polarized scanning tunneling microscope (SP-STM) (Fig. 13). This study identified that, due to charge transfer towards the molecule and the hybridization of molecular orbitals near the Fermi level with bulk electronic states of the electrodes, transport across the molecule is nearly resonant in the minority channel with an

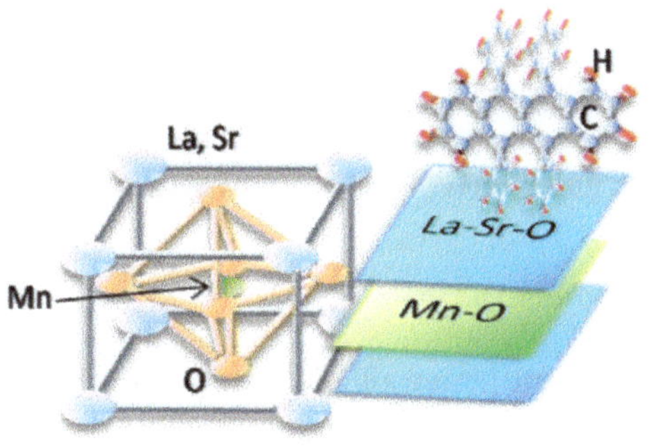

Fig 15. Schematic interpretation of the interface electronic structure between LSMO and rubrene molecules. Reprinted from Ref. 76 copyright (2015) American Chemical Society.

ultralow areal resistance product. It was shown that the molecules can sustain a large current density with substantial spin-polarization, thus validating the concept of nanoscale spintronic devices. Indeed, despite the absence of transition metal sites on the molecule that could contribute to spin filtering, a high differential GMR value of ~60% across the molecular junction was observed, with an areal resistance product of only 70 $\mu\Omega\ \mu m^2$ and conductance of $0.26G_0$, G_0 being the quantum of conductance. Ab initio theoretical analysis showed that this effect arises from a strong, generic, spin-dependent hybridization mechanism that can also dominate the magnetoresistance in the TMR regime.

Raman et al.[54] have recently shown that delocalized carbon-based radical species with unpaired spin, such as phenalenyl could be very promising for the construction of quantum memory registers. These phenalenyl moieties, which can be regarded as graphene fragments, are formed by the fusion of three benzene rings and belong to the class of open-shell systems. The spin structure of these molecules responds to external stimuli (such as light, and electric and magnetic fields), which provides novel schemes for performing spin memory and logic operations. A molecular device using such molecules as templates to engineer interfacial spin transfer resulting from hybridization and magnetic exchange interaction with the surface of a ferromagnet was fabricated and the device showed an interfacial magnetoresistance of more than 20% near room temperature (Fig. 14). Moreover, the formation of a nanoscale magnetic molecule with a well-defined magnetic hysteresis on FM surfaces was demonstrated. This finding suggests the use of chemically tunable phenalenyl-based molecules as a viable and scalable platform for building molecular-scale quantum spin memory and processors for room temperature molecular memory devices. Furthermore, it is predicted that the spin structure of the phenalenyl ligands can take the role of bus qubits in tailor-made molecular networks for realizing future molecular quantum information processing systems.

3. Ferromagnet – organic interface

3.1.1. *Interface spin polarization*

Interfaces between different materials and many unexpected and technologically important phenomenon such as reconstruction of charge, spin and orbital states occurring at those interfaces attracted attention of the scientists to study these interfaces in detail. Recent years saw a particular surge of interest on the organic-FM interface where a strongly spin-polarized interface termed as "spinterface" was experimentally observed. [55, 56] It was found out that adsorption of organic materials on a magnetic surface leads to bonding-specific electronic coupling at the interface affecting the magnitude and sign of the spin polarization. Due to the adsorption of the molecules, a chemical reaction takes place between the ferromagnet and the organic layer forming a hetero-interface. This hetero-interface state can have features that are completely different from the individual thin film surfaces and reports so far predicted promising effects occurring at such interfaces. Therefore, for improved spintronic components, it is important to investigate material properties at these hetero-interfaces. This section of the chapter will focus on the recent research progress of such interfaces. Since this type of interface not only determines the charge injection/ extraction but also the spin injection/detection, it critically affects the performance of organic spintronic devices. A series of reports using different techniques that can measure spin information are reviewed here, including X-ray magnetic circular dichroism (XMCD), spin-polarized photoemission spectroscopy (SPPES), spin-polarized metastable de-excitation spectroscopy (SPMDS), spin-polarized scanning tunneling microscope (SPSTM), two-photon photoemission etc.

Energy Level Alignment of FM Electrode/OSC: The energy level alignment of electrode/ OSC interfaces play a major role in organic electronic devices. The injection barrier due to the measured energy level alignment makes it possible to predict which interface is suitable for injecting electrons or holes into an electronic device. However, the energy levels at an interface often do not align with vacuum level, creating confusion regarding the device performance. Therefore in order to clarify the device properties in a comprehensive way, it is important to have a thorough knowledge of the interfaces. Based on the device fabrication process, there could be direct or indirect contact formations between the ferromagnet and the organic layers, as discussed in detail by Zhan et al. [57] Indirect contacts are formed between the OSC and the FM electrode when the interface formation is carried out under ambient conditions causing a native oxide and/or hydrocarbon surface layer on the FMs or when a tunnel barrier is intentionally inserted to separate the organic from the FM layer. In organic spin valve applications, typically Al_2O_3 or LiF are used as "engineered" barrier layers. These barrier layers (native or engineered) isolate the FM metal surface from the organic layer preventing any chemical interaction and therefore no chemical bonds can form between the OSC and the FM. This prevents any possible hybridization between the p-orbitals due to interaction with the FM continuum of states. Hence, spin polarization of π-conjugated orbitals of the OSC do not interact with the spin-polarized *d*-bands of the FM, at least not through direct overlap between wave functions across the interface such as interlayer exchange coupling. Due to indirect contacts, the work function of the FM electrode gets strongly modified due to the push-back effect [58] even when barrier layer is an insulator and has no intrinsic dipole (e.g., hydrocarbon contamination). Hence the work function values of atomically clean FM surfaces should not be used to estimate the energy level alignment. When the barrier layer (natural or engineered) is within tunneling limit, charge carriers can tunnel from the FM into the OSC (or vice versa) and the energy level alignment at the interface depend on the position of the Fermi level of the (modified) FM surface. Also it depends on the energies that correspond to the ionization potential and electron affinity (hole and electron polaron formation energies) of the OSC molecule at the interface that takes into account the intermolecular and intramolecular ordering of the molecules at the interface together with the electrostatic interaction with the substrate. The interface properties determine the energy needed to create a hole (or electron) in a particular molecular orbital of an

OSC molecule screening from its environment. The surface energy mismatch between the (modified) FM electrode and OSC interface layer can introduce an intermolecular order very different from that in the bulk of the OSC film. Depending on the local intermolecular order, the energies corresponding to ionization potential can vary even up to 1 eV. However, large variations in local intermolecular order might lead to a wide distribution of polaron energies that can be manifested in the injection and multistep hopping process, thus enabling low bias injection and transport even if the polaron energies of the majority intermolecular order distribution are far from the Fermi level of the spin valve device. In a direct contact, chemical interaction between metal surface and some specific atoms from organic molecule can lead to the reconstruction of molecular structure through covalent bonding to the surface and thus creation of hybrid states at the interfaces. A few experimental methods and results will be discussed in this section of the chapter.

Here it is important to point out that although the terminology of "spinterface" appeared recently, the high spin polarization at a FM/organic interface was reported already in 2002 using the spin polarized metastable de-excitation spectroscopy (SPMDS) studies. [59, 60] In these works, Suzuki et al. showed the electronic structure and spin asymmetry of metal phthalocyanines (Pcs) (metal) Mn, Fe, Cu, and Mg) and metal-free Pc thin films of 1-monolayer (ML) thickness on an Fe (100) substrate. The spin asymmetry of these interfaces were found positive on Fe (100). In SPMDS, positive (negative) spin asymmetry indicates negative (positive) spin polarization. Therefore, the spin polarization of both metal and metal-free phthalocyanines was found to be antiparallel to Fe (100) substrate. The surface density-of-states (SDOS) of the Pcs observed in SPMDS was almost identical, and thus it was concluded that, the Pc electronic structure is insensitive to the variation of the central metal atoms. On the other hand, the spin asymmetry in the SPMDS measurements exhibited noticeable differences among Pcs. These results were discussed from the viewpoint of the Pcs electronic structure. The fact that differences are found in the spin asymmetry but not in SDOS among Pcs indicated an important role of the central metal atoms for spin polarization. From a structural analysis using low-energy ion scattering spectroscopy, it was shown that Cu-phthalocyanine (CuPc) molecules adsorb on the Fe (100) surface with their planes parallel to the surface when the film thickness is one monolayer, while the structure of the film becomes disordered with an increase in the thickness. This thickness dependent structure of the CuPc film agrees with that determined from the SPMDS spectra.

In recent times, Djeghloul et al. [61] reported that phthalocyanine molecules on a Co single crystal surface clearly showed a highly efficient spinterface. Spin-polarized direct and inverse photoemission (SPPES) experiments were performed that reveal a high degree of spin polarization at room temperature at this interface. A magnetic moment on the molecule's nitrogen *p* orbitals was measured, which substantiates an ab-initio theoretical description of highly spin-polarized charge conduction across the interface due to differing spinterface formation mechanisms in each spin channel. Through this example, a recipe to engineer simple organic-inorganic interfaces with remarkable spintronic properties was proposed that can endure well above room temperature. Using spin-resolved photoelectron spectroscopy and ab initio calculations Lach et al.[62] studied possible injection channels in metal phthalocyanines (MPc). They reported selective hybrid interface states at the Fermi level of a FM-OSC hybrid junction only by changing the central metal atom of an MPc.

Gruber et al. [63] explicitly demonstrated how the magnetization of a FM layer (Co) promotes and controls, at room temperature, the magnetization of a molecular layer (MnPc) across a noble metal layer (Cu). The spin-polarization of quantum well states within the noble metal layer due to the underlying FM layer promotes magnetic moments on the molecular sites through adsorption-induced electronic hybridization. Concurrently, the resulting interlayer exchange coupling field aligns these molecular magnetic moments at room temperature. It is also shown that the sign and amplitude of spin-polarization is controllable by

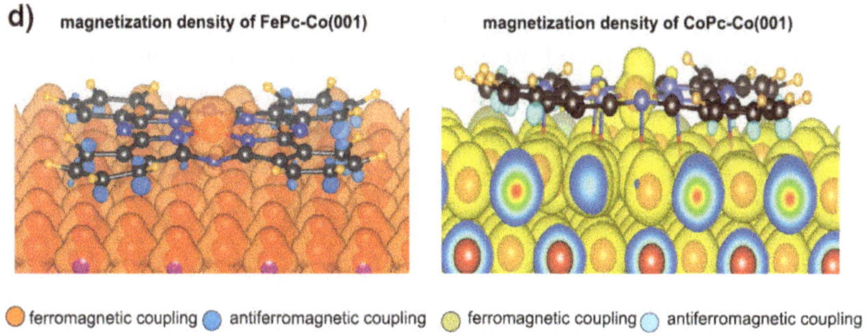

Fig 16. The magnetic coupling of FePc and CoPc to Co is shown on the left figure where the magnetization density is dominated by the iron and cobalt central atom at the molecular side in an out-of-plane geometry. Note the spin-polarization present on carbon and nitrogen atoms of the phenyl rings, which is created through interaction with the spin-polarized Co-d states of the substrate. Reprinted with permission from Ref. 62, copyright (2011) John Wiley & Sons.

nanoscale engineering of the spacer thickness. Furthermore, it is interesting to note that compared to a noble metal layer, when the FM surface was protected by a graphene layer, there was much lower magnetic coupling, and thus a less robust spin-polarized interface.

Steil et al. [64] showed via spin-resolved real-time pump–probe two-photon photoemission experiment (2PPE) that the prototypical interface between cobalt and the metalorganic complex Alq_3 can exhibit highly efficient spin-filtering properties. Femtosecond spin dynamics at the Co/Alq_3 interface was studied. To disentangle the microscopic origin of spin filtering, an optically generated transient spin polarization was created in a well-defined hybrid interface state that was followed with 2PPE measurements. It was found that the electrons are trapped at the interface in a spin-dependent manner for a surprisingly long time of the order of 0.5–1 ps. Therefore the authors concluded that FM metal/organic interfaces act as spin filters because electrons are trapped in hybrid interface states by spin-dependent confining potentials.

Another direct evidence for the spin polarization of Alq_3 sub-monolayers on Fe surfaces was reported by N K-edge X-ray magnetic circular dichroism (XMCD) measurement. [65] The authors concluded that hybridization and exchange coupling between π-conjugated orbitals in Alq_3 and the Fe 3*d* of the substrate was evident from the measurements. More recent publications including C K-edge XMCD measurements on C_{60}/Fe interfaces showed the mixture of C_{60} π/π^* orbitals at the FM metal/organic interfaces.[66] Also there are a few reports on the spin polarized ultraviolet photoemission spectroscopy (SPUPS) results, focusing mainly on metal phthalocyanines/FM metal interfaces.[67, 68] SPUPS measurements directly demonstrate the formation of hybrid interface states, induced by chemisorption of CuPc, CoPc, and FePc at the Co surface. Among these three interfaces, CuPc/Co interface show the strongest spin polarization directly at the Fermi level.

All the "spinterface" results mentioned above discussed mainly the spin polarization properties at the interface between OSCs and FM transition metals. However at the interfaces between OSC and the complex metal oxides such as LSMO or LCMO, even more versatile properties can be expected as the reconstruction of charge, spin and orbital states at the interface of a complex oxide with another oxide or an organic material can pave the way for many fascinating properties absent in the individual materials. In a recent report Grisolia et al. [69] have shown that reconstruction of charges at oxide interfaces can

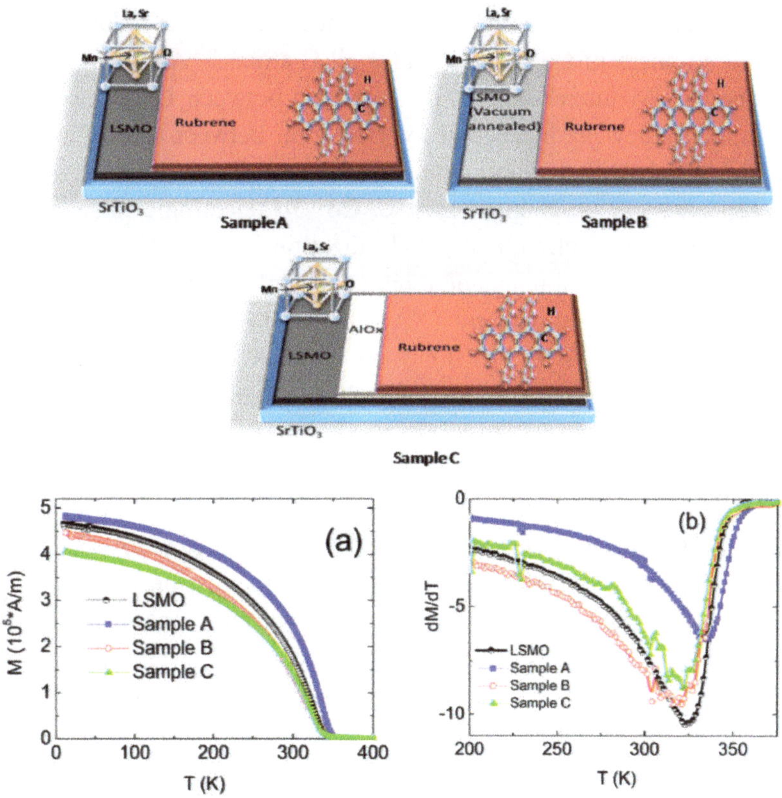

Fig 17. (Top) Schematic diagram of the architectures of the grown multilayers with LSMO crystal structure and chemical structure of the rubrene molecule. (Bottom) Field-cooled (FC) magnetization as a function of temperature (a) together with the first derivative of magnetization (b) for pure LSMO and the LSMO coated with rubrene as grown (sample A), surface treated (sample B), and one with an AlOx layer in between (sample C)). Reprinted with permission from Ref. 76, copyright (2015) American Chemical Society.

be engineered for novel phases at the interface that might be able to enhance the device responses and control them more accurately.

3.1.2. *Interface magnetism*

Previously, it has been shown by XMCD that adsorption of 1 monolayer (ML) of Mn-phthalocyanine (MnPc) molecules decreases the magnetic moment of Co from 1.73 μ_B to 1.67 μ_B (supplementary information of Ref. 61). From on-site local magnetization density maps, it was also concluded that a strong AFM coupling between Co and C benzene sites leads to a reduction of the total magnetic moment of Co by 0.22 μ_B for all C atoms.[61] Although Iacovita et al. [70] have shown FM interactions at the interface between CoPc molecules and Co nano-islands, they predicted that hybridization between Co 3*d* and N 2*p* states results in a reduced magnetic moment of 0.7 μ_B. Several other groups [71, 72] have also reported on a change in the Co 3*d* electronic distribution due to chemisorption and subsequently hybridization with phthalocyanine molecules leading to a suppressed magnetic moment of CoPc on Co or Fe substrates. Using X-ray absorption spectroscopy (XAS) and XMCD, it was shown that upon depositing sub-monolayer (ML) C_{60} on the $Fe_3O_4(001)$ surface, Fe^{3+}/Fe^{2+} ratio reduces due to electron donation by the C_{60} molecules. [73] Also C_{60} molecules on a Fe(001) surface shows hybridization that induces significant magnetic polarization of C_{60} $\pi*$ derived orbitals. [74]

The modified magnetic property at the interface between a FM half metallic oxide LSMO and two OSCs Cobalt Pthalocyanine (CoPc)[75] and rubrene [76] is recently reported. Also detailed X-ray photoemission spectroscopy (XPS) measurements were carried out which indicate chemical reaction between the LSMO and the organic layers resulting in a modified band diagram of the half-metallic oxide. For both the OSCs, increase of FM properties both in terms of increased saturation moment (M_S) and Curie temperatures were observed. The XPS studies show modified core level spectra of Sr and Mn of LSMO surface in contact with rubrene or CoPc which does not return to its normal state even after washing the OSCs away. This is an indicative of chemisorption of the organic molecules on the half-metallic oxide LSMO surface causing interface hybridization between π-orbitals of semiconductors and *d* electrons of ferromagnets. Magnetic and transport characterization of such interfaces indicate that this hybridization clearly improves FM properties of LSMO however at the same time the low field MR of the LSMO also increases. By modifying the interface with annealing treatments and insertion of additional dielectric layer of AlO_x, the interface magnetism can be modified significantly.[76] Also, it can be noted that after removal of rubrene molecules, the improved FM property of LSMO surface still exists although the M_S value decreases compared to the 7 nm rubrene coated LSMO. This indicates that there are two contributions present simultaneously. First, the rubrene molecule is chemically absorbed on the LSMO surface and changes its electronic and magnetic properties permanently. Secondly, in close contact with FM LSMO, a magnetic moment is induced in OSC rubrene molecules due to proximity effect that increases the saturation moment of the LSMO coated with rubrene. After removal of the rubrene layer from the LSMO surface, the real effect at the modified interface can be seen. It is clear that the saturation moment of LSMO is increased by 0.592 μ_B/formula unit (f. u.) because of rubrene coating. It is important to mention that this change in saturation moment was measured for a 20 nm thick LSMO film. Considering that the main change takes place in the top 1 nm layer (~2–3 unit cells) of the LSMO (in contact with rubrene molecules), the change in magnetization in the first 1 nm thickness of the film was calculated to be 11.84 μ_B/f.u., and the improvement in surface magnetization of LSMO is more than 230%.

Similar modified interface magnetic properties between Alq_3 and two different magnetic electrodes NiFe and CoFe has been reported by Morley et al. [77] using XPS, Kelvin Probe and magneto-optic Kerr effect (MOKE) magnetometry. Using MOKE magnetometry, the surface magnetization of the transition metal electrodes was studied. It was found that the surface magnetization of the NiFe electrode changed with the Alq_3 layer on top, which was considered to be due to the Alq_3 chemisorption onto the NiFe electrode that changed the electronic states at the interface. The XPS spectra and MOKE data showed that the Alq_3 strongly bonds to the surface of the transition metal electrode. This bonding was attributed to chemisorptions of the Alq_3 onto the metal surface, as traces of Alq_3 was seen by XPS on the surface after washing. Similar proof of chemisorption was shown previously also. [75, 76] This chemisorption changes the surface magnetization of the electrode, which according to the authors should change the spin polarization of the magnetic electrodes.

However it is important to mention here that whether this increased ferromagnetism at the FM/organic interface can be ascribed directly to increase in spin polarization, is not very clear. Increased magnetic moment can result from the electrons localized at the defect sites at the interface also that might not contribute to the tunneling current. However, recent SPMDS measurements at the LSMO/graphene indicated spin polarization at the interface improves compared to bare LSMO films. [78] Modification of the nature of molecules and the orientations of the molecules can affect this hybridization leading to a promising way of designing spin-polarized interfaces.

4. Major Challenges

Charge and spin transport in OSC: In order to understand the major hurdles and challenges in the field of organic spintronics, we have to carefully analyze different experimental and theoretical data on charge and spin transport in OSC. Performance of spin injection devices depend critically on spin polarized injection at the FM/OSC interface whereas the transport through the OSC layer depends on several issues. The charge transport depends on carrier mobility, defects, traps, grain boundaries etc. whereas for efficient spin polarized transport, additional factors like less spin relaxation within the OSC is expected. One of the major challenges for the OSC is their quite low mobilities restricting the achievement of long spin diffusion length in spite of their longer spin diffusion time. One possible step towards solving this challenge could be to focus more towards materials with better mobility and more single crystalline OSCs with less grain boundary related defects. However, the charge and spin transport can not be correlated very easily as increased density of charge carriers can easily destroy spin polarization of the injected charge carriers.

Already in the previous section of the article it was discussed how a spin polarized carriers can come out with any spin orientation if it hops into a trapped state which is already occupied by one charge carrier. At low temperature, variable range hopping (VRH) model has been employed to explain the charge transport in OSCs. However, at increased temperature, Yoo et al. suggested that VRH conduction could reduce efficient spin-polarized carrier transport.[79] Similar conclusion has been drawn from experiments by Majumdar et al. also. [21]

Although previously it was suggested that hyperfine interaction mainly dominate the spin relaxation in OSCs, subsequent reports pointed out that relaxation due to spin-orbit interaction also plays a major role. [80] Through the measurement of the electric-field dependence of λ_S in an organic spin-valve structure consisting of an Alq_3 spacer layer, together with the data on the temperature dependence of λ_S, a clear contradiction is pointed out that hyperfine interactions is the sole source of spin relaxation. Instead, the Elliott-Yafet (EY) mechanism, arising from spin-orbit interaction, is pointed out to be the more likely the dominant spin relaxing mechanism. It was pointed out that the rapid decrease of λ_S with temperature, reported in literature cannot be consistent with hyperfine interaction because in that mechanism, spin relaxation is caused by the magnetic field of the nuclear spins, which at best can have weak temperature dependence. To correctly identify the dominant spin relaxation mechanism, the electric-field dependence of the spin relaxation length and spin relaxation time in Alq_3 was investigated. It was found that the spin diffusion length λ_S monotonically decreases with increasing electric field. This result implied that the spin relaxation time decreases very rapidly with increasing electric field that contradicts the role of hyperfine interaction to be the main cause for spin relaxation. Similar conclusion was drawn from the study of RRP3HT and RRaP3HT based spin valve devices [22] where spin relaxation in less structured polymer RRaP3HT and more structured polymer RRP3HT had separate contributions from different effects. At lowest temperatures, almost temperature independent spin relaxation length of both the polymers were observed that indicated impurity scattering to be the most dominant mechanism in determining the spin relaxation time and rate. However, at elevated temperatures above 50 K, phonon scattering starts to become important and Dyakonov-Perel (DP) mechanism starts to dominate over the EY mechanism. However, non-linear dependence of spin relaxation rate on sample conductivity and hence mobility indicated that the observed effect is not purely DP mechanism but could be due to coexistence of both EY and DP mechanism. So in order to have improved spin diffusion length at elevated temperatures, it is important to suppress spin scattering by introducing structurally improved materials with minimum impurity and less spin orbit coupling.

Absence of Hanle effect: Also there are other significant issues both from fundamental understanding and practical device operation perspective that needs to be addressed. Practical issues like slow speed of operation, reproducibility, and low mobility of carriers inside an organic spacer remained a fundamental point of concern. From fundamental understanding point of view, absence of Hanle effect has created dilemma about the spin injection in organics. Hanle effect is a phenomenon where in a weak magnetic field, perpendicular to the spin polarization of the injected electrons, a spin precession is expected with a frequency proportional to the field intensity. This results in a modification of the spin component along the injection axis. If the duration of the precession is sufficiently long to produce a measurable deviation, the magnetoresistance will change. For example, in a 10 mT field, the rotation of the spin polarization by 45° would require less than 1 ns. As the average dwelling time of electrons in a typical 100-nm-long organic device is longer than 1 ns, Hanle effect should be measurable. Grunewald et al. [81] investigated the traces of Hanle precession in vertical organic spin valves. Due to the low mobility in organic semiconductors, the transit time of spin-polarized carriers was expected to allow the precession of the spins in perpendicular fields which in statistical average would quench the magnetoresistance. However, the authors failed to see any change in resistance while sweeping the perpendicular field, up to the point where the electrode's magnetization starts to reorient. They concluded that absence of Hanle-type effects is an indicative that the magnetoresistance is not based on the injection of spin-polarized electrons into the organic semiconductor but rather on tunneling through pinholes superimposed with tunneling anisotropic magnetoresistance. The authors concluded that even for clear spin-valve signatures, control experiments should be established which can confirm or disprove the electrical detection of injected spin-polarized carriers, similar to the Hanle precession in inorganic spin valves. Also Riminucci et al. [82] simultaneously reported the absence of Hanle effect in organic spin valves that remained a challenge. However, it is still not clear whether Hanle effect can be expected from a system where incoherent hopping of spin polarized carriers is the main conduction mechanism.

5. Conclusion and future perspective

In conclusion, it can be said that after a decade of intensive research effort, the field of organic spintronics is in a very interesting state now where some rapid development is expected in the direction of single molecular and two-dimensional device concepts with the newly found ideas of spin polarized interfaces. However, some major challenges in terms of yield, reproducibility, speed, large room temperature output in fabricated devices, still remained. Some promising new ideas also emerged that includes utilization of the optical control of spin-flipping of charge carriers in conjugated polymers for spintronic devices. [83] On the basis of the dynamics of polaron relaxation in conjugated polymers, it is theoretically found that the spin of an injected electron, which converts to a negative polaron in a conjugated polymer, is reversed by photoexcitation. This effect of photoinduced spin-flipping in polymers could be used to design a device of an optically controlled spin valve. Here the spin-flipping being driven by the electric dipole transition rather than the magnetic dipole transition, could lead to much higher efficiency. Also Velev et al. [84] theoretically predicted that using organic ferroelectric PVDF in a magnetic tunnel junction structure such as Co/PVDF/O/Co (0001) with asymmetric interfaces, both TMR and tunneling electroresistance (TER) can be obtained giving rise to four state memory devices which could lead to energy efficient memory components and hence could be extremely interesting.

Another interesting approach yet to be explored is the utilization of spin switching molecules that can show spin crossover between high and low spin states reversibly under external stimulus such as temperature, pressure, light etc. Control over the spin switching of these materials could open up unlimited possibilities for designing new functionalities and improved spintronic components with optimized performance in near future.

References

1. W. J. M. Naber, S Faez, W G van der Wiel, Organic Spintronics, *J. Phys. D: Appl. Phys.* **40**, R205 – R228 (2007).
2. V. Dediu, L. E. Hueso, I. Bergenti and C. Taliani, Spin routes in organic semiconductors, *Nat. Mater.* **8**, 707 – 716 (2009).
3. S. Majumdar, H. S. Majumdar and R. Österbacka, Organic Spintronics, *Comprehensive Nanoscience and Technology, Vol. 1* (Oxford Academic Press), pp. 109–142 (2011).
4. S. Sanvito, Molecular Spintronics, *Chem. Soc. Rev.*, **40**, 3336–3355 (2011).
5. D. Sun, E. Ehrenfreund, Z. Valy Vardeny. The first decade of organic spintronics research, *Chem. Commun.* **50**, 1781 – 1793 (2014).
6. M. N. Baibich, J. M. Broto, A. Fert, F. Nguyen Van Dau, F. Petroff, P. Etienne, G. Creuzet, A. Friederich, J. Chazelas, Giant Magnetoresistance of (001)Fe/(001)Cr Magnetic Superlattices. *Phys. Rev. Lett.* **61**, 2472 – 2475 (1988).
7. G. Binasch, P. Grünberg, F. Saurenbach, W. Zinn, Enhanced magnetoresistance in layered magnetic structures with antiferromagnetic interlayer exchange, *Phys. Rev. B* **39**, 4828 – 4830 (R) (1989).
8. P. M. Tedrow, R. Meservey, Spin-Dependent Tunneling into Ferromagnetic Nickel *Phys. Rev. Lett.* **26**, 192 – 195 (1971).
9. M. Julliere, Tunneling between ferromagnetic films, *Phys. Lett. A* **54**, 225 – 226 (1975).
10. J. S. Moodera, L. R. Kinder, T. M. Wong, and R. Meservey, Large Magnetoresistance at Room Temperature in Ferromagnetic Thin Film Tunnel Junctions, *Phys. Rev. Lett.* **74**, 3273 – 3276 (1995).
11. T. Miyazaki and N. Tezuka, Giant magnetic tunneling effect in Fe/Al_2O_3/Fe junction, *J. Magn. Magn. Mater.* **139**, L231 – L234 (1995).
12. I. Zutic, J. Fabian, S. Das Sharma, Spintronics: Fundamentals and applications, *Rev. Mod. Phys.* **76**, 323 – 410 (2004).
13. V. Dediu, M. Murgia, F. C. Matacotta, C. Taliani, S. Barbanera, Room temperature spin polarized injection in organic semiconductor, *Solid State Commun.* **122**, 181 – 184 (2002).
14. Z. H. Xiong, D. Wu, Z. V. Vardeny and J. Shi, Giant magnetoresistance in organic spin-valves, *Nature* 427, 821 – 824 (2004).
15. S. Majumdar, R. Laiho, P. Laukkanen, I. J. Väyrynen, Himadri S. Majumdar, R. Österbacka, Application of regioregular polythiophene in spintronic devices: Effect of interface, *Appl. Phys. Lett.* **89**, 122114–1 – 3 (2006).

16. V. Dediu, L. E. Hueso, I. Bergenti, A. Riminucci, F. Borgatti, P. Graziosi, C. Newby, F. Casoli, M. P. De Jong, C. Taliani, and Y. Zhan, Room-temperature spintronic effects in Alq_3-based hybrid devices, *Phys. Rev. B* **78**, 115203–1 – 6 (2008).
17. F. J. Wang, C.G. Yang and Z. Valy Vardeny, Spin response in organic spin valves based on LSMO electrodes, *Phys. Rev. B* **75**, 245324–245330 (2007).
18. S. Majumdar, H. Huhtinen, H. S. Majumdar, R. Laiho, R. Österbacka, Effect of $La_{0.67}Sr_{0.33}MnO_3$ electrodes on organic spin valves, *J. Appl. Phys.* 104, 033910 -1 – 6 (2008).
19. B. Chen, Y. Zhou, S. Wang, Y. J. Shi, H. F. Ding, D. Wu, Giant magnetoresistance enhancement at room-temperature in organic spin valves based on $La_{0.67}Sr_{0.33}MnO_3$ electrodes, *Appl. Phys. Lett.* **103**, 072402-1 – 3 (2013).
20. A. J. Drew et al., Direct measurement of the electronic spin diffusion length in a fully functional organic spin valve by low-energy muon spin rotation, *Nat. Mater.* **8**, 109 – 114 (2009).
21. S. Majumdar, H. S. Majumdar, On the origin of decay of spin current with temperature in organic spintronic devices, *Org. Electron.* **13**, 2653 – 2658 (2012).
22. S. Majumdar and H. S. Majumdar, Decay in spin diffusion length with temperature in organic semiconductors—An insight of possible mechanisms, *Synth. Met.* **173**, 26 – 30 (2013).
23. Theory for Spin Diffusion in Disordered Organic Semiconductors. P. A. Bobbert, W. Wagemans, F. W. A. van Oost, B. Koopmans, and M. Wohlgenannt, *Phys. Rev. Lett.* **102**, 156604-1 – 4 (2009).
24. S. Liang, R. Geng, B. Yang, W. Zhao, R. C. Subedi, X. Li, X. Han, T. D. Nguyen, Curvature-enhanced Spin-orbit Coupling and Spinterface Effect in Fullerene-based Spin Valves, *Sci. Rep.* **6**, 19461-1 – 9 (2016).
25. B. Dlubak et al., Highly efficient spin transport in epitaxial graphene on SiC, *Nat. Phys.* **8**, 557–561 (2012).
26. R. Lin, F. Wang, M. Wohlgenannt, C. He, X. Zhai, Y. Suzuki, Organic spin-valves based on fullerene C_{60}, *Synth. Met.* **161**, 553–557 (2011).
27. M. Gobbi, F. Golmar, R. Llopis, F. Casanova, L. E. Hueso, Room-Temperature Spin Transport in C_{60}-Based Spin Valves, *Adv. Mater.* **23**, 1609–1613 (2011).
28. X. Zhang, S. Mizukami, T. Kubota, Q. Ma, M. Oogane, H. Naganuma, Y. Ando, T. Miyazaki, Observation of a large spin-dependent transport length in organic spin valves at room temperature, *Nat. Commun.* **4**, 1392 (2012).
29. F. Li, T. Li, F. Chen, F. Zhang, Excellent spin transport in spin valves based on the conjugated polymer with high carrier mobility, *Sci. Rep.* **5**, 9355 (2015).
30. X. Sun, M. Gobbi, A. Bedoya-Pinto, O. Txoperena, F. Golmar, R. Llopis, A. Chuvilin, F. Casanova, L. E. Hueso, Room-temperature air-stable spin transport in bathocuproine-based spin valves, *Nat. Commun.* **4**, 2794 (2013).
31. B. Li, C.-Y. Kao, J.-W. Yoo, V. N. Prigodin, A. J. Epstein, Magnetoresistance in an All-Organic-Based Spin Valve, *Adv. Mater.* **23**, 3382–3386 (2011).
32. J. M. Manriquez, G. T. Yee, R. S. McLean, A. J. Epstein, J. S. Miller, A room-temperature molecular/organic-based magnet, *Science* **252**, 1415 – 1417 (1991).

33. V. Prigodin, N. Raju, K. Pokhodnya, J. Miller, A. Epstein, Spin-Driven Resistance in Organic-BasedMagnetic Semiconductor V[TCNE]$_x$, *Adv. Mater.* **14**, 1230 – 1233 (2002).
34. E. Ehrenfreund, Z. Valy Vardeny, Organic spin-valves: from unipolar to bipolar devices, *Phys. Chem. Chem. Phys.* **15**, 7967 – 7975 (2013).
35. T. D. Nguyen, E. Ehrenfreund, Z. Valy Vardeny, Spin-Polarized Light-Emitting Diode Based on an Organic Bipolar Spin Valve, *Science* **337**, 204 – 209 (2012).
36. S. Majumdar, H. S. Majumdar, R. Laiho, R. Österbacka, Organic spin valves: effect of magnetic impurities on the spin transport properties of polymer spacers, *New J. Phys.* **11**, 013022 – 1 – 11 (2009).
37. H. Vinzelberg, J. Schumann, D. Elefant, R. B. Gangineni, J. Thomas, B. Büchner, Low temperature tunneling magnetoresistance on LSMO/Co junctions with organic spacer layers, *J. Appl. Phys.* **103**, 093720 (2008).
38. Z. Pang, L. Lin, F. Wang, P. Yao, T. Yang · S. Han, Positive Giant Magnetoresistance in $La_{0.67}Sr_{0.33}MnO_3/Alq_3/Co$ Organic Spin-Valves, *J. Supercond. Nov. Magn.* **23**, 859 – 861 (2010).
39. B. Chen, S. –W. Jiang, H. –F. Ding, Z. –S. Jiang, D. Wu, The basis of organic spintronics: Fabrication of organic spin valves, *Chin. Phys. B* **23**, 018104 (2014).
40. S. Majumdar, H. S. Majumdar, R. Laiho, R. Österbacka, Comparing small molecules and polymer for future organic spin-valves, *J. Alloys Comp.* **423**, 169–171 (2006).
41. C. Barraud, P. Seneor, R. Mattana, S. Fusil, K. Bouzehouane, C. Deranlot, P. Graziosi, L. Hueso, I. Bergenti, V. Dediu, F. Petroff and A. Fert, Unravelling the role of the interface for spin injection into organic semiconductors, *Nat. Phys.* **6**, 615–620 (2010).
42. S. Sanvito, Molecular spintronics: The rise of spinterface science, *Nat. Phys.* **8**, 562–564 (2010).
43. D. Ciudad, M. Gobbi, C. J. Kinane, M. Eich, J. S. Moodera, L. E. Hueso Sign Control of Magnetoresistance Through Chemically Engineered Interfaces, *Adv. Mater.* **26**, 7561–7567 (2014).
44. D. Sun, M. Fang, X. Xu, L. Jiang, H. Guo, Y. Wang, W. Yang, L. Yin, P. C. Snijders, T. Z. Ward, Z. Gai, X.-G. Zhang, H. Nyung Lee, J. Shen, Active control of magnetoresistance of organic spin valves using ferroelectricity, *Nat. Commun.* **5**, 4396 (2014).
45. G. Szulczewski, S. Sanvito and J. M. D. Coey, A spin of their own, *Nat. Mater.* **8**, 693–695 (2009).
46. J. J. H. M. Schoonus, P. G. E. Lumens, W. Wagemans, J. T. Kohlhepp, P. A. Bobbert, H. J. M. Swagten, and B. Koopmans, Magnetoresistance in Hybrid Organic Spin Valves at the Onset of Multiple-Step Tunneling, *Phys. Rev. Lett.* **103**, 146601-1 – 4 (2009).
47. T. S. Santos, J. S. Lee, P. Migdal, I. C. Lekshmi, B. Satpati, and J. S. Moodera, Room-Temperature Tunnel Magnetoresistance and Spin-Polarized Tunneling through an Organic Semiconductor Barrier, *Phys. Rev. Lett.* 98, 016601-1 – 4 (2007).
48. J. H. Shim, K. V. Raman, Y. J. Park, T. S. Santos, G. X. Miao, B. Satpati, J. S. Moodera, Large Spin Diffusion Length in an Amorphous Organic Semiconductor, *Phys. Rev. Lett.* **100**, 226603 -1 - 4 (2008).
49. J.-W. Yoo, C.-Y. Chen, H. W. Jang, C. W. Bark, V. N. Prigodin, C. B. Eom, A. J. Epstein, Spin injection/detection using an organic-based magnetic semiconductor, *Nat. Mater.* **9**, 638 – 642 (2010).

50. R. Meservey and P. M. Tedrow, Spin-polarized electron tunneling, *Phys. Rep.* **238**, 173 – 243 (1994).
51. A. Thomas, J. S. Moodera, and B. Satpati, Evidence for positive spin polarization in Co with $SrTiO_3$ barriers, *J. Appl. Phys.* **97**, 10C908-1 – 3 (1999).
52. M. Münzenberg and J. S. Moodera, Superconductor-ferromagnet tunneling measurements indicate sp-spin and d-spin currents, *Phys. Rev. B* **70**, 060402(R) -1 – 4 (2004).
53. S. Schmaus, A. Bagrets, Y. Nahas, T. K. Yamada, A. Bork, M. Bowen, E. Beaurepaire, F. Evers, W. Wulfhekel, Giant magnetoresistance through a single molecule *Nat. Nanotech.* **6**, 185–189 (2011).
54. K. V. Raman, A. M. Kamerbeek, A. Mukherjee, N. Atodiresei, T. K. Sen, P. Lazić, V. Caciuc, R. Michel, D. Stalke, S. K. Mandal, S. Blügel, M. Münzenberg, J. S. Moodera, Interface-engineered templates for molecular spin memory devices, *Nature* **493**, 509–513 (2013).
55. M. Galbiati, S. Tataya, C. Barraud, A. V. Dediu, F. Petroff, R. Mattana, P. Seneor, Spinterface: Crafting spintronics at the molecular scale, *MRS Bulletin* **39**, 602-607 (2014).
56. R. J. Brooke, C. Jin, D. S. Szumski, R. J. Nichols, B. –W. Mao, K. S. Thygesen, W. Schwarzacher, Single-Molecule Electrochemical Transistor Utilizing a Nickel-Pyridyl Spinterface, *Nano Lett.* **15**, 275–280 (2015).
57. Y. Zhan, M. Fahlman. The Study of Organic Semiconductor/Ferromagnet Interfaces in Organic Spintronics: A Short Review of Recent Progress, *J. Pol. Sc. B: Pol. Phys.* **50**, 1453–1462 (2012).
58. X. Crispin, V. Geskin, A. Crispin, J. Cornil, R. Lazzaroni, W. R. Salaneck, J. L. Bredas, Characterization of the Interface Dipole at Organic/ Metal Interfaces, *J. Am. Chem. Soc.* **124**, 8131–8141 (2002).
59. T. Suzuki, M. Kurahashi, X. Ju, Y. Yamauchi, Spin Polarization of Metal (Mn, Fe, Cu, and Mg) and Metal-Free Phthalocyanines on an Fe(100) Substrate, *J Phys. Chem. B* **106**, 11553-11556 (2002).
60. T. Suzuki, M. Kurahashi, Y. Yamauchi, Spin Polarization in Molecular Orbitals of Copper-Phthalocyanine Deposited on a Magnetized Fe(100) Substrate, *J. Phys. Chem. B* **106**, 7643-7646 (2002).
61. F. Djeghloul et al., Direct observation of a highly spin-polarized organic spinterface at room temperature, *Sci. Rep.* **3,** 1272 (2013).
62. S. Lach, A. Altenhof, K. Tarafder, F. Schmitt, Md. E. Ali, M. Vogel, J. Sauther, P. M. Oppeneer, C. Ziegler, Metal–Organic Hybrid Interface States of A Ferromagnet/Organic Semiconductor Hybrid Junction as Basis For Engineering Spin Injection in Organic Spintronics, *Adv. Funct. Mater.* **22**, 989–997 (2012).
63. M. Gruber et al., Spin-Dependent Hybridization between Molecule and Metal at Room Temperature through Interlayer Exchange Coupling, *Nano Lett.* **15**, 7921–7926 (2015).
64. S. Steil, N. Großmann, M. Laux, A. Ruffing, D. Steil, M. Wiesenmayer, S. Mathias, O. L. A. Monti, M. Cinchetti, M. Aeschlimann, Spin-dependent trapping of electrons at spinterfaces, *Nat. Phys.* **9**, 242–247 (2013).
65. Y. Zhan, E. Holmstrom, R. Lizaarraga, O. Eriksson, X. Liu, F. Li, E. Carlegrim, S. Stafstrom, M. Fahlman, Efficient Spin Injection Through Exchange Coupling at Organic Semiconductor/Ferromagnet Heterojunctions, *Adv. Mater.* **22**, 1626–1630 (2010).

66. T. L. A. Tran, T. Q. Le, J. G. M. Sanderink, W. G. van der Wiel, M. P. de Jong, Hybridization-induced oscillatory magnetic polarization of C60 orbitals at the C_{60}/Fe(001) interface, *Appl. Phys. Lett.* **98**, 222505-1 – 3 (2011).
67. T. Methfessel, S. Steil, N. Baadji, N. Großmann, K. Koffler, S. Sanvito, M. Aeschlimann, M. Cinchetti, H. J. Elmers, Spin scattering and spin-polarized hybrid interface states at a metal-organic interface, *Phys. Rev. B* **84**, 224403-1 – 7 (2011).
68. S. Steil, K. Goedel, A. Ruffing, I. Sarkar, M. Cinchetti, M. Aeschlimann, Investigation of the spin-dependent properties of electron doped cobalt–CuPc interfaces, *Synth. Met.* **161**, 570–574 (2011).
69. M. N. Grisolia et al., Hybridization-controlled charge transfer and induced magnetism at correlated oxide interfaces, *Nat. Phys.* (2016). DOI: 10.1038/NPHYS3627.
70. C. Iacovita, M. V. Rastei, B. W. Heinrich, T. Brumme, J. Kortus, L. Limot, and J. P. Bucher, Visualizing the Spin of Individual Cobalt-Phthalocyanine Molecules, *Phys. Rev. Lett.* **101**, 116602-1 – 4 (2008).
71. X. Chen and M. Alouani, Effect of metallic surfaces on the electronic structure, magnetism, and transport properties of Co-phthalocyanine molecules, *Phys. Rev. B* **82**, 094443-1 – 11 (2010).
72. J. Brede, N. Atodiresei, S. Kuck, P. Lazić, V. Caciuc, Y. Morikawa, G. Hoffmann, S. Blügel, R. Wiesendanger, Spin- and Energy-Dependent Tunneling through a Single Molecule with Intramolecular Spatial Resolution, *Phys. Rev. Lett.* **105**, 047204-1 – 4 (2010).
73. P. K. J. Wong, W. Zhanga K. Wang, G. van der Laan, Y. Xu, W. G. van der Wiel, M.1 P. de Jong, Electronic and magnetic structure of C_{60}/Fe_3O_4(001): a hybrid interface for organic spintronics, *J. Mater. Chem. C* **1**, 1197-1202 (2013).
74. T. L. A. Tran, P. K. J. Wong, M. P. de Jong, W. G. van der Wiel, Y. Q. Zhan and M. Fahlman, Hybridization-induced oscillatory magnetic polarization of C60 orbitals at the C_{60}/Fe(001) interface *Appl. Phys. Lett.* **98**, 222505-1 – 3 (2011).
75. S. Majumdar, S. Dey, H. Huhtinen, J. Dahl, M. Tuominen, P. Lukkanen, S. van Dijken, H. S. Majumdar, Comparative Study of Spin Injection and Transport in Alq3 and CoPc based organic spin valves. *Spin* **4**, 1440009 (2014).
76. S. Majumdar, K. Grochowska, M. Sawczak, G. Śliwiński, H. Huhtinen, J. Dahl, M. Tuominen, P. Laukkanen, H. S. Majumdar, Interfacial Properties of Organic Semiconductor–Inorganic Magnetic Oxide Hybrid Spintronic Systems Fabricated Using Pulsed Laser Deposition, *ACS Appl. Mater. Interfaces* **7**, 22228–22237 (2015).
77. N. A. Morley, A. J. Drew, H. Zhang, K. Scott, S. Hudziak, D. J. Morgan, Study of the magnetic–Alq_3 interface in organic spin-valves, *Appl. Surf. Sc.* **313**, 850–857 (2014).
78. S. Sakai, S. Majumdar, Z. Popov, P. V. Avramov, S. Entani, Y. Hasegawa, Y. Yamada, H. Huhtinen, H. Naramoto, P. B. Sorokin, Y. Yamauchi, Proximity-induced spin polarization of graphene in contact with half-metallic manganite. *ACS Nano* **10**, 7532–7541 (2016).
79. Yoo, J.-W.; Jang, H. W.; Prigodin, V. N.; Kao, C.; Eom, C. B.; Epstein, A. J., Tunneling vs. giant magnetoresistance in organic spin valve, *Synth. Met.* **160**, 216–222 (2010).
80. Dominant spin relaxation mechanism in compound organic semiconductors, S. Bandyopadhyay, *Phys. Rev. B* **81**, 153202-1 – 4 (2010).

81. M. Grünewald, R. Göckeritz, N. Homonnay, F. Würthner, L. W. Molenkamp, G. Schmidt, Vertical organic spin valves in perpendicular magnetic fields, *Phys. Rev. B* **88**, 085319 -1 – 11 (2013).
82. A. Riminucci et al., Hanle effect missing in a prototypical organic spintronic device, Appl. Phys. Lett. 102, 092407-1 – 3 (2013).
83. B. Di, S. Yang, Y. Zhang, Z. An, X. Sun, Optically Controlled Spin-Flipping of Charge Carriers in Conjugated Polymers, *J. Phys. Chem.* **C 117**, 18675–18680 (2013).
84. J. P. Velev, J. M. Lopez-Encarnacion, J. D. Burton, E. Y. Tsymbal, Multiferroic tunnel junctions with poly(vinylidene fluoride), *Phys. Rev. B* **85**, 125103 (2012).

Chapter 9

Spintronics with Graphene and van der Waals Heterostructures

M. Venkata Kamalakar[*,†,‡], *André Dankert*[*] *and Saroj P. Dash*[*,§]

[*]Department of Microtechnology and Nanoscience, Chalmers University of Technology, SE-41296, Göteborg, Sweden.
[†]Department of Physics and Astronomy, Uppsala University, Box 516, 75120, Uppsala, Sweden
[‡]*venkata.mutta@physics.uu.se;* [§]*saroj.dash@chalmers.se*

The outstanding spin transport properties of graphene make it an ideal candidate for chip scale spin communication devices. Motivated by this prospect, over the past decade, remarkable progress has been made in enhancing the spin transport parameters in graphene. Apart from simple graphene devices, van der Waals heterostructures of graphene have been fabricated by laminating other two-dimensional crystals with graphene. Such heterostructures of graphene with insulating hexagonal boron nitride (h-BN) as a substrate and gate dielectric or as spin tunnel barrier have been used to achieve efficient spin injection, large spin coherence time and diffusion length in graphene. In this chapter, we present two important advancements in the field of graphene spintronics: First, the recent achievement of long distance spin communication in large scale chemical vapor deposited graphene, and second, the demonstration of enhanced spin injection and spin filtering effects in ferromagnet/h-BN-graphene van der Waals heterostructures. We discuss how these results feature in the present state-of-the art and open new avenues for future developments.

1. Introduction

Future spintronic devices are expected to perform information storage and processing using the spin degree of freedom of electrons instead of their charge. Three basic requirements for such devices are efficient spin injection, detection, transportation and manipulation. Importantly, the viability of chip scale spintronic devices relies on how far spin polarized electrons or spin currents can be communicated without losing their spin coherence. This requires materials that can efficiently transport spin currents over long distances at room temperature. Such spin transport carries huge potential for all spin-logic[1], and integrated spin assisted logic and memory devices on the same chip. A major challenge in this direction lies in discovering a suitable material for spin transport, where the spin polarization of charge carriers can be maintained for a sufficiently long time period and distance at room temperature. In this context, most of the materials showed little promise until the arrival of graphene.

Graphene is a two dimensional (2D) crystal consisting of a single layer of carbon atoms arranged in a honeycomb lattice[2,3]. It is a material with superlative electrical, mechanical, optical, and spintronic properties. The uniqueness of graphene arises from its zero-band gap with its valence and conduction bands meeting at the Dirac point. It has a linear energy-momentum dispersion relation, which leads to massless carriers following the Dirac equation instead of the Schrödinger equation. Such a unique electronic structure leads to a very high electron velocity and tunable charge carrier concentration presenting an enormous potential for many electronic applications. Graphene is also proposed to be an ideal material for spintronic applications as the spin of an electron is expected to live longer and travel much further than in other known materials, because of the relatively weaker coupling of the spin to the crystal lattice. The performance of a spin transport material mainly depends on the spin relaxation mechanisms in the material channel. The low atomic weight of carbon atoms and the symmetric hexagonal crystal structure of graphene results in a relatively weak spin-orbit coupling. Furthermore, the hyperfine interaction between the nuclear and electron spins in graphene is negligible, due to the predominance of the ^{12}C isotope with zero-nuclear spin. Considering these factors, a long spin lifetime of $\tau_s \sim 1$ μs and a diffusion length of $\lambda_s \sim 100$ μm have been theoretically predicted in graphene at room temperature[4,5], making it the best medium for spin transport.

In spin transport experiments, ferromagnetic tunnel contacts can be used to inject and detect spins into graphene. The spin relaxation time is typically measured based on Hanle spin precession experiments on graphene spin valve devices. In 2007, Tombros et al.[16] reported measurements on spin transport in graphene spin-valve devices at room temperature with a spin relaxation time $\tau_s \sim 150$ ps and a diffusion length $\lambda_s \sim 1.5$ μm. These values are orders of magnitude lower than theoretical predictions. Since this first report, experimental and theoretical research focused on understanding spin relaxation processes in graphene and improvements in the spin parameters[6–12]. Two main factors have been identified to be responsible for such low values of experimental spin parameters in graphene spintronic devices: the quality of tunnel barrier, and the graphene channel properties. A tunnel barrier is an ultrathin layer of an insulating material, which enhances the spin injection and detection efficiency when introduced between the graphene and ferromagnetic contacts. However, roughness and pin-holes in the tunnel contact can introduce additional spin relaxation in graphene due to inhomogeneous magnetic fringe fields at the interface. The back-diffusion of spins through pin-holes and relaxation in the ferromagnet can also severely affect measured spin parameters in the graphene. Therefore, it is essential to develop homogeneous and crystalline ferromagnetic tunnel contacts having atomically smooth interface with graphene. At the same time, it is also necessary to improve the graphene channel properties, in a device environment the graphene is highly sensitive to disorder at the interfaces with substrates and contaminations during the fabrication process. These extrinsic factors can reduce the spin polarization and limit the actual spin lifetime in graphene.

Major efforts in addressing these above discussed issues have consistently resulted in improving values of the spin parameters in graphene[6,9–12]. In this context, van der Waals heterostructures (vdWh) of graphene with other two dimensional (2D) crystals are considered to be very important materials. A vdWh is formed by directly stacking two or more layers of 2D crystals, where the interactions between the layers are purely van der Waals forces (as depicted in Fig. 1). One example is 2D hexagonal boron nitride (h-BN), which is an insulating isomorph of graphene and has been used successfully as substrate[13], gate dielectric[11] and spin tunnel barrier[10,14,15] in graphene spintronic devices. The absence of dangling bonds on the surface of vdWh materials helps in the creation of high quality hetero-interfaces. In this chapter, we review the recent progress in graphene spintronics with focus on large scale chemical vapor deposited (CVD) graphene, followed by vdWh of graphene with h-BN tunnel barriers for efficient spin injection and transport.

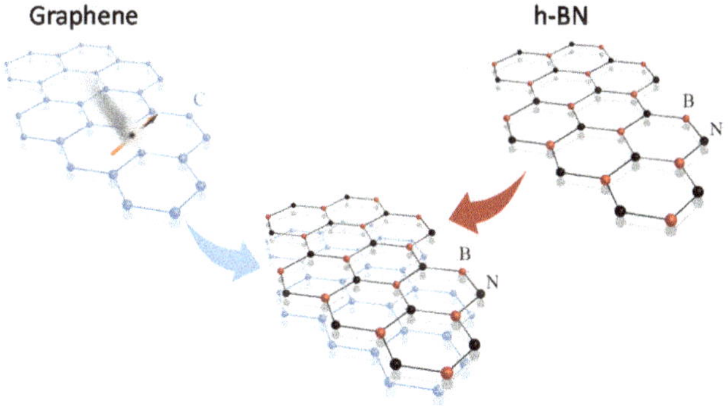

Figure 1. A representation of structures of graphene, hexagonal boron nitride (h-BN) and their van der Waals heterostructure.

2. Graphene spintronics: state of the art

The theoretical prediction of large spin lifetime and diffusion length in graphene ignited the interest in lateral spin transport experiments using multi-terminal spintronic devices. The first demonstration by Tombros et al. employed a nonlocal measurement geometry to measure pure spin transport and precession in graphene[16]. A schematic representation of such a nonlocal electrical spin injection and detection method in a graphene device is illustrated in Fig. 2. The electrical injection of spin polarized electrons into graphene is achieved by driving an electric current I from a ferromagnetic tunnel contact with an intrinsic spin polarization. The spin accumulation thus created in graphene results into a diffusive spin current across the graphene channel and a nonlocal spin signal $V=V_{NL}$ can be sensed by a ferromagnetic detector placed at a distance L within the spin diffusion length in graphene. Such a measurement configuration is called non-local measurement technique[16] due to the isolation of the current and voltage measurement circuits, where the spin signal is quantified by a nonlocal resistance $R_{\mathrm{NL}}=V_{\mathrm{NL}}/\mathrm{I}$. This method results in the detection of pure spin signal devoid of other spurious magnetoresistive effects. A spin accumulation producing an imbalance in chemical potential $\Delta\mu = \mu_\uparrow - \mu_\downarrow$ between up (↑) and down (↓) spins can be measured by changing the magnetization of injection and detection ferromagnetic electrodes from parallel (↑↑ or ↓↓) to anti-parallel (↑↓ or ↓↑) configuration by the application of a longitudinal magnetic field $\mathbf{B}_\parallel$. This measurement configuration is called spin-valve geometry. In addition, the spins entering the graphene channel from the injector can be made to precess with Larmor frequency about an out of plane magnetic field ($\mathbf{B}_\perp$), resulting in a modulation of the detected spin signal, the technique being called the Hanle spin precession measurement. The modulation of the nonlocal resistance R_{NL} with application of $\mathbf{B}_\perp$ leads to a quantification of the spin lifetime τ_s and diffusion length λ_s in the graphene channel. The magnitude of the spin valve or Hanle signal is also utilized to estimate the spin polarization of the injector and detector electrodes.

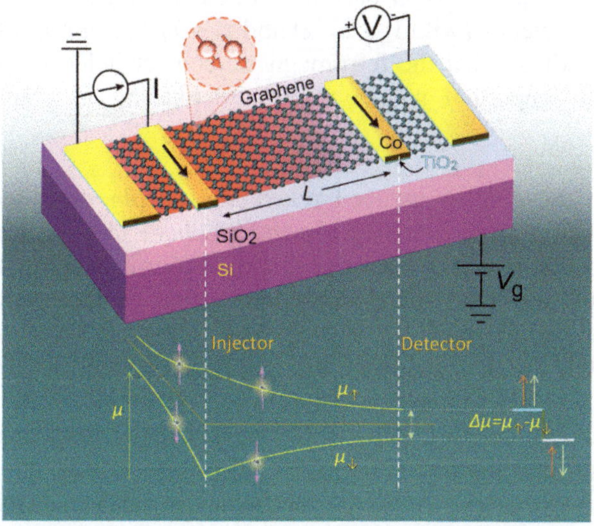

Figure 2. Scheme of spin injection and detection in graphene using ferromagnetic tunnel contacts in a non-local measurement geometry, together with the spin accumulation and variation of spin chemical potential profile across the graphene channel presented below the device scheme. (Adapted from Ref. 29)

For the realization of graphene spintronic devices, the fabrication of high quality ferromagnetic tunnel contacts on graphene with sharp interfaces is one of the basic requirements. Although, the direct spin injection from a ferromagnet into graphene without a tunnel barrier is possible, it results in a very low spin polarization (P ~ 1% or less) and low τ_s ~ 50 ps [10,17]. The spin injection efficiency of such direct contacts is limited mainly due to the conductivity mismatch problem between graphene and ferromagnet[18,19]. Furthermore, the low spin lifetime in such devices has been attributed to doping of the graphene under the ferromagnetic contacts, producing additional spin scattering centers. Introducing an ultrathin Al_2O_3 tunnel barrier can yield τ_s ~150 ps and λ_s ~ 1.5 μm with P ~ 10 % at room temperature[16]. Even though this discovery has been a remarkable achievement, these spin parameters are much lower than the theoretically predicted values[4]. The discrepancy is attributed to the possible pin-holes in the barrier, due to which the spins injected into the graphene channel can get back into the ferromagnet, where they experience a faster relaxation effectively reducing the measured spin lifetime. Additionally, the interface roughness can produce local magnetostatic fields which can severely suppress the spin accumulation and dynamics in the graphene under the contacts. Using a TiO_2 tunnel barrier, spin lifetime of τ_s ~150 ps was achieved[17], which is similar to τ_s obtained with Al_2O_3 barrier. Further improvements in tunnel barrier properties by depositing a TiO_2 seeded MgO barrier resulted in a significant enhancement of spin injection efficiency (with P ~ 27%) and $\tau_s \sim 770$ ps at room temperature[20,21]. These reports indicate that the ferromagnetic contact induced spin-relaxation in graphene is an important source of spin dephasing in graphene spintronic devices[21].

In addition to the contact induced relaxation, a spin–orbit coupling induced relaxation can arise from impurities in the graphene channel introduced during fabrication process, lattice deformations, and electronic coupling to the substrate. Experiments to overcome these limitations using cleaner samples with high mobility suspended graphene[22,23] and epitaxial graphene[24] have shown that τ_s is indeed controlled by the quality of the graphene channel. Recently, significant improvements in spin parameters with $\tau_s \sim 2$ ns and $\lambda_s \sim 22$ μm have been reported in cleaner devices using h-BN encapsulated graphene channels[9,11,12]. Most of these developments for spin transport studies in graphene were performed on micrometer size flakes, mechanically exfoliated from bulk graphite crystals. Even though CVD method can deliver large scale graphene which can be useful for practical applications, so far, there are only few reports on spin transport in such graphene[7,25,26].

3. Spin transport in chemical vapor deposited graphene

Chemical vapor deposited (CVD) graphene is the most technologically prospective form of graphene as it can be grown over a large area and can be practically transferred to any substrate. However, there are challenges in CVD graphene device processing that led to spin parameters limited to values of $\tau_s \sim$ 200 ps and $\lambda_s \sim 2$ μm on Si/SiO_2 substrates at room temperature[7,26–28]. Graphene prepared by CVD method is polycrystalline with grains boundaries, wrinkles, ripples and contaminations, which significantly affect its charge and spin transport properties. In spite of these challenges, in a recent development, long distance spin transport up to 16 μm and $\tau_s \sim 1.2$ ns in CVD graphene on Si/SiO_2 substrate has been demonstrated at room temperature[29].

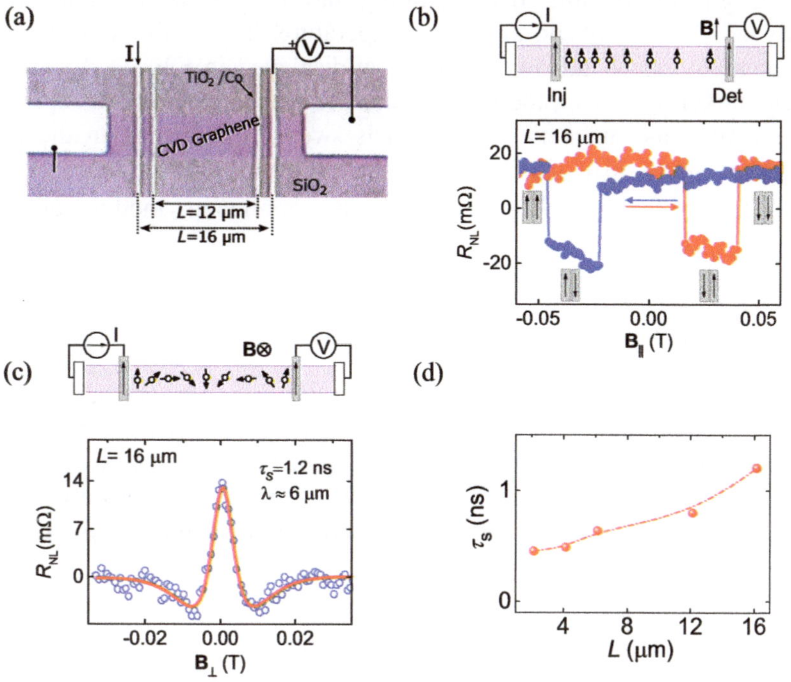

Figure 3. Long distance spin transport in CVD graphene. (a) Optical microscope image of a fabricated long channel CVD graphene device on Si/SiO_2 substrate with multiple ferromagnetic tunnel contacts with non-local measurement connections. (b) Non-local spin valve measurement, and (c) Hanle spin precession signal on a 16 μm long graphene channel yielding a τ_s ~ 1.2 ns and λ_s ~ 6 μm at room temperature. (d) Enhancement in the value of spin lifetime for longer graphene channel lengths. Adapted from [Ref. 29].

Figure 3a shows a typical multi-terminal CVD graphene spintronic device along with non-local measurement configuration used in the study. For device fabrication, a commercial CVD graphene layer was first transferred onto a Si/SiO_2 substrate by a wet chemical process[10,29], which was then patterned into stripes via optical lithography and oxygen plasma etching. Followed by this, electron-beam lithography assisted patterning and metal evaporation were performed for preparing the TiO_2/Co ferromagnetic tunnel contacts onto graphene. The detailed structural and electrical characterization of the CVD graphene revealed grains with size ~ 2 - 10 μm and a carrier mobility of ~2000 $cm^2\ V^{-1}\ s^{-1}$ at room temperature with good wafer scale uniformity.

The measured nonlocal spin-valve switching signal is shown in Fig. 3b for a channel length of 16 μm, demonstrating the long distance spin transport capability of CVD graphene at room temperature. The Hanle spin precession measurement in the 16 μm long channel is displayed in Fig. 3c. In the Hanle measurement configuration, the injector and detector electrodes are kept parallel and an out of plane magnetic field ($\mathbf{B}_\perp$) is swept. This results in the Larmor precession of spins diffusing in the graphene channel with a frequency $\omega_L = \frac{g\mu_B}{\hbar}\mathbf{B}_\perp$ (Landé g-factor g = 2) and results in a continuous modulation of spin signal R_{NL}, given by

$$\Delta R_{NL} \propto \int_0^\infty \frac{1}{\sqrt{4\pi Dt}}\, e^{-\frac{L^2}{4Dt}}\, \cos(\omega_L t)\, e^{-(t/\tau_s)}\, dt, \quad (1)$$

where L is the length of the channel, D is the diffusion constant, with the expression containing the spin diffusion, precession and dephasing contributions [16,30,31]. In Fig. 3c, the red curve represents fitting of Eq. (1) to the raw data, which gives a value of spin lifetime of $\tau_s \sim 1.23 \pm 0.10$ ns and spin diffusion constant of $D = 0.0264 \pm 0.002$ m^2s^{-1} resulting in a spin diffusion length of $\lambda_s \sim 6$ μm ($\lambda_s = \sqrt{D\tau_s}$). These are the highest values of spin parameters for any form of pristine graphene on normal oxidized silicon/silicon (SiO_2/Si) substrates at room temperature. The long distance spin transport in CVD graphene obtained on Si/SiO_2 substrate is remarkable as compared to similar demonstration in exfoliated graphene on atomically smooth h-BN substrate. In addition, a systematic dependence of spin parameter on the length of the CVD graphene channel has been observed, where longer channels leading to longer spin lifetimes (shown in Fig. 3d). This clearly indicates a larger role of contacts in spin relaxation and shows that a longer diffusion channel for spins leads to better estimation of the actual spin parameters in graphene. This demonstration is a significant step towards large scale graphene spintronics, since it has been achieved without suspension or encapsulation of graphene. Considering the wafer scale uniformity of CVD graphene and envisioning the possibilities of cleaner and dry transfer processes, it has the ability to guide the future developments of graphene spintronics towards practical applications.

4. Spin transport in graphene/h-BN van der Waals heterostructures

Van der Waals hetereostructures (vdWh)[32] fabricated by stacking different kinds of 2D crystals have demonstrated abilities to enhance the efficiency for graphene devices. For spintronic devices with such structures, long distance spin transport has been demonstrated in exfoliated graphene on h-BN substrate[13]. Recently, h-BN encapsulated graphene channel devices showed an enhanced spin lifetime and diffusion length up to 2 ns and 22 μm, respectively[9,11,12]. The atomically thin layers of h-BN and other 2D crystals such as graphene, fluorographene and h-BN have also been demonstrated to be ideal for spin based applications[10,14,15,28,33–36].

This has raised profound interest in utilizing insulating 2D crystals in a more active role as spin tunnel barriers, apart from utilization as a substrate or for encapsulation.

The structural compatibility of the h-BN and graphene (with a lattice mismatch of only 1.7 %) makes them ideal partners for vdWh. The insulating property of h-BN with a band gap ~ 6 eV and atomically thin nature makes it an ideal tunnel barrier for spin injection into graphene[37]. Especially when compared to conventional metal-oxide barriers such as Al_2O_3, MgO, TiO_2, the h-BN barrier can avoid the possible doping problems during the deposition process. Large scale h-BN can be grown by CVD methods and can be transferred on to graphene efficiently[38]. In this part of thc chapter, we discuss the use of large scale CVD h-BN as a spin tunnel barrier, which has been recently shown to enhance spin parameters in graphene[10,33], with efficient spin injection and related novel observations[39].

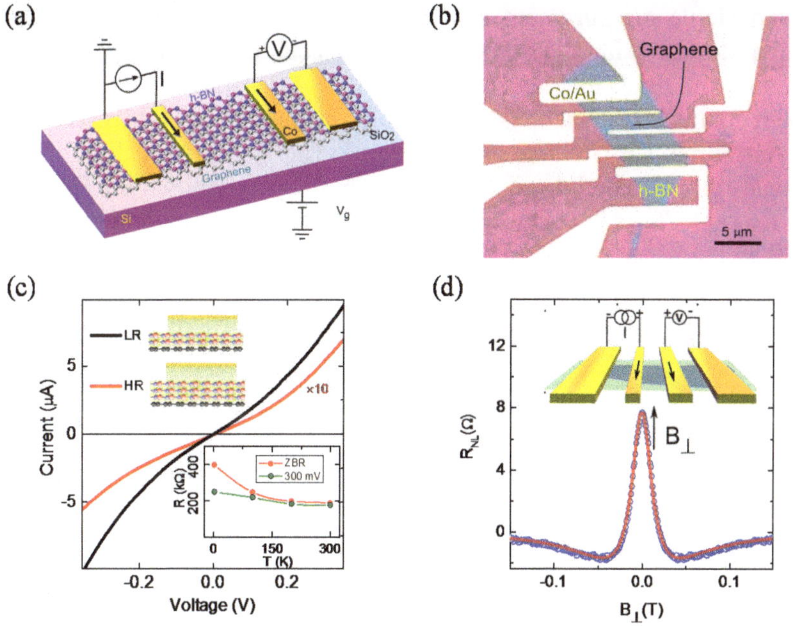

Figure 4. Spin transport in graphene/h-BN van der Waals heterostructures. (a) Schematic of a graphene/h-BN van der Waals heterostructure device with nonlocal spin transport measurement configuration. (b) Optical microscope image of a fabricated device. (c) Current-Voltage characteristic curves obtained on a high resistance (HR) and low resistance (LR) h-BN contacts. Inset: Temperature dependence of resistances at of zero and 300 mV bias voltages. (d) Hanle spin signal of a graphene|h-BN van der Waals device measured in nonlocal geometry. Adapted from [Ref. 10, 33, 39].

In these experiments, commercially procured CVD h-BN grown over copper substrates was employed. The h-BN was transferred onto graphene on Si/SiO_2 substrates by a well optimized wet chemical process. Devices were fabricated by standard electron beam assisted lithography and metal evaporation. A schematic and micrograph of a device with graphene/h-BN vdWh and cobalt (Co) contacts is shown in Fig. 4a and 4b, respectively. One unique feature of CVD h-BN is its spatial variation of thickness between 1-3 layers as confirmed by atomic force microscopy[10,26,33]. This leads to a variation in the resistance area product (RA) of the contacts as presented in Fig. 4c, where the current-voltage (IV) characteristic curves for a high resistance (HR) ~100 kΩ μm^2 and a low resistance (LR) ~10 kΩ μm^2 contacts are shown. The IV characteristics are of non-linear tunneling nature and showed a weak temperature dependence, which is typical for good quality tunnel barriers (inset in Fig. 4c for the HR contact).

Spin transport measurement were performed in these devices in non-local spin valve and Hanle measurement configurations. In Fig. 4d, Hanle measurement data of a device with Co|h-BN tunnel contacts with RA ~10 kΩ μm^2 are displayed. From the Hanle data (fitting with Eq. 1), a spin lifetime of τ_s ~ 500 ps has been obtained, which is a ten-fold enhancement when compared to transparent direct Co contacts on graphene[10]. The spin polarization of Co/h-BN contacts were evaluated from the spin valve signal (shown in Fig. 5a) with $\Delta R_{NL} = R_{NL}(\uparrow\uparrow) - R_{NL}(\uparrow\downarrow)$ using *Eq. 2*. On the devices with a contact resistance of RA ~10 kΩ μm^2, a spin polarization of 11-14% was obtained for the Co/h-BN/graphene junctions.

$$\Delta R_{NL} = \frac{P^2 \lambda_s}{2W\sigma} \exp(-L/\lambda_s), \quad (2)$$

where P is the spin polarization of the contacts, λ_s is the spin diffusion length, σ is the channel conductivity, L and W are the length and width of the graphene channel respectively.

The thickness variations of CVD h-BN barriers allowed further investigations of the dependence of spin signal and spin lifetime as a function of increasing contact resistance. In particular, these studies revealed a systematic enhancement of τ_s and ΔR_{NL} with increasing the contact resistance. Specifically, a unique inversion of the spin signal has been observed for devices consisting of a high resistance HR (~100 kΩ.μm^2) injector and a low resistance LR (~10 kΩ.μm^2) detector combinations. A comparison of such an inverted spin signal for a HR-LR device along with a normal spin valve signal obtained for an LR-LR device is present in Fig. 5(a). This indicates opposite signs of spin polarization in the HR and LR contact. In Fig. 5(b), the spin polarization of several devices is presented as a function of effective tunnel resistance, showing an enhancement in magnitude of spin polarization for high resistive h-BN barriers. This is consistent with the fact that the HR h-BN barriers are in the optimum range for efficient spin injection into graphene and lead to an enhanced spin polarization. From the HR-LR h-BN injector-detector combination devices, an effective nonlocal spin polarization $P = \sqrt{|P_i P_d|} \sim 31\%$ is obtained. Considering the fact that the LR h-BN contacts show a polarization ~15%, the HR h-BN contacts result in a large opposite (negative) spin polarization up to 65%. These observations indicate a possible selective spin filtering in HR Co/h-BN/graphene junctions[39]. These results show that h-BN is not only a novel tunnel barrier for charge and spin based devices but also exhibits a unique possible spin filtering effect in van der Waals heterostructures with graphene.

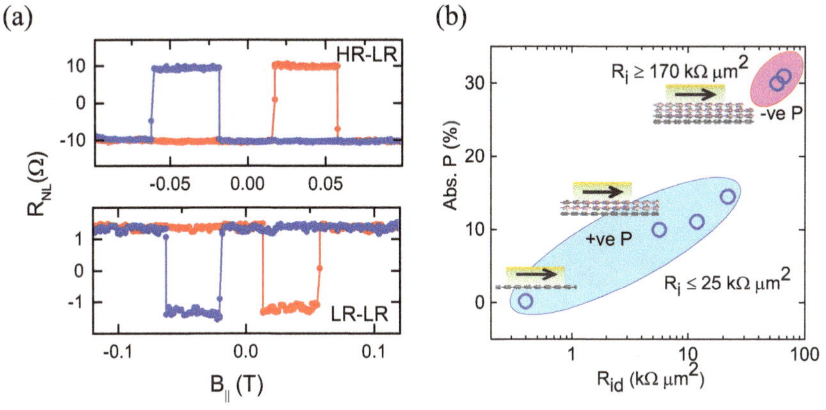

Figure 5. Sign inversion and spin filtering in Co/h-BN/graphene van der Waals heterostructures. (a) Inverted spin valve signal obtained for HR-LR injector-detector combination (top panel) vs. normal signal obtained on devices with LR-LR injector-detector combination (lower panel). (b) Spin polarization dependence on the h-BN tunnel contact resistance $R_{id} = \sqrt{R_i R_d}$, where R_i and R_d are the injector and detector resistances. Adapted from [Ref. 39].

5. Spin relaxation in graphene and practical challenges

Understanding spin relaxation in graphene is a key issue in the field of graphene spintronics. Primarily, there are four mechanisms that are responsible for spin relaxation in solids: Elliot-Yafet (EY), Dyakonov-Perel (DP), Bir–Aronov–Pikus (BAP) and hyperfine interaction. In graphene, the spin relaxation is primarily attributed to the EY and DP mechanism, originating from spin-orbit coupling and momentum scattering. In the EY mechanism, spin-flip scattering occurs during momentum scattering events. Thus the spin relaxation is proportional to the momentum relaxation in EY mechanism with $\tau_s \propto \tau_p$, where τ_p is the momentum relaxation time. In the DP mechanism the spin precession and relaxation occurs in between momentum scattering events due to local spin-orbit field as a consequence of broken space inversion symmetry. Consequently, the spin relaxation is inversely related to momentum relaxation and $\tau_s \propto 1/\tau_p$ in DP mechanism.

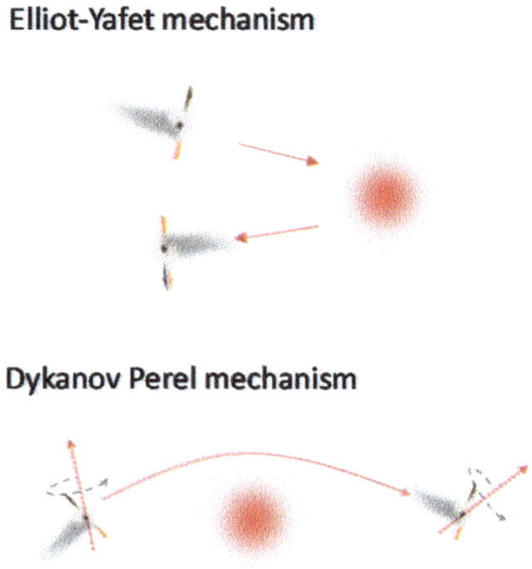

Figure 6: Schematic illustration of the Elliot-Yafet (EY) and Dyakonov-Perel (DP) spin relaxation mechanisms.

A schematic representation of EY and DP mechanisms is shown in Fig. 6. Due to the low spin-orbit coupling, the theoretical calculations considering both the EY and DP mechanisms estimate a spin lifetime $\tau_s \sim 1\,\mu s$ in graphene. This is several orders of magnitude higher than experimentally measured values, leaving a huge discrepancy between theory and experiments.

In order to understand the spin relaxation in graphene spintronic devices, significant experimental efforts were made to improve the ferromagnetic tunnel contacts and graphene channel properties. The use of metal oxide tunnel barriers (Al_2O_3, TiO_2 and MgO) and recent 2D tunnel barriers (such as flourographene and h-BN) have shown promising enhancement of spin lifetime up to 0.7 ns in graphene with highest estimated tunnel spin polarization up to 65%[25,39,40]. Devices made to retain the integrity and limit extrinsic effects in the graphene channel, either via suspension[22,23], atomically flat h-BN substrate[13] or h-BN encapsulation[9,11] have also yielded spin lifetime up to several nanoseconds.

Similar order of spin lifetime has also been reported in CVD graphene with nominal carrier mobility (~2000 $cm^2V^{-1}s^{-1}$ on conventional SiO_2 substrates[29]). To be noted, large variations in spin parameters from device to device and from different laboratories have also been observed due to fabrication related issues[8]. Even though these studies identified the signatures of both the EY and DP mechanisms for the spin relaxation from the dependence on τ_s and τ_p, the discrepancy of about three orders from the theoretical spin lifetime (τ_s in the μs range)[4,5,41] remains an unaddressed issue.

Recently, it has been theoretically proposed that scattering from spin-flip exchange field of magnetic impurities, adatoms or vacancies can result in resonant scattering leading to spin lifetime values similar to the ones observed in experiments[42,43]. In addition, spin-orbit induced spin-pseudo spin entanglement and relaxation in the ballistic regime has also been proposed[44]. Recent experiments using clean graphene channels encapsulated with thick h-BN layers have demonstrated spin lifetimes of several nanoseconds and spin diffusion length up to 22 μm at higher carrier concentration. While these recent theories and experiments seem to converge, there are also issues related to the effect of contacts. Especially in the recent report on long distance spin transport in CVD graphene, it has been observed that longer channels lead to longer spin lifetimes, confirming a greater role of the contacts in spin relaxation. These studies suggest that a more explicit role of ferromagnetic contacts and the device geometry has to be considered in theoretical expressions employed for the evaluation of Hanle spin lifetime. In future, the large area CVD graphene and their vdWh with CVD h-BN not only have the potential for application of graphene spintronic devices, but also present robust systems for systematic experiments to unlock the different contributions of spin relaxation mechanisms in graphene.

6. Summary & Outlook

Graphene spintronics has made remarkable progress in the past decade. Since the first report of spin transport in graphene at room temperature, several important contributions were made to improve the spin parameters in graphene and understand the spin relaxation in graphene. These experiments were focused on improvements in ferromagnetic tunnel contacts and graphene channel properties. Better quality metal oxide tunnel barriers have improved spin injection efficiency and spin lifetime in graphene. More recently, atomically thin h-BN based spin tunnel contacts and its van der Waals heterostructures with graphene have been employed to demonstrate efficient spin injection, enhancements in spin lifetime and spin filtering effects. In order to reduce the graphene channel induced spin relaxation, cleaner samples with h-BN encapsulated graphene channel were successfully used to achieve several nanosecond spin lifetimes. More recently, the work on large area CVD graphene devices demonstrated spin lifetime up to 1.2 ns and long distance spin transport up to 16 μm, highest for any form of pristine graphene on conventional oxidized Si wafers.

The experimental results and theoretical studies of spin relaxation mechanisms in graphene suggest that careful engineering of contacts and channel properties (controlling the nature of impurities) in devices lead to enhanced spin parameters. High quality interfaces with h-BN tunnel contacts and h-BN encapsulated graphene structures appear to be promising candidates for further advancements. The recent demonstration of long distance spin transport capabilities of CVD graphene is a key development, which combined with h-BN tunnel barrier and large scale encapsulation schemes has the potential to play a promising role in future development of graphene spintronic devices with potential applications.

References

1. Behin-Aein, B., Datta, D., Salahuddin, S. & Datta, S. Proposal for an all-spin logic device with built-in memory. *Nat. Nanotechnol.* **5,** 266–270 (2010).

2. Novoselov, K. S. *et al.* Electric field effect in atomically thin carbon films. *Science* **306**(5696), 666–669 (2004).

3. Geim, A. K. & Novoselov, K. S. The rise of graphene. *Nat. Mater.* **6,** 183–191 (2007).

4. Huertas-Hernando, D., Guinea, F. & Brataas, A. Spin-orbit coupling in curved graphene, fullerenes, nanotubes, and nanotube caps. *Phys. Rev. B* **74,** 155426 (2006).

5. Huertas-Hernando, D., Guinea, F. & Brataas, A. Spin-Orbit-Mediated Spin Relaxation in Graphene. *Phys. Rev. Lett.* **103,** 146801 (2009).

6. Han, W. *et al.* Tunneling Spin Injection into Single Layer Graphene. *Phys. Rev. Lett.* **105,** 167202 (2010).

7. Avsar, A. *et al.* Towards wafer scale fabrication of graphene based spin valve devices. *Nano Lett.* **11,** 2363–2368 (2011).

8. Volmer, F. *et al.* Role of MgO barriers for spin and charge transport in Co/MgO/graphene nonlocal spin-valve devices. *Phys. Rev. B* **88,** 161405(R) (2013).

9. Drögeler, M. *et al.* Nanosecond spin lifetimes in single- and few-layer graphene-hBN heterostructures at room temperature. *Nano Lett.* **14,** 6050–6055 (2014).

10. Kamalakar, M. V., Dankert, A., Bergsten, J., Ive, T. & Dash, S. P. Enhanced Tunnel Spin Injection into Graphene using Chemical Vapor Deposited Hexagonal Boron Nitride. *Sci. Rep.* **4,** 6146 (2014).

11. Guimarães, M. H. D. *et al.* Controlling Spin Relaxation in Hexagonal BN-Encapsulated Graphene with a Transverse Electric Field. *Phys. Rev. Lett.* **113,** 086602 (2014).

12. Ingla-Aynés, J., Guimarães, M. H. D., Meijerink, R. J., Zomer, P. J. & van Wees, B. J. 24 − μ m spin relaxation length in boron nitride encapsulated bilayer graphene. *Phys. Rev. B* **92,** 201410 (2015).

13. Zomer, P. J., Guimarães, M. H. D., Tombros, N. & van Wees, B. J. Long-distance spin transport in high-mobility graphene on hexagonal boron nitride. *Phys. Rev. B* **86,** 161416 (2012).

14. Fu, W., Makk, P., Bräuninger, M., Maurand, R. & Schönenberger, C. Large-scale fabrication of BN tunnel barriers for graphene spintronics. *J. Appl. Phys.* **116,** 074306 (2014).

15. Yamaguchi, T. *et al.* Electrical Spin Injection into Graphene through Monolayer Hexagonal Boron Nitride. *Appl. Phys. Express* **6,** 073001 (2013).

16. Tombros, N., Jozsa, C., Popinciuc, M., Jonkman, H. T. & van Wees, B. J. Electronic spin transport and spin precession in single graphene layers at room temperature. *Nature* **448,** 571–574 (2007).

17. Dankert, A., Kamalakar, M. V., Bergsten, J. & Dash, S. P. Spin transport and precession in graphene measured by nonlocal and three-terminal methods. *Appl. Phys. Lett.* **104,** 192403 (2014).

18. Fert, A. & Jaffrès, H. Conditions for efficient spin injection from a ferromagnetic metal into a semiconductor. *Phys. Rev. B* **64,** 184420 (2001).

19. Rashba, E. I. Theory of electrical spin injection: Tunnel contacts as a solution of the conductivity mismatch problem. *Phys. Rev. B* **62,** R16267–R16270 (2000).

20. Wang, W. H. *et al.* Growth of atomically smooth MgO films on graphene by molecular beam epitaxy. *Appl. Phys. Lett.* **93,** 183107 (2008).

21. Han, W. *et al.* Spin transport and relaxation in graphene. *J. Magn. Magn. Mater.* **324,** 369–381 (2012).

22. Guimarães, M. H. D. *et al.* Spin transport in high-quality suspended graphene devices. *Nano Lett.* **12**(7), 3512–3517 (2012).

23. Neumann, I. *et al.* Electrical detection of spin precession in freely suspended graphene spin valves on cross-linked poly(methyl methacrylate). *Small* **9,** 156–60 (2013).

24. Dlubak, B. *et al.* Highly efficient spin transport in epitaxial graphene on SiC. *Nat. Phys.* **8,** 557–561 (2012).

25. Friedman, A. L., van 't Erve, O. M. J., Li, C. H., Robinson, J. T. & Jonker, B. T. Homoepitaxial tunnel barriers with functionalized graphene-on-graphene for charge and spin transport. *Nat. Commun.* **5,** 3161 (2014).

26. Fu, W., Makk, P., Maurand, R., Bräuninger, M. & Schönenberger, C. Large-scale BN tunnel barriers for graphene spintronics. *J. Appl. Phys.* **116,** 074306 (2014).

27. Wojtaszek, M., Vera-Marun, I. J., Whiteway, E., Hilke, M. & van Wees, B. J. Absence of hyperfine effects in C-13 graphene spin-valve devices. *Phys. Rev. B* **89,** 035417 (2014).

28. A. L. Friedman, O. M. J. Van Erve, J. T. Robinson, et al., ACS Nano **9,** 6747 (2015).

29. Kamalakar, M. V., Groenveld, C., Dankert, A. & Dash, S. P. Long Distance Spin Communication in Chemical Vapour deposited Graphene. *Nat. Commun.* **6,** 6766 (2015).

30. Han, W. & Kawakami, R. K. Spin Relaxation in Single-Layer and Bilayer Graphene. *Phys. Rev. Lett.* **107,** 047207 (2011).

31. Shiraishi, M. *et al.* Robustness of Spin Polarization in Graphene-Based Spin Valves. *Adv. Funct. Mater.* **19,** 3711–3716 (2009).

32. Geim, A. K. & Grigorieva, I. V. Van der Waals heterostructures. *Nature* **499,** 419–425 (2013).

33. Kamalakar, M. V., Dankert, A., Bergsten, J., Ive, T. & Dash, S. P. Spintronics with graphene-hexagonal boron nitride van der Waals heterostructures. *Appl. Phys. Lett.* **105,** 212405 (2014).

34. Godel, F. *et al.* Voltage-controlled inversion of tunnel magnetoresistance in epitaxial nickel/graphene/MgO/cobalt junctions. *Appl. Phys. Lett.* **105,** 152407 (2014).

35. Dankert, A., Kamalakar, M. V., Wajid, A., Patel, R. S. & Dash, S. P. Tunnel Magnetoresistance with Atomically Thin Two-Dimensional Hexagonal Boron Nitride Barriers. *Nano Res*. **8,** 1357 (2015). doi:10.1007/s12274-014-0627-4

36. Dlubak, B. *et al.* Graphene-passivated nickel as an oxidation-resistant electrode for spintronics. *ACS Nano* **6,** 10930–4 (2012).

37. Britnell, L. *et al.* Electron tunneling through ultrathin boron nitride crystalline barriers. *Nano Lett.* **12,** 1707–1710 (2012).

38. Song, L. *et al.* Large scale growth and characterization of atomic hexagonal boron nitride layers. *Nano Lett.* **10,** 3209–15 (2010).

39. Kamalakar, M. V., Dankert, A., Kelly, P. J. & Dash, S. P. Inversion of Spin Signal and Spin Filtering in Ferromagnet|Hexagonal Boron Nitride-Graphene van der Waals Heterostructures. *Sci. Rep.* **6,** 21168 (2016).

40. Han, W., Kawakami, R. K., Gmitra, M. & Fabian, J. Graphene spintronics. *Nat. Nanotechnol.* **9,** 324–340 (2014).

41. Ochoa, H., Castro Neto, a. H. & Guinea, F. Elliot-Yafet Mechanism in Graphene. *Phys. Rev. Lett.* **108,** 206808 (2012).

42. Kochan, D., Gmitra, M. & Fabian, J. Spin Relaxation Mechanism in Graphene: Resonant Scattering by Magnetic Impurities. *Phys. Rev. Lett.* **112,** 116602 (2014).

43. Lundeberg, M., Yang, R., Renard, J. & Folk, J. Defect-Mediated Spin Relaxation and Dephasing in Graphene. *Phys. Rev. Lett.* **110,** 156601 (2013).

44. Tuan, D. Van, Ortmann, F., Soriano, D., Valenzuela, S. O. & Roche, S. Pseudospin-driven spin relaxation mechanism in graphene. *Nat. Phys.* **10,** 857–863 (2014).

Index

www.ingramcontent.com/pod-product-compliance
Ingram Content Group UK Ltd.
Pitfield, Milton Keynes, MK11 3LW, UK
UKHW022104170926
8624IPUK00024B/56

9 789813 149816